First
ASTRONOMY
DICTIONARY

FIREFLY BOOKS

A FIREFLY BOOK

SECOND AND THIRD EDITIONS: REVISED AND UPDATED BY
John Woodruff

FIRST EDITION: EDITORIAL CONSULTANTS
Ian Ridpath and John Woodruff

Published by Firefly Books Ltd. 2003

© Philip's 2003

First printing

National Library of Canada Cataloguing in Publication

    Firefly astronomy dictionary. – 1st North American ed.
First ed. published London : G. Philip Ltd., 1995 under title:
Philip's astronomy dictionary.
First ed. edited by Ian Ridpath and John Woodruff.
2nd and 3rd eds. revised and updated by John Woodruff.
ISBN 1-55297-837-0 (pbk.)
1. Astronomy–Dictionaries. I. Woodruff, John
II. Title: Philip's astronomy dictionary.

QB14.F56 2003   520'.3   C2003-901918-7

Publisher Cataloging-in-Publication Data
(Library of Congress standards)

    Firefly astronomy dictionary. – 1st American ed.
[256] p. : ill. (some col.), charts, ports. (some col.), maps ; cm.
Originally published: Philip's Astronomy Dictionary. London:
Philip, 1999.
Summary: Reference guide to the major terms and theories
used in astronomy, with over 1000 up-to-date entries,
extensively cross-referenced, plus concise coverage of new
developments in space exploration.
ISBN 1-55297-837-0 (pbk.)
1. Astronomy – Dictionaries. I. Title.
520/ .3 21   QB14.F57   2003

Published in Canada in 2003 by
Firefly Books Ltd.
3680 Victoria Park Avenue
Toronto, Ontario, M2H 3K1

Published in the United States in 2003 by
Firefly Books (U.S.) Inc.
P.O. Box 1338, Ellicott Station
Buffalo, New York 14205

Produced in Great Britain in 2003 by Philip's,
a division of Octopus Publishing Group Ltd,
2–4 Heron Quays, London E14 4JP

Printed in China

Front cover (*clockwise from top left*):
Io *NASA/JPL*; X-ray view of Milky Way
center *NASA/U Mass./D. Wang, et al*;
reflecting telescope *Philip's*; Lagoon
Nebula *NOAO/AURA/NSF*; Saturn V
rocket carrying Apollo 11 *NASA*.

Back cover: Neptune and Earth
compared *NASA/JPL*; Pioneer 10
Philip's.

Title page: Hubble Space Telescope
European Space Agency.

Using this book

The *Firefly Astronomy Dictionary* contains more than 1,000 articles and over 180 color illustrations covering a wide range of topics. The book has been designed to be easy to use by readers of all levels.

The alphabetical order of headwords is letter by letter, ignoring the spaces between words. So, for example, "**G star**" comes between "**Grus**" and "**Gum Nebula**." The exception is that "**H α**," "**H I region**," and "**H II region**" appear at the beginning of the "**H**" section. Terms beginning with a number are alphabetized as if the number were spelled out; an example is "**Sixty-one Cygni**."

Cross-references are used extensively, and appear in **bold type**. A cross-reference usually indicates that the entry referred to provides secondary information on the topic under discussion. Sometimes it means that the term in **bold type** is defined or discussed at greater length in its own entry. An entry which is just a cross-reference means that the term is defined in another entry, where it will be found in *italic type*. A heading or cross-reference followed by a number, for example "**aberration** (2)," is used whenever the same term has more than one entry.

Most entries are fairly concise, but a number of subjects are treated at greater length, for example "**spectral classification**" and, in particular, the entries for the major planets. The term "planetary body" is used in this dictionary to refer to large satellites and asteroids in addition to the major planets. For the purposes of the values given for the diameters of the giant planets, their "surfaces" have been taken to be the levels in their atmospheres at which the pressure is 1 bar, i.e. the average pressure at the surface of the Earth.

In the past few years many new satellites of the outer planets, particularly of Jupiter, have been discovered, and several have been named. None of these have been given entries; all previously named satellites are, however, included in this dictionary.

Particularly in tables, large numbers are sometimes given in exponential form. Thus 10^3 is a thousand, 2×10^6 is two million, and so on. "Billion" means a thousand million, or 10^9.

Dates in this book have been expressed in the order month, day, year, although it is customary in astronomy to show dates in the order year, month, day.

In biographical entries, if the first forename of a person is not the one by which they were most commonly known, it is enclosed in parentheses.

A

aberration (1) The small apparent displacement of a star's position produced by the combination of the motion of the Earth and the finite velocity of light.

Annual aberration is the displacement resulting from the motion of the Earth in its orbit around the Sun. This is detected by an observer as an apparent motion of a star in an ellipse over the course of the year. The maximum displacement of the star, given by the semimajor axis of this ellipse, is just over 20 seconds of arc. This angle (α) is the constant of annual aberration and is determined from the relationship

$$\tan \alpha = v/c$$

where v is the Earth's orbital velocity and c is the velocity of light. For a star at the pole of the ecliptic the ellipse becomes a circle; for a star on the ecliptic the ellipse becomes a straight line.

Diurnal aberration is a very much smaller displacement of a star, to the east, resulting from the rotation of the Earth on its axis. It reaches a maximum of 0.32 seconds of arc, for an observer at the equator, and is zero at the poles. The Earth's rotation carries an observer in a circle with a velocity that is greatest at the equator and zero at the poles.

The aberration of starlight was discovered by James **Bradley**, who published his findings in 1729. In demonstrating that the Earth is not fixed in space – thus furnishing the first observational evidence to support Copernicus's theory – and that light travels at a finite velocity, the discovery is important in the development of cosmological ideas.

aberration (2) A defect, such as blurring, distortion, or false coloration that can occur in the image produced by a lens or curved mirror. The major aberrations are **chromatic aberration**, spherical aberration, **coma** (1), and **astigmatism**. Spherical aberration results from the unequal deflection of light rays by different zones of a spherical lens or mirror. The rays are not brought to the same focus, causing a lack of definition about the edges of the image. Spherical aberration in lenses can be minimized by using a suitable combination of elements, and in reflecting telescopes it is overcome by the use of a mirror in the shape of a **paraboloid**.

ablation Wearing away of the surface of an object. An example is the erosion of a meteoroid during its high-speed passage through the Earth's atmosphere: atmospheric molecules striking the meteoroid vaporize its surface layers.

absolute magnitude (symbol M) The apparent magnitude a star would have if it were at a standard distance of 10 parsecs from the Earth. The absolute magnitude may be derived from the **apparent magnitude** (m) and the **parallax** (π) by the formula

$$M = m + 5 + 5 \log \pi$$

where π is in seconds of arc. The absolute magnitudes of most stars lie between -5 and $+15$. The Sun has an absolute magnitude of $+4.8$. SEE ALSO **magnitude**.

absolute temperature Temperature as measured on a scale whose zero point is *absolute zero*, the point at which almost all motion at the molecular level ceases. The unit of absolute temperature is the degree kelvin; an interval of one kelvin (K) is equivalent to an interval of one degree Celsius (°C). The relationship between a Kelvin temperature (T) and the same temperature as measured on the Celsius scale (t) is

$$T = t + 273.15$$

absolute zero being $-273.15°C$. Absolute temperatures are widely used in astronomy. (SEE ALSO the table of conversion factors on page 255.)

absorption line A dark line or band in a **continuous spectrum**. Absorption lines are produced when particular wavelengths of radiation from a hot source are absorbed by a cooler, intervening medium. They are typical of the spectra of stars – radiation from the hot interior is absorbed by a star's cooler outer layers, producing absorption lines. The absorption lines in the Sun's spectrum are called **Fraunhofer lines**. The wavelengths at which absorption lines occur are characteristic of the chemical elements present in the radiating source. This enables a star's chemical composition to be determined from its spectrum.

accretion The coalescence of small particles in space as a result of collisions, and the gradual building up of larger bodies from smaller ones by gravitational attraction; also, the accumulation of matter by a star or other celestial object. Matter can also be transferred from one component of a close binary system to another by an accretion process, via an **accretion disk**. Accretion is an important factor in the evolution of stars, planets, and comets.

accretion disk A disk of gas in orbit around an object, formed by inflowing matter. In a binary star system in which the two component stars are very close, gas can be transferred from the larger to the smaller star, creating an accretion disk. If the companion has a strong gravitational field, as it will have if it is a neutron star or black hole, friction will heat the orbiting gas to millions of degrees, hot enough to emit X-rays. Accretion disks also occur at the center of galaxies.

Achernar The star Alpha Eridani, magnitude 0.46, distance 144 light years, luminosity 1150 times that of the Sun. It is a main-sequence star of spectral type B3 and the ninth-brightest star in the sky.

Achilles Asteroid no. 588, one of the **Trojan asteroids** sharing Jupiter's orbit, and the first to be discovered, by Max Wolf in 1906. It has a diameter of 147 km (91 miles).

achondrite A type of **stony meteorite**, consisting of rock that was once molten, and which therefore was once part of a planetary body that underwent **differentiation**. Like all stony meteorites, achondrites contain no uncombined metallic elements, and they also lack chondrules – millimeter-sized globules of silicate rock. The principal types are *eucrites*, *diogenites*, and *howardites*, differing in chemical make-up and structure, but all with properties (such as radiation damage) that suggest they were once part of a planetary surface. A main candidate is the asteroid Vesta. Achondrites include a handful of meteorites known to have originated on the Moon, and the **SNC mete-**

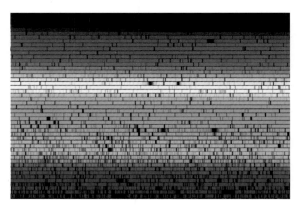

▶ **absorption line** The visible light spectrum of the cool giant star Arcturus (α Boötis) is shown here. The dark vertical lines in the spectrum are caused by atoms in the star's atmosphere absorbing radiation. Because each element absorbs radiation at characteristic wavelengths, the spectrum of a star can be used to determine which elements are present.

◀ **active optics** The active optics actuators on the reverse of the primary mirror of the WIYN telescope at Kitt Peak National Observatory allow the mirror to be flexed continually to compensate for the effects of gravity as the telescope moves. This system means that far thinner mirrors can be used without risking distortion of the images and data obtained.

orites, believed to have originated on Mars. SEE ALSO **chondrite**.

achromatic lens (achromat) A lens designed to reduce **chromatic aberration**, cutting down the amount of false color in the image. This is usually achieved by using two lens elements made from different types of glass, for example one element of crown glass and the other of flint glass. The combination is chosen such that two selected wavelengths of light come to the same focus, and at the same time the residual aberration is reduced to a minimum. SEE ALSO **apochromat**.

active galactic nucleus (AGN) The region, at the heart of an *active galaxy*, in which huge amounts of energy are generated by processes other than those that power stars. The energy is believed to be released as matter is accreted by a supermassive **black hole**, of up to a few billion solar masses, at the galaxy's center. **Jets** are often emitted by AGNs. Active galaxies show a range of behavior, depending on the amount of energy generated in their nuclei: SEE **Seyfert galaxy**, **BL Lac object**, **blazar**, **quasar**.

active optics A technique for maintaining the primary mirror of large telescopes in accurate shape and alignment, despite the distortions caused by temperature changes or tilting the telescope. The mirror rests on numerous supports which are automatically adjusted by computer to keep its figure smooth under all conditions. The technique can also compensate for any slight imperfections left after the manufacturing of the mirror. Mirrors with active supports can be made thinner, and thus lighter and cheaper, than ordinary telescope mirrors. The term is sometimes used to refer to the more complicated technique more properly known as **adaptive optics**.

Adams, John Couch (1819–92) English astronomer. While still an undergraduate at Cambridge, he set out to analyze the motion of Uranus. The planet's observed path was not in agreement with its calculated orbit, and Adams believed that the discrepancies could be accounted for by the gravitational influence of an undiscovered planet. He calculated an orbit for the new planet, but no search was mounted. It was Johann **Galle** who located Neptune, as it was subsequently called, near a position predicted by the French astronomer Urbain **Le Verrier**. After a lengthy priority dispute, the honors of predicting the existence of Neptune came to be shared between Adams and Le Verrier. However, it now seems that neither Adams's nor Le Verrier's calculations were that accurate, and that the two predicted orbits and the actual orbit roughly coincided at the point where Neptune happened to be when Galle found it.

adaptive optics A technique to counteract the blurring of images caused by **seeing** conditions in the Earth's atmosphere. Sensors measure the distortions in a telescope's image caused by atmospheric effects. The image falls on to a small, wafer-thin subsidiary mirror, which can be flexed in a controlled fashion hundreds of times a second to cancel out the imperfections in the image. The resulting image is almost as sharp as if the telescope were in space.

Adhara The star Epsilon Canis Majoris, magnitude 1.50, 431 light years away, and 3700 times as luminous as the Sun. It is a blue giant of spectral type B2.

Adrastea One of the small inner **satellites** of Jupiter discovered in 1979 during the Voyager missions. It is irregular in shape, measuring about 25 × 20 × 15 km (15 × 12 × 9 miles), and orbits near the outer edge of Jupiter's main ring.

AGN ABBREVIATION FOR **active galactic nucleus**.

airglow A faint glow in the Earth's atmosphere. Molecules in the tenuous upper atmosphere, where the solar radiation is strong, are ionized (split up into charged particles) by ultraviolet radiation or by collision with other charged particles. These particles are short-lived and soon recombine to form the original molecules, emitting a faint light as they do so. The recombination of oxygen molecules is a principal source of emission, which continues long after sunset. From a dark site on a clear, moonless night the airglow is visible as a gray luminous background to the stars, but for many observers **light pollution** makes the airglow impossible to see. The **ashen light** observed on Venus's night side may be a similar phenomenon.

Airy, George Biddell (1801–92) English astronomer and geophysicist. He became Astronomer Royal and Director of the Royal Greenwich Observatory, and re-established the Observatory's standards and authority. Airy discovered irregularities in the orbits of Venus and the Earth, and determined the mass of the Earth by taking gravity measurements at the top and bottom of a mineshaft.

Airy disk The central spot in the image of a star formed in a telescope. The effects of **diffraction** make it impossible for the image to be formed as a point. Instead, it is spread out into an Airy disk, surrounded by diffraction rings and spikes. (After George Biddell Airy.)

al-Battānī, Muḥammad ibn Jābīr (Latinized name **Albategnius**) (858–929) Arab astronomer. His 40 years of observations yielded a catalog of star positions more accurate than those given in Ptolemy's *Almagest*, and introduced trigonometry into Arab astronomy. He made accurate determinations of the precession of the equinoxes and the obliquity of the ecliptic, and showed that the Earth's distance from the Sun varies over the course of the year.

► **adaptive optics** The Very Large Telescope (VLT) uses adaptive optics. Here the light from a close binary pair has been reflected from the primary mirror on to a subsidiary mirror, which is continually adjusted to compensate for variations in the Earth's atmosphere; it is then computer processed.

PSF   Raw   Processed

albedo A measure of the reflectivity, or brightness, of a material or body. The albedo scale goes from 1, for a perfectly reflecting, white, surface, to 0, for a perfectly absorbing, black, surface. In astronomy, albedo is used to indicate the fraction of sunlight reflected by Solar System bodies. Rocky bodies such as Mercury and the Moon have low albedoes, while cloud-covered bodies like Saturn or those consisting of a high proportion of water ice, like many of the satellites of the outer Solar System, and comets, have high albedoes. Albedo can be defined in various ways. The one most used now is *geometric albedo* – the ratio of the reflectance of the body to that of a flat white surface of the same surface area at the same position.

Aldebaran The star Alpha Tauri, magnitude 0.87 (but slightly variable, of irregular type), 65 light years distant, luminosity about 150 times that of the Sun. Aldebaran is a giant of type K5. It appears to lie in the Hyades cluster, but is actually a foreground object at about half the cluster's distance.

Algol The star Beta Persei, the prototype of a class of eclipsing binaries known as Algol-type variables. The first recorded observation of Algol's variability was made by Geminiano Montanari of Bologna in 1669. John **Goodricke** in 1782 was the first to suggest that the variability resulted from two stars in binary motion, the fainter and larger one from time to time passing in front of the brighter one. There are two stars, of spectral types B8 and G, orbiting each other in 2.867 days. The B8 star is a dwarf and is the visible component; the fainter star is a subgiant. During eclipses, Algol's brightness drops from 2.1 to 3.4. There is also a third star, but it does not take part in the eclipses. Algol lies 93 light years away.

Algonquin Radio Observatory A radio astronomy observatory at Lake Traverse, Ontario, run by the National Research Council of Canada. Its principal instrument is a 46-m (150-ft) fully steerable radio dish, which has been in operation since 1966.

Almagest An encyclopedia of astronomy compiled by **Ptolemy** in about AD 140. It is thought to be based mainly on the work of **Hipparchus**, whose star catalog it incorporates. It was the ancient world's standard accurate description of the heavens and remained authoritative until the middle of the 16th century, as nothing else was available.

almanac A yearbook containing predicted positions of celestial objects and other astronomical and calendrical data. For astronomical and navigational purposes the leading publication is *The Astronomical Almanac*, produced jointly each year by Her Majesty's Nautical Almanac Office and the US Naval Observatory.

Alpha Centauri The brightest star in the constellation Centaurus, and the third-

◀ **albedo** This albedo map of Mars was produced by NASA's Mars Global Surveyor. Red areas are bright and show where there is dust. Blue areas show where the underlying, darker rocks have been exposed.

brightest star in the sky, magnitude −0.28; also known as Rigil Kentaurus. It is a visual binary, with G2 and K1 main-sequence components of magnitudes −0.01 and 1.35, orbital period 80 years, distance 4.4 light years. Associated with it is the much fainter star **Proxima Centauri.**

al-Sūfi, Abu'l-Husain (903–86) Arab astronomer. In his *The Book of the Fixed Stars* he presented a detailed revision of the star positions in Ptolemy's *Almagest*, based on his own observations. It contains the first recorded reference to the Andromeda Galaxy. He measured the length of the year, and attempted to determine the length of a degree of the meridian.

Altair The star Alpha Aquilae, magnitude 0.76, distance 17 light years, luminosity 11 times that of the Sun. It is a main-sequence star of type A7.

altazimuth A mounting for an instrument such as a telescope that can move in both altitude and azimuth – that is, about both horizontal and vertical axes. The disadvantage of this mounting for an amateur telescope is the problem of moving it simultaneously in azimuth and altitude in order to correct the Earth's diurnal motion, and so keep the object being observed in the field of view. This is made easier with the **equatorial telescope**, and simply ignored with the **Dobsonian telescope**. Since the introduction of computer-controlled drive mechanisms, the altazimuth has been the preferred mounting for all large professional instruments.

altitude The angular distance of a celestial body above the observer's horizon. It is measured in degrees from 0 (on the horizon) to 90 (at the zenith) along the **great circle** passing through the body and the zenith. If the object is below the horizon, the altitude is negative. SEE ALSO **coordinates.**

Amalthea The largest of Jupiter's small inner **satellites**, discovered in 1892 by E. E. Barnard. It is an irregular body, meas-

▲ **altazimuth mounting** This simple form of telescope mount allows free movement in both horizontal and vertical axes. An altazimuth mounting is not suitable for use with motordrives, however, unless they are computer controlled.

uring 270 × 165 × 150 km (170 × 100 × 90 miles). Amalthea is dark red in color, which may indicate that its surface is covered with sulfur from the volcanoes of Io, the innermost of the large satellites.

Amor asteroid One of a class of small **asteroids** whose **perihelion** distances lie inside the orbit of Mars (specifically, those with perihelion between 1.017 and 1.3 AU). They are named for the 1-km (0.6-mile) diameter asteroid 1221 Amor. The first to be discovered, the most studied, and the largest is 433 **Eros**. Over a thousand were known by early 2003.

Some of these small bodies may have been perturbed out of the main asteroid belt by Mars, while others may be the nuclei of extinct comets. Close encounters with Mars or Earth can further perturb Amor asteroids into Earth-crossing **Apollo asteroid** orbits; the two groups are often classified together as *Apollo–Amor objects.* Like other small bodies in the inner Solar System, Amor asteroids have limited lifetimes, typically about 10 million years, and many of them will eventually impact with the Earth or Mars. SEE ALSO **Aten asteroid.**

▲ **analemma** A plot of the Sun's apparent position from 52°N, looking south at midday, at 5-day intervals throughout the course of a year. The Sun is at the top of the figure 8 at the summer solstice and at the bottom at the winter solstice.

analemma A long, thin figure-of-eight formed by plotting the position of the Sun in the sky at the same time of day throughout the year. The variation in the Sun's position along the long axis of the "8" – highest at the summer **solstice**, lowest at the winter solstice – is caused by the Earth's axial inclination. The variation across the short axis is a result of the eccentricity of the Earth's orbit.

Ananke One of Jupiter's four small outermost **satellites**, discovered in 1951 by Seth Nicholson. It is in a retrograde orbit – SEE **retrograde motion** (1) – and may well be a captured asteroid.

anastigmatic lens (anastigmat) A type of lens designed to compensate for **astigmatism**.

Andromeda A large constellation of the northern hemisphere, adjoining the Square of Pegasus. The main stars lie in a line leading away from Pegasus, and the star Alpha Andromedae (Sirrah or Alpheratz, mag. 2.1) actually forms one corner of the Square. In mythology, Andromeda was the princess rescued by Perseus. Gamma (Almach or Alamak) is a fine double of magnitudes 2.3 and 4.8, spectral types K3 and B9. The open cluster NGC 752 is an easy binocular object, and NGC 7662 is a planetary nebula for small telescopes. The most famous object in the constellation is the **Andromeda Galaxy**.

Andromeda Galaxy A spiral galaxy 2.4 million light years away in the constellation Andromeda, the most distant object visible to the naked eye, also known as M31 or NGC 224. The Andromeda Galaxy has a mass of over 300,000 million Suns. Cepheid variable stars and novae have been observed in it. To the naked eye it appears as a smudge of magnitude 3.4; through binoculars and small telescopes it is seen to have an elongated structure some $1° × 3°$ in extent, and its plane is tilted at 13° to our line of sight. Its true diameter is about 150,000 light years, somewhat larger than our own Galaxy. The Andromeda Galaxy is accompanied by two elliptical galaxies, M32 (NGC 221) and M110 (NGC 205).

Andromedids A meteor shower which gave some dramatic displays in the 19th century, following the breakup in the 1840s of its parent comet, **Biela's Comet**. The shower, also known as the *Bielids*, had its radiant in the constellation Andromeda, near the star Gamma. The Andromedid meteor stream no longer intersects the Earth's orbit.

Anglo-Australian Observatory An observatory at Siding Spring Mountain, near Coonabarabran, New South Wales, jointly funded by the UK and Australian governments. It contains the 3.9-m (150-inch) Anglo-Australian Telescope and the 1.2-m (48-inch) UK Schmidt Telescope, and has unique facilities for wide-field observations.

angstrom (angstrom unit) (symbol Å) A unit of length, equal to 10^{-10} m, formerly widely used in spectroscopy for wavelengths of light. It is no longer recommended for use in astronomy, having been superseded by the nanometer, symbol nm (1 nm = 10^{-9} m, 1 Å = 0.1 nm). It is named for Swedish physicist and astronomer Anders Ångström (1814–74), who introduced it in his atlas of the solar spectrum, published in 1868.

angular diameter The apparent diameter of a celestial body expressed in angular measure (usually in **arc minutes** and **seconds**). It is the angle subtended at the observer by the true diameter of the body under observation. If the distance of the celestial body from the observer is known, its true diameter can be calculated.

annual parallax The angle subtended at a celestial object by the radius of the Earth's orbit, which is 1 astronomical unit. It is measured by determining the semimajor axis of the parallactic ellipse traced by the star on the celestial sphere. The reciprocal of the annual parallax in seconds of arc is the distance of the object in **parsecs**. SEE ALSO **parallax**.

annular eclipse A type of **solar eclipse** in which the Sun's disk is not completely hidden by the Moon, a thin ring of light (*annulus* is Latin for "ring") remaining visible around the dark face of the Moon. A solar eclipse is annular if it occurs when the Moon is close to **apogee**, at its farthest point from Earth in its elliptical orbit. The tip of its shadow-cone does not reach down to the Earth's surface, and its apparent diameter is a little less than the Sun's.

anomalistic month One revolution of the Moon around the Earth relative to its **perigee**. It is equal to 27.55455 days of **mean solar time**.

anomalistic year One revolution of the Earth around the Sun relative to its **perihelion**. It is equal to 365.25964 days of **mean solar time**. This is about 4m 43.5s longer than a sidereal year, because the perihelion point moves eastward.

anomaly An angular measurement used in determining the position of a body in an elliptical orbit. It is reckoned from periapsis (SEE **apsides**), in the direction of the object's motion. There are three types of anomaly. *True anomaly*, in the case of a planet, is the angle between the planet's radius vector and its perihelion. In the diagram it is the angle PSX. *Mean anomaly* is the same angle, but for an imaginary planet moving along its orbit at constant velocity, rather than at nonuniform velocity as in reality. *Eccentric anomaly* relates to the motion of a point along a circle, the diameter of which is the long axis of the planet's orbit. In the diagram, the eccentric anomaly is the angle PCX'. The point X' is defined by producing a perpendicular from the long axis through the planet to the circumscribing circle.

ansae (singular **ansa**) The outer parts of Saturn's rings, which through a telescope can appear like handles projecting from the planet's disk. The word is Latin for "handles."

antapex SEE **apex**.

Antares The star Alpha Scorpii, a supergiant of type M1 that varies irregularly from about magnitude 0.9 to 1.2. Its distance is 604 light years, and it is more than

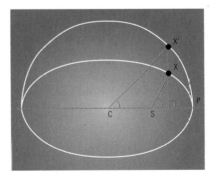

▲ **anomaly** The true anomaly is described by the angle PSX in the diagram here. The eccentric anomaly, PCX', is given by projecting the body's position on to a circumscribing circle, and can be used as a matter of convenience in orbital calculations. C, center of ellipse; S, Sun; P, perihelion; X, position of planet.

10,000 times as luminous as the Sun. Antares has a main-sequence companion star of magnitude 5.4, type B2, orbiting it with a period of about 900 years.

anthropic principle An argument that life in the Universe is somehow related to the physical properties of the Universe. Its simplest form is the *weak anthropic principle*, which states that the Universe we inhabit has properties that have made it possible for life to develop. This is not controversial, but the *strong anthropic principle* goes further. It argues that the laws of physics that have shaped the Universe are such that the emergence of intelligent life is an inevitable outcome.

Antlia A small southern constellation representing an air pump, added to the sky by **Lacaille**. It has no star brighter than magnitude 4.3.

Antoniadi, Eugène Michael (1870–1944) French astronomer, born in Turkey of Greek parents. He was a highly skilled observer, particularly of the inner planets. His maps of the surface features of Mars and Mercury were not bettered until those planets were visited by spacecraft. The **canals** of Mars were shown by Antoniadi to be illusory. He devised a system known as the **Antoniadi scale**, used to record seeing conditions.

Antoniadi scale The scale devised by Eugène **Antoniadi**, widely used by amateur astronomers to record the **seeing** conditions (the steadiness of the atmosphere) under which lunar and planetary observations are made. It distinguishes five gradations of seeing:

I   Perfect seeing, without a quiver
II Slight undulations, with periods of calm lasting several seconds
III Moderate seeing, with larger air tremors
IV Poor seeing, with constant troublesome undulations
V   Very bad seeing, scarcely allowing a rough sketch to be made.

ap-, apo- Prefixes referring to the farthest point in an object's orbit around another body, as in **aphelion** and **apogee**.

Apache Point Observatory An observatory in the Sacramento Mountains of New Mexico. Its main instrument is a 3.5-m (138-inch) reflector, and the Sloan Digital Sky Survey's 2.5-m (100-inch) reflector is also sited there. The Observatory is operated by a group of US universities called the Astrophysical Research Consortium.

apastron The point in the orbit of one component of a binary system at which it is farthest from the other star.

aperture The clear diameter of an optical telescope's objective lens or primary mirror. More generally, for other types of telescope, the aperture is the size of the principal radiation-collecting element – for example, the antenna of a radio telescope. Aperture is the most significant parameter of any telescope, for the amount of radiation it can collect from distant objects is all-important.

aperture synthesis A technique in radio astronomy in which an array of radio dishes is used to synthesize the resolution of a much larger telescope. The technique involves electronically combining the output from numerous small antennae, which are usually arranged in a line many kilometers long. As the Earth rotates, the array sweeps out an area equivalent to a single large dish with a diameter equal to the maximum length of the line. An example is the **Very Large Array**. SEE ALSO **radio interferometry**.

apex The point on the celestial sphere toward which the Sun and the entire Solar System appear to be moving at a velocity of 19–20 km/s (12 mile/s) relative to the nearby stars; it is also known as the *solar apex*. The apex lies in the constellation Hercules at about RA 18h, dec. +30°. The point on the celestial sphere opposite the apex is the *antapex*, in the constellation Columba.

aphelion The point in the orbit of a planet or a comet at which it is at its farthest distance from the Sun.

apochromat A lens especially corrected to a high degree to eliminate chromatic **aberration**. By using three elements, all of different types of optical glass, a greater amount of correction is possible than with the two-element **achromat**.

apogee The point in the orbit of a body around the Earth at which it is at its farthest distance from the Earth.

Apollo asteroid One of a class of small **asteroids** whose **perihelion** distances lie inside the Earth's orbit (specifically, those with perihelion less than 1.017 AU and **semimajor axis** greater than 1 AU). They are named for the 2-km (1.25-mile) diameter object 1862 Apollo. Over a thousand were known by early 2003.

Like the **Amor asteroids**, with which they are often grouped as *Apollo–Amor objects*, some may be extinct cometary nuclei, but others may have been perturbed inward by Jupiter from one of the **Kirkwood gaps** in the main asteroid belt. Many Apollo asteroids could impact the Earth within a few million years. SEE ALSO **Aten asteroid**.

Apollo program A US space project that landed astronauts on the Moon. The Apollo spacecraft had three main parts: a pressurized *command module* for the three crew members; a *service module* containing oxygen and electrical supplies and the main maneuvering engine; and the *lunar module* in which two astronauts made the actual Moon landing. The whole affair was launched by a Saturn V rocket. After four preparatory manned flights, Apollo 11 made the first Moon landing in July 1969. Neil Armstrong and Edwin Aldrin became the first men to step on the Moon, remaining on the surface for nearly a day and bringing back 22 kg of lunar rock and dust. Subsequent Apollos extended the exploration of the Moon's surface, bringing back further samples and leaving behind instruments which transmitted data about moonquakes. The last of the series, Apollo 17, was in December 1972.

apparent magnitude (symbol *m*) The brightness of an object as seen from Earth. The apparent magnitude differs from the **absolute magnitude** because of the effect of the object's distance from us. SEE ALSO **magnitude**.

apparent solar time Local time, based upon the apparent daily movement of the Sun. It is the time shown on a sundial. Noon occurs when the Sun reaches its maximum altitude, i.e. when it crosses the observer's meridian. Time reckoned in this way is subject to considerable variations because of the Sun's non-uniform motion, a result of both the elliptical shape

▶ **Apollo program** The Lunar Roving Vehicles allowed the astronauts on the final three Apollo lunar missions to range farther from the landers and so collect more varied samples from the Moon's surface. Here, James B. Irwin works next to the Vehicle during the Apollo 15 mission. The photograph was taken by David R. Scott.

of the Earth's orbit and the fact that the Sun moves along the ecliptic, not the celestial equator. The time kept by clocks is **mean solar time**, based on the motion of a hypothetical mean Sun.

apparition The appearance in the sky of a comet, or a period during which a planet is best observed.

appulse The apparent close approach of two celestial bodies, such as a planet and a star, whose directions of motion as seen on the celestial sphere converge (though the bodies in question are in reality remote from each other). SEE ALSO **occultation**.

apsides (singular **apsis** or **apse**) The two points in an orbit that are respectively closest to (*periapsis*) and farthest from (*apapsis*) the primary body. For example, the apsides of the Earth's orbit are its perihelion and aphelion; in the Moon's orbit the apsides are its perigee and apogee. The *line of apsides* is the line connecting these points.

Apus A far-southern constellation representing the bird of paradise, first shown in Johann **Bayer**'s atlas of 1603. Its brightest star is of magnitude 3.8.

▲ **Arecibo Observatory** As well as its more famous role in the SETI project, the radio telescope is a powerful radar transmitter and receiver, and is used to map small bodies such as asteroids and to examine the Earth's upper atmosphere.

Aquarids Either of two **meteor showers**, the **Delta Aquarids** and the **Eta Aquarids**.

Aquarius A constellation of the zodiac, representing a figure pouring water from a jar. Alpha (Sadalmelik, magnitude 3.0) and Beta (Sadalsuud, 2.9) are the brightest stars in the constellation. Zeta is a fine binary with a period of 856 years, and components of magnitudes 4.4 and 4.6. R Aquarii is a **symbiotic variable**, with a range from 5.8 to 12.4 and a mean period of 387 days. There are two important planetary nebulae: the **Saturn Nebula** and the **Helix Nebula**. Aquarius also contains two globular clusters, M2 and M72.

Aquila A distinctive constellation, led by first-magnitude **Altair**; most of it lies in the northern hemisphere of the sky, although the celestial equator crosses it. Mythologically it represents the eagle sent by Zeus to collect the shepherd-boy Ganymede, who was to be cup-bearer to the gods. Gamma (Tarazed), magnitude 2.7, and Beta (Alshain), 3.7, stand either side of Altair. Eta, a Cepheid variable, has a period of 7.18 days and a range from 3.5 to 4.4; it lies between Theta (3.2) and Delta (3.4), which make good comparison stars. The Mira variable R Aquilae has a range from 5.5 to 12.0 and a period of 284 days. The Milky Way runs through the constellation and is very rich in this region.

Ara A southern constellation between Scorpius and Triangulum Australe, representing the altar of the gods. The brightest stars are Alpha and Beta, both magnitude 2.8. The brightest variable in the constellation is the eclipsing binary R Arae (6.0 to 6.9; period 4.43 days, Algol type). There is also a prominent open cluster, NGC 6193, and a scattered globular, NGC 6397.

arc minute, second (symbols ′, ″) Units of angular measure, amounting to 1/60 of a degree and 1/3600 of a degree, respectively. These units are used to express the apparent diameter or separation of astronomical objects and the resolving power of a telescope. Often abbreviated to arcmin, arcsec.

Arcturus The star Alpha Boötis, the fourth-brightest in the sky, magnitude −0.05, distance 37 light years, luminosity 100 times that of the Sun. It is a giant of type K2.

Arecibo Observatory The radio and radar telescope near Arecibo in Puerto Rico, diameter 305 m (1000 ft), the largest single-bowl type of instrument in operation. It is built into a natural hollow in the limestone terrain and cannot be steered. However, an ingenious feed system enables sources as far as 20° from the zenith to be tracked. It was opened in 1963 and is operated by Cornell University. An upgrade completed in 1998 corrected the telescope's spherical aberration.

areo- Prefix referring to the planet Mars, as in *areography*, the description and mapping of the surface features of Mars. (From *Areos*, Greek for Mars.)

Argelander, Friedrich Wilhelm August (1799–1875) German astronomer. In 1852 he began the vast undertaking of preparing an atlas and catalog of all stars down to magnitude 9.5 in the northern hemisphere. This immense work, covering over 300,000 stars, was published as the ***Bonner Durchmusterung***. Argelander introduced the system of nomenclature for variable stars by which they are assigned one or two capital roman letters, beginning with R and RR, and also the subdivision of magnitudes into tenths.

Ariel One of the five main **satellites** of Uranus, diameter 1158 km (720 miles), discovered in 1851 by William Lassell. It is composed largely of a mixture of rock and water ice, with a density of 1.6 g/cm^3. Ariel has been geologically active in the past. Its characteristic features are wide, steep-sided troughs like Korrigan Chasma that wind across a landscape made up of cratered terrain, ridged terrain, and plains.

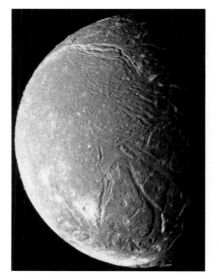

▲ **Ariel** Voyager 2 captured this image of Uranus's satellite Ariel in 1986. Ancient faults and valleys are overlain by abundant impact craters, indicating that the moon has not been geologically active for billions of years.

Tectonics accounts for the ridges, and volcanic flooding for the plains; the troughs formed when the water interior froze and the satellite expanded, splitting the crust.

Aries The first constellation of the zodiac although, since **precession** has now shifted the **vernal equinox** into Pisces, Aries has technically lost this distinction. In mythology, it represents the lamb with the golden fleece. Its brightest stars are Alpha (Hamal), magnitude 2.0, and Beta (Sheratan), magnitude 2.6.

Aristarchus of Samos (3rd century BC) Greek mathematician and astronomer. He tried to calculate the distances of the Sun and Moon from the Earth, as well as their sizes. His method was sound but the results were inaccurate. He explained the immobility of the fixed stars by their great distance from the Earth's orbit. He put the Sun at the center of the Universe, making him the first to propose a heliocentric theory, but the idea was not taken up because it did not seem to make the calculation of planetary positions any easier.

Aristotle (384–322 BC) Greek philosopher and encyclopedist. He developed a view of the Universe based on the four "elements" (earth, air, fire, and water) and the system of concentric spheres proposed by Eudoxus of Cnidus upon which various celestial bodies were carried. Aristotle added more spheres so as to account for the motion of all celestial bodies. The celestial bodies were, he said, made of a substance called aether, and are perfect and incorruptible, unlike the Earth. He attempted to estimate the size of the Earth, which he maintained was spherical, but unmoving – the center of the Universe. This "Aristotelian" view of the world remained almost unchallenged until Copernicus.

Arizona Meteor Crater SEE **Meteor Crater**.

array An arrangement of antennae used in radio astronomy. SEE **radio telescope**.

artificial satellite A man-made object moving in orbit around the Earth or some other body, in particular one that is uncrewed. The first artificial satellite, Sputnik 1, was successfully put into orbit by the USSR on October 4, 1957.

ascending node The point at which an orbit crosses from south to north of, for example, the celestial equator or the ecliptic. SEE **node**.

ashen light A faint luminosity on the night side of Venus when the planet appears as a thin crescent, close to inferior conjunction. The ashen light is occasionally reported by observers, but it has never been photographed. Its appearance is very similar to **earthshine** on the Moon, although that is produced in a different way. If it is a real phenomenon, and not illusory, it is probably the result of electrical disturbance in the planet's ionosphere,

and may therefore be similar in nature to the **airglow** in the Earth's atmosphere.

aspect The position of a planet or the Moon relative to the Sun as seen from Earth. SEE **conjunction**, **elongation**, **opposition**, **quadrature**.

association, stellar A group of stars which have formed together, but which are more loosely linked than a star cluster. Many are young and define portions of the spiral arms of our Galaxy. SEE **OB association**, **T association**.

A star A star of spectral type A, characterized by the great strength of the hydrogen absorption lines in its spectrum. On the main sequence A stars range in surface temperature from 7500 K at type A9 up to 9900 K at type A0. Their corresponding masses and radii are 1.8 and 1.4 times the Sun's respectively at A9, increasing to 3.2 and 2.5 at A0. Examples are Sirius, Vega, Altair, and Deneb. Roughly 10% of A stars have strong magnetic fields. In these, certain chemical elements become concentrated in their atmospheres and produce unusually strong spectral lines. These are classified as Ap (for peculiar) stars. Another kind of spectral anomaly produces the metallic-line (Am) stars, of which Sirius is an example.

asterism A group of easily recognizable stars, part of a constellation. Examples are the seven stars that form the shape of the Big Dipper or Plough in Ursa Major, and the "Teapot" in Sagittarius. The term is also used for distinctive groupings of stars visible in binoculars or small telescopes, some of which have been given descriptive names by observers, such as the Coathanger (in Vulpecula) and Kemble's Cascade (in Camelopardalis).

asteroid (minor planet) A small Solar System body in an independent orbit around the Sun. The majority move between the orbits of Mars and Jupiter. The largest asteroid, and the first one to be discovered, on January 1, 1801, by

Giuseppe **Piazzi**, is **Ceres**, which has a diameter of 934 km (580 miles). The smallest measured are the so-called **near-Earth asteroids**, some of which are just a few meters across. There are thought to be a million asteroids with a diameter greater than 1 km (0.6 mile); below this, they extend in size down to dust particles. Objects less than about 10 m (30 ft) in diameter are best classed as **meteoroids**. Some of the very small objects have found their way to Earth as **meteorites**.

The second asteroid to be found was **Pallas**, in 1802, followed by **Juno** in 1804 and **Vesta** in 1807. It was not until 1845 that the fifth asteroid, **Astraea**, was discovered, but from then on the pace of discovery quickened. In 1891 Max **Wolf** made the first asteroid discovery by photography, and went on to find hundreds more. Since then the introduction of a new technique has often increased the rate of discovery; the most recent example is the CCD. By early 2003 over 50,000 asteroids had been cataloged and had their orbits calculated, and this figure is now increasing by several thousand a year. Tens of thousands more have been observed, but not often enough for an accurate orbit to be calculated. Some of the larger asteroids are spherical, but most are irregularly shaped. Since the 1970s, polarization, infrared and radar techniques have allowed improved measurements to be made of asteroids' albedos, sizes, and shapes.

Most asteroids orbit within the main *asteroid belt*, which is a torus-shaped (like a ring doughnut) region of space between about 2.2 and 3.3 AU from the Sun. However, within the main belt are the **Kirkwood gaps** at distances from the Sun at which **resonances** rule out stable orbits. The typical asteroid orbit is elliptical but more eccentric than those of the major planets, and with a greater inclination to the ecliptic. Some asteroids have orbits that bring them into the inner Solar System (SEE **Amor asteroid**, **Apollo asteroid**, and **Aten asteroid**), while others, such as **Hidalgo** and **Chiron**, have orbits that lie partially or entirely in the outer Solar System.

The great majority of asteroids almost certainly originate from small **protoplanets** left over from the formation of the Solar System, prevented by Jupiter's gravita-

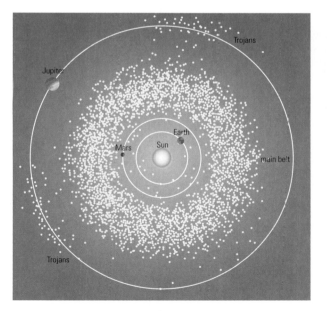

▶ **asteroid** Plots of the larger known asteroids: most of them lie in the main belt between the orbits of Mars and Jupiter, while the Trojans move in the same orbit as Jupiter, 60° in front of or behind it.

ASTEROIDS						
Name	Number	Discovery	Diameter (km)	Av. distance from Sun (million km)	Orbital period (years)	Rotation period (hours)
The first discovered						
Ceres	1	1801, Piazzi	934	413.6	4.60	9.08
Pallas	2	1802, Olbers	525	414.0	4.61	7.81
Juno	3	1804, Harding	268	399.3	4.36	7.21
Vesta	4	1807, Olbers	510	353.2	3.63	5.34
Astraea	5	1845, Hencke	120	385.2	4.13	16.81
Hebe	6	1847, Hencke	186	362.8	3.78	7.27
Iris	7	1847, Hind	200	357.0	3.69	7.14
Flora	8	1847, Hind	162	329.4	3.27	13.6
Metis	9	1848, Graham	190	357.1	3.69	5.08
Hygeia	10	1849, De Gasparis	408	469.2	5.55	27.6
Other large asteroids						
Davida	511	1903, Dugan	326	475.4	5.67	5.13
Interamnia	704	1910, Cerulli	316	458.1	5.36	8.72
Europa	52	1858, Goldschmidt	302	463.3	5.46	5.63
Sylvia	87	1866, Pogson	260	521.5	6.52	5.19
Eunomia	15	1851, De Gasparis	256	395.5	4.30	6.08
Euphrosyne	31	1854, Ferguson	256	472.1	5.58	5.53
Psyche	16	1852, De Gasparis	254	437.1	5.00	4.20
Some Trojan asteroids						
Hektor	624	1907, Kopff	225	775.3	11.80	6.92
Agamemnon	911	1919, Reinmuth	175	780.3	11.91	—
Patroclus	617	1906, Kopff	149	782.7	11.97	—
Achilles	588	1906, Max Wolf	147	774.4	11.78	—
Some Amor asteroids						
Ganymed	1036	1924, Baade	41	398.7	4.35	—
Eros	433	1898, Witt	21	218.6	1.76	5.27
Betulia	1580	1950, Johnson	6.5	328.3	3.25	6.13
Amor	1221	1932, Delporte	1	287.2	2.66	—
Some Apollo asteroids						
Toro	1685	1948, Wirtanen	12.2	204.5	1.60	10.20
Phaethon	3200	1983, Green	5.0	190.2	1.43	4
Daedalus	1864	1971, Gehrels	3.3	218.5	1.77	8.57
Apollo	1862	1932, Reinmuth	2.0	220.1	1.78	3.07
Some Aten asteroids						
Ra-Shalom	2100	1978, Helin	3.5	124.5	0.76	19.79
Aten	2062	1976, Helin	1	144.6	0.95	—
Hathor	2340	1976, Kowal	0.5	126.2	0.78	—

Diameters and rotation periods of asteroids are often uncertain, and different measurement techniques give different results. For irregular asteroids, the diameter given is an average.

tional influence from forming into a major planet. A few may be the extinct nuclei of **comets**. In the early days of the Solar System Jupiter's gravitation diverted many asteroids into planet-crossing orbits, causing a period of intense cratering of plane-tary and satellite surfaces, and ejected many other asteroids from the Solar System altogether. The larger of those remaining underwent **differentiation**, acquiring metallic cores and rocky outer layers; many of these were broken up by later collisions.

This, and the different distances from the Sun at which the original protoplanets formed, accounts for the wide range of compositional types observed today. The present asteroid population does not represent the remnants of a large planet that disintegrated, as was once thought: put together, they would make up a body only about 1500 km (950 miles) in diameter.

New discoveries are designated by the year of discovery, and a pair of letters and subscripted numbers to denote the month and order of discovery. A permanent number is given when the orbit has been accurately calculated, and the discoverer may then add a name to the number (as with, for example, 1707 Chantal), subject to the approval of the International Astronomical Union.

SEE ALSO **Centaur, Hirayama family, Kuiper Belt, Trojan asteroid**.

astigmatism An **aberration** (2) in lenses and mirrors that affects light from objects away from the center of the observer's field of view. Instead of being bought to a point focus, these oblique rays of light form an image which is either an ellipse or a straight line, depending on the distance from the lens at which it is viewed. A lens designed so as to minimize astigmatism is called an *anastigmat*.

Astraea Asteroid no. 5, diameter 120 km (75 miles), discovered by Karl Hencke in 1845.

astrobiology Another name for **exobiology**.

astrograph A telescope used to take wide-angle photographs of star fields, mainly for **astrometry**. In particular, the term is used for the instruments designed for the *Carte du Ciel* photographic survey of the sky – equatorially mounted refractors having an aperture of 330 mm and a focal length of 3.4 m, with objectives corrected for color and coma. The work of astrographs is now largely carried out with **Schmidt telescopes**, which can be made with larger apertures.

astrolabe An early astronomical instrument for showing the appearance of the celestial sphere at a given moment, and for determining the altitude of celestial bodies. The basic form consisted of two concentric disks – one with a star map and one with a scale of angles around its rim – joined and pivoted at their centers (rather like a modern **planisphere**), with a sighting device attached. Astrolabes were used from the time of the ancient Greeks down to the 17th century for navigation, measuring time, and terrestrial measurement of height and angles.

astrology A pseudo-science professing to assess people's personality traits and to predict events in their lives and future trends in general from aspects of the heavens, in particular the positions of the planets. Astrology is based on ideas which are scientifically unsound and which the great majority of rational people dismiss, but horoscopes are still published in the popular press, though mainly for their entertainment value. In many ancient cultures it was for astrological reasons that observational records first came to be kept. Such records from ancient China have proved of great historical importance in research into past eclipses, novae, and comets. Even in later times, casting horoscopes provided astronomers such as Johann **Kepler** with the livelihood to sustain them in their astronomical work.

astrometric binary A star which is recognized as a **binary star** from its wavelike **proper motion**, caused by its orbital motion with an unseen companion.

astrometry The branch of astronomy that is concerned with the measurement of precise positions of celestial objects, usually determined from photographic or CCD images obtained specially for the purpose. **Parallaxes** and **proper motions** of stars are derived from positions obtained at widely differing times. The most accurate positional measurements have been made by the astrometry satellite **Hipparcos**.

Astronomer Royal An additional title, until 1972, of the Director of the Royal Greenwich Observatory, but since then an honorary title bestowed upon a prominent British astronomer. The first, John Flamsteed, was appointed by Charles II when he founded the Greenwich Observatory.

astronomical twilight The dawn and dusk periods when the Sun is less than 18° below the horizon. SEE ALSO **twilight**.

astronomical unit (symbol AU) The mean distance between the Earth and the Sun, used as a fundamental unit of distance, particularly for distances in the Solar System. It is equal to 149,598,000 km (92,956,000 miles). There is a more complicated "official" definition, which gives a value that differs from the mean Earth–Sun distance by a few kilometers.

astrophotography The application of photography to astronomy. Photography makes it possible to obtain a permanent record of the positions and appearance of objects, while long exposures show faint objects such as remote galaxies invisible to the naked eye. In 1850 the first star images, of Vega and Capella, were obtained by W. C. Bond at Harvard College Observatory. Henry Draper obtained a spectrogram of Vega in 1872 and photographed the Orion Nebula in 1880. The 1880s marked photography's breakthrough in astronomy, as techniques improved and plates became more sensitive. The first photographic atlases of the sky were made, and libraries of stellar spectra were built up. In the 20th century astrophotography rapidly supplanted visual observation for many purposes, but has itself now been superseded by electronic detectors such as CCDs. Photography using conventional film is now practiced only by amateur astronomers, and many of those are now using CCDs to image astronomical objects.

Astrophotography differs from normal picture-taking in that exposures may be very long, often an hour or more, and the camera or telescope must accurately track the object as the Earth rotates during the exposure. The objects on an astrophotograph have an enormous range of brightness, perhaps a million to one, compared with 100 to 1 or less for objects in normal terrestrial photography. There is no "correct" exposure; the longer it is, the fainter the objects recorded, although a practical limit is imposed by the build-up of background sky fog.

Special emulsions are designed to be most efficient for long exposures. The speeds of these emulsions can be improved by between 10 and 30 times by *hypersensitizing* them before use, which involves baking them in nitrogen or in a nitrogen/hydrogen mixture. Astronomical emulsions are available in a variety of spectral sensitivities, chosen to isolate well-defined parts of the spectrum so that measurements of star colors, for example, can be compared between observatories. The astrophotographer can also choose between emulsions on the basis of contrast, resolution and granularity. Modern emulsions have low photographic speeds and very high contrast. The high contrast ensures that objects which are fainter than the night sky will be visible on the photograph, while the fine, uniform grain structure gives good resolution. Various post-processing techniques, such as *contact copying* and *unsharp masking*, can be used to enhance faint features. Images taken

ASTRONOMERS ROYAL	
Name	Held office
John **Flamsteed** (1646–1719)	1675–1719
Edmond **Halley** (1656–1742)	1720–42
James **Bradley** (1693–1762)	1742–62
Nathaniel Bliss (1700–64)	1762–64
Nevil **Maskelyne** (1732–1811)	1765–1811
John Pond (1767–1836)	1811–35
George Biddell **Airy** (1801–92)	1835–81
William Christie (1845–1923)	1881–1910
Frank Watson Dyson (1868–1939)	1910–33
Harold Spencer **Jones** (1890–1960)	1933–55
Richard Woolley (1906–86)	1956–71
Martin **Ryle** (1918–84)	1972–82
Francis Graham-Smith (1923–)	1982–90
Arnold Wolfendale (1927–)	1991–94
Martin Rees (1942–)	1995–

through color filters can be combined in the darkroom to give true-color pictures.

astrophysics The study of the physical properties of celestial bodies. It is based mainly on the study of radiation from these bodies, and has developed since the 19th century through the application of photography, photometry, and spectroscopy. New observing techniques at wavelengths from radio and infrared to X-rays and gamma-rays enormously increased the scope of astrophysics in the second half of the 20th century. Theoretical astrophysics applies mathematical methods and physical laws to explain or predict the behavior of the Universe and objects within it.

Aten asteroid One of a class of small **asteroids** whose orbits lie largely within the Earth's (specifically, those with **semi-major axis** less than 1 AU). They are named for the 1-km (0.6-mile) diameter object Aten. Nearly 200 were known by early 2003. SEE ALSO **Amor asteroid**, **Apollo asteroid**.

Atlas One of the small inner **satellites** of Saturn, discovered in 1980 during the Voyager missions. Atlas orbits just outside Saturn's A Ring, and seems to act as the **shepherd moon** to the whole bright ring system.

atmosphere The envelope of gases around a star, planet or satellite. The ability of a planet or satellite to retain an atmosphere depends on its outer temperature, determined by its distance from the Sun, and its escape velocity, determined by its mass. Smaller bodies in the inner Solar System, such as Mercury and the Moon, have almost no atmosphere or none at all, while the giant planets have deep, massive atmospheres. Far from the Sun, even small worlds like Pluto and Triton have clung on to a tenuous atmosphere. Venus, Earth, and Mars were too small to hang on to the lighter gases, such as hydrogen, from their original atmospheres, which had the composition of the solar nebula from which the Solar System formed. Subsequent geolog-

ical processes such as **volcanism** have modified their atmospheres considerably, and they now contain carbon dioxide, oxygen, and nitrogen (although in very different proportions). The giant planets have their original, largely hydrogen, atmospheres. Saturn's largest moon, **Titan**, is the only satellite with a substantial atmosphere, of nitrogen.

The outer layers of a star, which is where the lines in the star's spectrum originate, are referred to as its atmosphere.

atmospheric extinction The reduction in the brightness of light from celestial bodies when it passes through the Earth's atmosphere. The effect is more pronounced for objects at low altitude, for their light has then had to pass through a greater volume of atmosphere. Observations of, for example, a variable star have to be corrected for extinction, which can dim the star's light by half a magnitude when viewed at an altitude of 20°, and by a whole magnitude at 10°. The main cause of atmospheric extinction is the **scattering** of light by molecules in the atmosphere. As scattering affects wavelengths at the blue end of the visible spectrum more than it does the red, the light becomes reddened, as is most obvious at sunrise or sunset.

atmospheric refraction The small apparent change in altitude of a celestial object, seen by an observer at the Earth's surface, caused by its light passing through the Earth's atmosphere. Just as light is refracted when it passes from air into glass, so it is refracted when it passes from the vacuum of space into the atmosphere. Atmospheric refraction makes a body appear to be at a higher altitude than it actually is. The effect is greater the closer the body is to the horizon, and a body can even become visible before it has risen. All astronomical observations have to be corrected for atmospheric refraction in order to obtain the true position from the observed apparent position.

AU ABBREVIATION FOR **astronomical unit**.

Auriga A large northern constellation, containing the first-magnitude star **Capella**. Mythologically it represents Erichthonius, a king of Athens, who invented the four-horse chariot. Beta (Menkalinan), magnitude 1.9, is an eclipsing variable of very small range. The stars Eta and Zeta are popularly called the Haedi or Kids. **Epsilon Aurigae** is a remarkable eclipsing binary. Zeta is also an eclipsing binary, consisting of a K5 giant orbited by a B7 main-sequence star with a period of 972 days; the visual range is magnitude 3.7 to 4.0. There are three prominent open clusters: M36, M37, and M38.

aurora (plural **aurorae**) An illumination of the night sky, also known as the *aurora borealis* in the northern hemisphere and as the *aurora australis* in the southern. An aurora can take many forms, for example a horizontal arc with rays radiating upward, or a hanging curtain exhibiting beautiful, flickering colors, usually red and green. Aurorae occur at altitudes between 20 and 1000 km (12 and 600 miles), but most typically at 100 km (60 miles). The phenomenon is produced when fast-moving particles arriving in the **solar wind**, attracted by the Earth's magnetic field, collide with atoms of oxygen and nitrogen, which then emit energy at red and green wavelengths. Auroral displays are commonest at times of high solar activity, when solar **flares** emit powerful bursts of charged particles. Aurorae have been observed on all the giant planets, which have strong magnetic fields.

Australia Telescope National Facility A large **aperture synthesis** telescope in New South Wales, the most powerful radio astronomy instrument in the southern hemisphere. With three separate parts, it is an example of **very long baseline interferometry**. One consists of a 6 km line of six dish aerials, each 22 m (72 ft) in diameter, at Culgoora. These can be linked to another 22 m dish, 100 km to the south at Mopra, and also the large radio dish 200 km farther south at **Parkes Observatory**. The Australia Telescope came into operation in 1988.

autumnal equinox SEE **equinox**.

averted vision Looking slightly to one side of a faint object in order to see it better. The most sensitive part of the retina in the human eye is not right at the center, but in a small region surrounding the center. Using averted vision when observing causes the light from the object being observed to fall in this region, making it easier to see.

axis (plural **axes**) The imaginary line, joining the north and south poles of a celestial body, about which it rotates. The *magnetic axis* of a body possessing a magnetic field is the line joining the north and south poles of the field. A body's rotational and magnetic axes do not necessarily coincide; with **Uranus** and **Neptune**, for example, they are markedly different.

azimuth The angle measured westward from north in a horizontal plane to the vertical circle (the meridian) that runs through a celestial object. SEE ALSO **coordinates**.

B

Baade, (Wilhelm Heinrich) Walter (1893–1960) German–American astronomer. From Mount Wilson Observatory in the 1943 wartime blackout, he was able to observe individual stars in the Andromeda Galaxy and distinguish the younger, bluer **Population I** stars from the older, redder **Population II** stars. He went on to improve the use of **Cepheid variable** stars as distance indicators, and showed that the Universe was older and larger than had been thought.

background radiation SEE **cosmic microwave background**.

Baily's beads A very short-lived phenomenon seen during a solar eclipse, just before or after totality, in which the extreme edge of the Sun's disk appears to break into a string of bright lights. It is caused by rays of light from the Sun shining through the valleys on the Moon's limb, while other rays are blocked off by mountains on the limb. It was first described by English astronomer Francis Baily (1774–1844), who observed it in 1836. SEE ALSO **diamond-ring effect**.

Balmer series The distinctive series of lines in the **hydrogen spectrum**. The Balmer series is a dominant characteristic of the spectra of A stars. The first Balmer line is called the **hydrogen alpha line** (Hα), and is at 656.3 nm; the second line is Hβ, at 486.1 nm; the third is Hγ, at 434.2 nm; and so on. The formula governing the wavelengths of lines in the spectrum of hydrogen was worked out in 1885 by Swiss mathematics teacher Johann Balmer (1825–98).

Barlow lens An extra lens used in conjunction with a telescope's eyepiece in order to increase the magnification, usually by a factor of 2. It can also improve definition. (After English physicist Peter Barlow, 1776–1872.)

Barnard, Edward Emerson (1857–1923) American astronomer. He was a skilled visual observer, and also a pioneer of photographic astronomy. Barnard discovered many comets, the fifth satellite of Jupiter (Amalthea), and **Barnard's Star**. He speculated (correctly) that what appear to be regions containing few stars are dark nebulae, and he published a catalog of them.

Barnard's Loop A large but faint emission nebula, spanning 14°, in the form of an arc centered on the Orion Nebula. It was discovered by E. E. Barnard; it may be a supernova remnant.

Barnard's Star A red dwarf star 6 light years away in the constellation Ophiuchus. It has the largest known **proper motion**, at 10.4 arc seconds per year, and is the closest star to the Sun after the Alpha Centauri system. It was discovered in 1916 by E. E. Barnard. Barnard's Star is magnitude 9.5, with a luminosity 0.0004 times the Sun's, and type M4. Irregularities in its proper motion were once attributed to the presence of planets similar to Jupiter and Saturn around it, but are now thought to have been artifacts of instrumental error.

barred spiral galaxy A spiral **galaxy** in which the arms extend from the ends of a bar through the main plane of the system. Most galaxies – including our own – are thought to possess a bar, which plays an important role in their evolution and in triggering star formation.

Barringer Crater SEE **Meteor Crater**.

barycenter The **center of mass** of two or more celestial bodies – it is the point around which they revolve. The term is used in particular for the Earth–Moon system. As the Earth is so much more massive than the Moon, the barycenter lies within the Earth.

Bayer, Johann (1572–1625) German astronomer. In 1603 he published a star atlas, *Uranometria*, in which he introduced 12 new constellations. In it he allocated letters of the Greek alphabet (SEE the table on page 255) to the stars in each constellation: alpha (α), beta (β), gamma (γ), and so on, generally in order of brightness. This system of what came to be known as *Bayer letters* is still in use.

BD ABBREVIATION FOR *Bonner Durchmusterung*.

Beijing Astronomical Observatory
An observatory with its main optical/infrared observing site at Xinglong, about 150 km (95 miles) northeast of Beijing. It houses China's largest telescope, a 2.16-m (85-inch) reflector. The Large Sky Area Multi-Object Fiber Spectroscopic Telescope (LAMOST) is due to open in 2004.

Belinda One of the small inner **satellites** of Uranus discovered in 1986 during the Voyager 2 mission.

Bell Burnell, (Susan) Jocelyn (1943–) English astronomer. As a gradu-

ate student, she ran a survey of radio galaxies which led to the discovery of the first **pulsar**. She is now generally credited, with her then supervisor, Antony **Hewish**, as the codiscoverer of pulsars, but she received no share of the Nobel prize that was subsequently awarded.

Bennett, Comet One of the brightest comets of the 20th century, discovered by South African amateur astronomer John Caister Bennett (1914–90) in 1969, and a first-magnitude object in 1970. Its tail extended for 20°.

Bessel, Friedrich Wilhelm (1784–1846) German mathematician and astronomer. He was a pioneer of exact measurements in positional astronomy. His catalog of 63,000 star positions, based on refinements of previous observations as well as his own, marks the beginning of modern astrometry. In 1838 Bessel became the first to announce that he had determined the parallax of a star (61 Cygni). From the "wobbles" in the motions of Sirius and of Procyon he deduced the existence of their companion stars, which were not detected until many

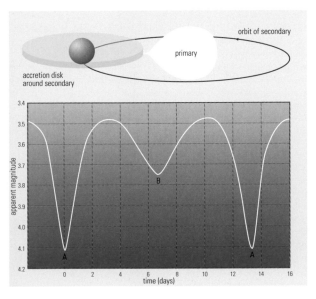

◀ **Beta Lyrae star** The continuous light curve of Beta Lyrae stars is believed to be caused by an accretion disk of material which surrounds the secondary star and eclipses the primary. This theory also explains why the secondary star appears to be dimmer than the primary, despite being far more massive.

years after his death. His name is associated with mathematical functions (*Bessel functions*) which were used in the calculation of planetary orbits.

Beta Centauri The second-brightest star in the constellation Centaurus, magnitude 0.61 (slightly variable), also known as Hadar or Agena. It is a giant of spectral type B1, and lies 525 light years away.

Beta Lyrae star A type of **eclipsing binary** in which gas escapes from a bloated primary star and falls on to an **accretion disk** that surrounds the secondary star. Beta Lyrae stars are members of a broad class of double stars called *semidetached* binaries; they are said to be semidetached because only one of the two stars fills its **Roche lobe**. In Beta Lyrae itself, the primary star is overflowing its Roche lobe and gases stream on to the disk at the rate of 0.00001 solar masses per year; some gas escapes from the system altogether. The lower half of the diagram shows the light curve of Beta Lyrae; the main minimum (A) is when the secondary star passes in front of the brighter primary, and the shallower minimum (B) is when the opposite occurs.

Beta Pictoris SEE **Pictor**.

Betelgeuse The star Alpha Orionis, and the second-brightest star in Orion. It is an orange-red supergiant of type M2 and a semiregular variable that ranges from magnitude 0 to 1.3 with a period of about 6½ years. Its luminosity is about 10,000 times that of the Sun, and it lies 427 light years away. Betelgeuse has a diameter about 500 times that of the Sun (about 700 million km/400 million miles).

Bethe, Hans Albrecht (1906–) German–American physicist. In 1938 he and Carl von **Weizsäcker** independently proposed a detailed theory to account for the production of energy in the Sun and other stars (SEE **carbon–nitrogen cycle**) in which hydrogen is converted into helium by nuclear fusion; this is essentially still

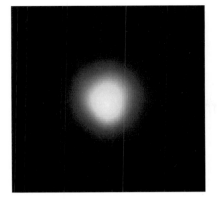

▲ **Betelgeuse** The Hubble Space Telescope's first direct image of a star other than our Sun revealed that Betelgeuse (α Orionis) has a bright spot more than 2000 K hotter than the rest of the star's surface. The exact nature of the spot is not yet understood.

accepted today. In recognition, he was awarded the 1967 Nobel Prize for Physics. SEE ALSO George **Gamow**.

Bianca One of the small inner **satellites** of Uranus discovered in 1986 during the Voyager 2 mission.

Biela's Comet A now disintegrated comet of period 6.62 years, discovered by Wilhelm von Biela (1782–1856) in 1826. At its return in 1846 it was observed to have split into two parts, the smaller of which increased in brightness. At the next return in 1852 the two portions were about 2.3 million km (3.7 million miles) apart. The comet has not been seen since, but in 1872 there was a magnificent display of the **Andromedid** meteor shower, at precisely the time when the Earth crossed the orbital path of the vanished comet. This provided dramatic proof that the particles which produce meteors are the debris of disintegrated comets.

Big Bang A theory advanced to explain the origin of the Universe. It was developed in the 1940s by George **Gamow** from the ideas of Georges **Lemaître**. According to the Big Bang theory a giant explo-

◀ **Big Bear Solar Observatory** The suite of telescopes in the dome of the Big Bear Solar Observatory observe the Sun in visible, near-infrared and ultraviolet light, thus giving astronomers detailed information about different parts of the solar surface and atmosphere. This information helps them to monitor and understand the way in which features on the Sun's surface form and evolve.

sion 10 to 20 thousand million years ago began the expansion of the Universe, which still continues.

Everything in the Universe once constituted an exceedingly hot and compressed gas with a temperature exceeding 10,000 million degrees K. When the Universe was only a few minutes old its temperature would have been 1000 million degrees K. As it cooled, nuclear reactions would have taken place that would have led to the material emerging from the fireball consisting of about 75% hydrogen and 25% helium by mass, the composition of the Universe as we observe it today. There were fluctuations from place to place in the density or expansion rate. Slightly denser regions of gas whose expansion rate lagged behind the mean value collapsed to form galaxies when the Universe was perhaps 10% of its present age.

The **cosmic microwave background** radiation detected in 1965 is considered to be the residual radiation of the Big Bang explosion. The *ripples* in this background radiation discovered by the **Cosmic Background Explorer** satellite in 1992 are evidence of the initial density variations from which galaxies formed.

We can trace the Universe back at least to an age of 10^{-5} seconds after the Big Bang, when it contained matter, antimatter, and radiation at a temperature of around 10^{12} K. In the standard Big Bang theory we cannot extrapolate back to the moment of creation itself because the the-

ory predicts the existence of a singularity, where the laws of physics break down. Suggestions for an origin have arisen from progress in the unification of the fundamental forces of physics, known as grand unified theories, or GUTs. These lead to inflationary cosmology, in which a finite region, of which the Universe is a visible part, starts off as a small, hot, expanding bubble of matter in a form that can exist only at a temperature above 10^{27} K. The bubble cools and expands rapidly as it changes into normal matter. The inflationary model would take the Universe back to 10^{-35} or 10^{-45} seconds, depending on whether gravity is included in the unification of forces, but it may be more correct to imagine that time itself begins at this point.

However, this leads back to one of the three main problems with the Big Bang theory: the so-called causality problem, of why the Universe did expand from a singularity. The other two are the problem of the missing mass, and the age problem: the Universe appears to be younger than some of the objects it contains.

Big Bear Solar Observatory A solar observatory built on an artificial island in Big Bear Lake, in the San Bernardino Mountains, California. The surrounding expanse of water ensures that there is none of the bad seeing caused by heat haze over land. The main instrument is a 0.65-m (26-inch) reflector, opened in 1973.

Big Dipper Popular name for the shape formed by the seven main stars in the constellation Ursa Major: Alpha (Dubhe), Beta (Merak), Gamma (Phekda), Delta (Megrez), Epsilon (Alioth), Zeta (**Mizar**), and Eta (Alkaid). In the UK, the stars are known as the *Plough*.

binary pulsar A **pulsar** orbiting another star, forming a binary. The first to be discovered was PSR 1913+16 in 1974; its companion is also thought to be a neutron star, but one that does not emit pulses. The orbital period is nearly 8 hours. Observations show that the pulsar's orbit is gradually contracting, evidently due to the emission of energy in the form of **gravitational waves** as predicted by Einstein's General Theory of Relativity. American radio astronomers Joseph Taylor and Russell Hulse were awarded the 1993 Nobel Prize for Physics for the discovery of PSR 1913+16.

binary star Two stars in orbit around a common **center of mass**. They are usually classified as visual, spectroscopic or eclipsing binaries, according to the means by which they are observed. Their orbital periods vary enormously, from hours to years for spectroscopic and eclipsing binaries, and from a decade to many centuries for visual binaries. A star with more than two components is called a *multiple star*, **Castor** being an example.

In a *visual binary*, of which about 50,000 are known, the components are sufficiently far apart, with an angular separation greater than about 0.1 arc seconds, for the two to be seen separately through a telescope. The first visual binary discovered was **Mizar** (Zeta Ursae Majoris) in 1650 by Giovanni Battista Riccioli, who saw it had a close companion (as distinct from its wider companion, Alcor).

In an **eclipsing binary**, one star periodically passes in front of the other, so that the total light output appears to fluctuate. Most eclipsing binaries are also spectroscopic binaries. **Algol** (Beta Persei) was the first eclipsing binary to be identified, by John **Goodricke** in 1782.

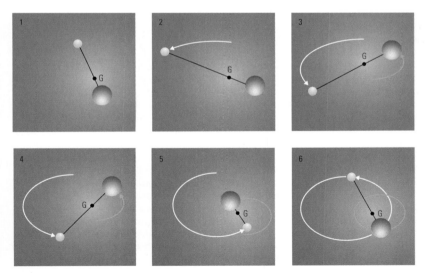

▲ **binary star** The two components of a binary system move in elliptical orbits around their common center of gravity (G), which is not half way between them but nearer to the more massive component. In accordance with Kepler's second law of motion, they do not travel at uniform rates. In (2) they are greatly separated and moving slowly. In (6) they are closer and moving more rapidly.

A **spectroscopic binary** is a system whose components are too close for their separation to be measured visually. The binary nature is revealed by a Doppler shift in the absorption lines of the stars' spectra as they move around their orbits. The brighter component of Mizar was the first star shown to be a spectroscopic binary, by Edward **Pickering** in 1889.

The study of irregularities in the **proper motions** of stars has in some cases revealed the existence of invisible companions; such a star is known as an *astrometric binary*.

binoculars Two small telescopes joined by a hinge, giving a low-magnification view. The hinge allows the binoculars to be adjusted for the distance between the observer's eyes, and there is usually a means of focusing the eyepieces independently. Compactness is achieved by using prisms to fold the light-paths within the instrument. Binoculars have designations such as 8 × 40, where 8 is the magnification and 40 is the diameter of each object lens in millimeters.

bipolar flow The flow of material from a star in two streams in opposite directions. A star that generates bipolar flow is usually surrounded by a dusty envelope that is oriented perpendicularly to its rotation axis. The interplay of stellar rotation and mass outflow is thought to create bipolar flows.

black-body radiation The radiation that would be emitted by a *black body*, a hypothetical body which absorbs all thermal radiation (heat) falling upon it, and is a perfect emitter of thermal radiation. The spectrum of the radiation emitted by a black body is a **continuous spectrum**, and the intensity of the radiation is greatest at a wavelength which depends only on the body's temperature. Although the black body is only a theoretical object, its importance in astronomy is that the spectra of stars can be interpreted by assuming that they approximate to black bodies. SEE ALSO **effective temperature**.

black drop An optical effect, visible at **second contact** or **third contact** during a transit of Mercury or Venus across the Sun's disk, in which a dark ligament – the "drop" – appears to connect the limb of the planet with the limb of the Sun. The phenomenon has to be allowed for in the timing of transits.

Black Eye Galaxy A spiral galaxy (M64, NGC 4826) in the constellation Coma Berenices that contains a dark cloud of dust near the nucleus, giving the "black eye" effect. The Black Eye Galaxy lies 24 million light years away.

black hole A localized region of space from which neither matter nor radiation can escape – in other words, the escape velocity exceeds the velocity of light. The boundary of this region is called the *event horizon*. Its radius, the **Schwarzschild radius**, depends on the amount of matter that has fallen into the region and it increases linearly as the mass increases.

A black hole of stellar mass is thought to form when a massive star undergoes total gravitational collapse. For stars up to about 1.4 solar masses, gravitational collapse can be physically halted to produce a **white dwarf**. A slightly more massive object will collapse to form a **neutron star**. If, however, the mass exceeds about 3 solar masses, even after a supernova explosion has blasted away the outer layers of the star, the collapse continues beyond even the neutron star stage. As it contracts below its Schwarzschild radius, the object becomes a black hole and effectively disappears. A star of 3 solar masses has a Schwarzschild radius of about 9 km. Inside the event horizon of the black hole, space and time are highly distorted and the stellar matter is increasingly compressed until it forms an infinitely dense **singularity**.

Black holes can have an immense range of sizes. *Supermassive black holes* with up to a billion solar masses are believed to be the source of energy in **quasars** and other **active galactic nuclei**. At the other end of the scale, some primordial black holes (SEE below) could be truly microscopic.

Since no light or other radiation can escape from black holes, they are extremely difficult to detect. Any matter encountered by the black hole will most likely go into orbit first rather than being drawn directly into it. A rapidly spinning disk of matter, known as an **accretion disk**, forms around the object and heats up through friction to such high temperatures that it emits X-rays. Black holes may therefore appear as X-ray sources in binary stars. The most famous candidate is **Cygnus X-1**. Other promising candidates include A0620-00 in Monoceros, the recurrent nova V404 Cygni, and LMC-3 in the Large Magellanic Cloud.

Not all black holes result from stellar collapse. During the Big Bang, some regions of space might have become so compressed that they formed so-called *primordial black holes*. Such black holes would not be completely black, because radiation could still "tunnel out" of the event horizon at steady rate, leading to the evaporation of the hole. (Such small-scale effects are not important for larger black holes.) Primordial black holes could thus be very hot.

blazar A term compounded from **BL Lacertae object** and **quasar**, referring to galaxies with exceptionally **active galactic nuclei**. Blazars show variable optical brightness, strong and variable optical polarization, and strong radio emission. The optical variations can be on timescales as short as days. Much of the activity is related to jets of gas expelled from their central regions at relativistic velocities. One explanation of the exceptional activity in blazars is that in these galaxies we are viewing the jets end-on.

blink microscope SEE **comparator**.

BL Lacertae object (BL Lac object) A highly luminous **active galactic nucleus**, the prototype of which, BL Lacertae, was originally classified as a variable star. BL Lacertae objects are variable at all wavelengths from radio to X-rays, sometimes over a timescale of just a few hours, and their optical spectra are unusual in being completely featureless, containing neither absorption nor emission lines. This is consistent with them being jets from **quasars** or **radio galaxies** emitted along our line of sight.

blue moon In everyday speech, something rare – as in the phrase "once in a blue moon." On rare occasions the Moon does appear blue, when, for example, a large volcanic eruption or an extensive forest fire injects many particles into the atmosphere. The particles scatter light, making it appear bluer. The term has also come to mean a second full Moon in the same calendar month, which happens about seven times in every 19 years.

blue straggler A bright, blue star in a globular cluster that remains on the **main sequence** long after stars of similar or lower mass have evolved into red giants. It is thought that they "straggle behind" because they are close binaries in which mass has been transferred from one to the other, or, in some cases, stars that have merged together.

Bode, Johann Elert (1747–1826) German astronomer. He published the theoretical relationship between the distances of the then known planets previously pointed out by Johann Titius (1729–96). On the basis of this relationship (SEE **Bode's law**), Bode suggested the existence of an undiscovered planet between Mars and Jupiter. He published a catalog of stars and nebulae.

Bode's law More properly called the *Titius–Bode law* (SEE **Bode**, Johann), a numerical sequence announced by Bode in 1772 which matches the distances from the Sun of the then known planets. It is formed by adding 4 to each number in the sequence 0, 3, 6, 12, 24, 48, 96, and 192, giving 4, 7, 10, 16, 28, 52, 100, and 196. Taking the Earth's distance from the Sun as 10, Mercury, Venus, and Mars fell into place reasonably well (at 3.9, 7.2 and 15.2), as did Jupiter and Saturn (52 and 95). The discovery in 1781 of Uranus at a distance corresponding to 192 prompted a search for a major planet to fill the gap at 28, but led instead to the discovery of the asteroids. The next number in the sequence is 388, but although Neptune (301) does not fit the bill, Pluto (395) does.

Although, in the form in which it was originally proposed, Bode's law has no real basis, it does give an indication of the way in which **resonances** and tidal interactions (SEE **tides**) between bodies in the Solar System can result in them having orbits which are **commensurable**.

Bok, Bartholomeus Jan ("Bart")

(1906–83) Dutch–American astronomer. He made extensive studies of the Milky Way, particularly of its star-forming regions (SEE **Bok globule**). He showed that stellar **associations** consist of young stars.

Bok globule A round, dark cloud of gas and dust which is a likely precursor of a **protostar**; named for Bart **Bok**. Small globules have diameters of 0.1 light year (6000 AU) and can be seen in front of bright nebulae. Larger globules, up to 3 light years across, are seen as dark patches against the stellar background of the Milky Way. Estimated globule masses range from about 0.1 solar masses for the

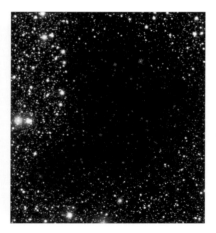

▲ **Bok globule** The dark cloud Barnard 68 (B68), shown here in an image obtained by the Very Large Telescope, is a Bok globule. It is 410 light years away.

smallest to about 2000 solar masses for those in the Rho Ophiuchi dark cloud. Globules undergo gravitational contraction to form firstly protostars, and eventually normal main-sequence stars.

bolide An exceptionally bright meteor – a major **fireball** – that is accompanied by a sonic boom. Objects whose passage through the atmosphere produce bolides may survive to reach the surface as meteorites.

bolometer An instrument for measuring all the electromagnetic radiation from a source, used in astronomy to measure the total radiation from a star. There are many types of bolometer, but nearly all work by registering the change in electrical resistance that results when radiation from a star falls on a thin conductor.

bolometric magnitude (symbol m_{bol}) A theoretical figure which measures the total amount of energy radiated by a body at all wavelengths. It can be calculated from the visual magnitude.

Bondi, Hermann (1919–) British cosmologist and mathematician, born in Austria. With Thomas **Gold** and Fred **Hoyle** he developed the **steady-state theory**, in which the continuous creation of matter drives the expansion of the Universe. He also studied the constitution of stars and general relativity.

Bonner Durchmusterung (BD) A catalog containing data on over 300,000 stars, published in 1859–62 by F. W. A. **Argelander** and extended in 1886 to over 450,000 stars by Eduard Schönfeld. The stars are numbered in declination zones from +90° to −23°, and are cited in the form "BD +52° 1638."

Boötes A prominent northern constellation containing **Arcturus**, the brightest star north of the celestial equator. In mythology Boötes represents a herdsman. Epsilon (Izar or Pulcherrima) is a beautiful double star, with K0 and A0 compo-

▶ **Brahe, Tycho** In the Tychonian model of the Solar System, there is a static Earth at the center. The Moon and Sun orbit the Earth, and the other planets move around the Sun.

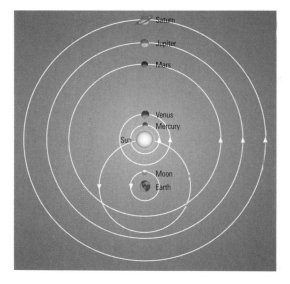

nents, magnitudes 2.6 and 4.8, separation 2.9 arc seconds. Mu (Alkalurops) is a wide double, magnitudes 4.3 and 6.5.

bow shock The boundary around a planet within which the planet's **magnetosphere** predominates over the surrounding **solar wind**. The stream of particles making up the solar wind is deflected around the bow shock, rather like the effect set up in front of a ship's bow as it moves through the water. SEE ALSO **heliosphere**.

Bradley, James (1693–1762) English astronomer. In 1742 he succeeded Edmond Halley as Astronomer Royal. He discovered the **aberration** (1) of light – the first observational evidence for Copernicus's heliocentric theory – and the **nutation** of the Earth's axis. His catalog of 3000 stars was published posthumously, and formed the basis of Friedrich **Bessel's** catalog.

Brahe, Tycho (1546–1601) Danish astronomer. In 1572 he observed the supernova in Cassiopeia, and his report of it, *De nova stella*, soon made him famous. Under the patronage of King Frederick II of Denmark he built and equipped two observatories – Uraniborg and Stjerneborg – on the island of Hven in the Baltic. He became the most skilled observer of the pretelescope era, expert in making accurate naked-eye measurements of the positions of stars and planets. He calculated the orbit of the comet seen in 1577, and this, together with his study of the supernova, showed that **Aristotle**'s picture of the unchanging heavens was wrong. He could not, however, accept the world system put forward by **Copernicus**. In his own planetary theory (the *Tychonian system*), the planets move round the Sun, and the Sun itself, like the Moon, moves round the stationary Earth. In 1597 he left Denmark and settled in Prague, where Johann **Kepler** became his assistant. The accurate series of planetary observations made by Tycho Brahe were used by Kepler in deriving his laws of planetary motion.

breccia A type of rock made up of coarse, angular fragments set in a finer-grained material. Impacts produced the fragments, which were later incorporated in a younger rock formation. Most of the lunar rock samples returned by the Apollo missions were breccias, and the surface rocks of many other cratered, geologically inactive bodies in the Solar System are expected to be made up largely of breccias.

bremsstrahlung (German, meaning "brake radiation") Electromagnetic radiation produced when energetic electrons are decelerated, for example when approaching the nucleus of an atom. It is generated in ionized gas clouds.

Bright Star Catalogue A catalog of naked-eye stars (above magnitude 6.5) first published by Yale University in 1930. The current (5th) edition is in electronic form.

brown dwarf An object with mass less than 0.08 solar masses, in which the core temperature does not rise high enough to initiate thermonuclear reactions. Such a *substellar object* is, however, luminous, for as it slowly shrinks in size it radiates away its gravitational energy. The surface temperature of a brown dwarf is below the 2500 K lower limit of a **red dwarf**. Over a hundred had been found by 2003, both singly and in orbit around larger stars. One of the coolest at 900 K is the first brown dwarf to be discovered, designated Gliese 229B. SEE ALSO **L star**.

B star A star of spectral type B, characterized by strong neutral helium absorp-

tion lines and the presence of hydrogen lines (although not as strong as in the A stars). On the **main sequence**, the temperatures of B stars range from 10,500 K at B9 to 31,000 K at B0. Their masses and sizes range correspondingly from 2.3 to 14 solar masses, and from 2.5 to 10 solar radii. At their hottest, B stars have 25,000 times the luminosity of the Sun. Examples of B stars are Achernar, Regulus, Rigel, and Spica. A B9 star stays on the main sequence for about 800 million years, but a B0 star remains there for only 11 million years. The hottest B stars (and the even hotter **O stars**) form loose groupings known as **OB associations**. O and B stars evolve to become supergiants with luminosities up to 100,000 times that of the Sun.

butterfly diagram A diagram illustrating the changing distribution of sunspots in solar latitude over the course of the 11-year cycle. At the start of a cycle there are spots at latitudes of up to around 30° north and south, but very few near the equator. As the cycle progresses, spots occur nearer the equator. When all the spots in a cycle are plotted on the diagram, a characteristic pattern in the shape of a butterfly appears. The diagram was devised by Edward Maunder in 1904. SEE ALSO **Maunder minimum**.

▲ **butterfly diagram** During the solar cycle, the latitude at which sunspots form changes. This is related to changes in the Sun's magnetic field during its cycle. Cycles usually overlap, with the first spots of a new cycle forming concurrently with the last spots of the old cycle.

C

Caelum A constellation of the southern sky, adjoining Columba, introduced by **Lacaille**, representing an engraving tool. It has no star above magnitude 4.4. One variable, R Caeli, of Mira type, ranges from magnitude 6.7 to 13.7 with a period of 391 days.

Calar Alto Observatory A German–Spanish observatory near Almeria in southern Spain. Its main telescope is a 3.5-m (138-inch) reflector, opened in 1984.

calendar A list of the days of the year, grouped into months, and based on the apparent motions of the Sun or Moon, or both. The ancient Egyptians used a calendar based on a solar year. The Babylonians (and the modern Hebrews and Muslims) used a lunar year of 12 months, which is 11 days shorter than a solar year, so an extra month is added every third year.

Our present calendar is based on that of the Romans, which originally had only 10 months. In 45 BC Julius Caesar introduced a revised calendar, because the Roman calendar had become badly out of phase with the seasons. In this revised calendar, known as the *Julian calendar*, the year was divided into 12 months. Every fourth year an extra day (leap day) was added to the usual year of 365 days. But the average length of the Julian calendar year, 365.25 days, was longer than a **tropical year** (the interval between two successive passages of the Sun through the vernal equinox) by over 11 minutes. The Julian calendar year thus slowly drifted out of phase with the seasons, by over 18 hours a century.

By the 16th century there was a discrepancy of 10 days in the date of the vernal equinox given by the calendar. In 1582 Pope Gregory XIII introduced a revised calendar, called the *Gregorian calendar*. This revision meant dropping 10 days from the calendar, so October 5, 1582, became October 15. The average length of the calendar year was reduced by inserting a leap day every four years, except in century years which are not divisible by 400. Thus the years 1600 and 2000 are leap years, but 1700, 1800, and 1900 are not. The Gregorian calendar was adopted in Great Britain and its territories in 1752, when September 2 was followed by September 14. The Gregorian calendar year of 365.2425 days is accurate to one day in 3300 years.

In about AD 525 Dionysius Exiguus, a Scythian scholar, introduced the method of reckoning the years by reference to the Christian era. He fixed the beginning of the present calendar so that Christ was born in December AD 1. However, it is believed that he made a slight error in his calculations, so that the birth of Christ actually occurred a few years before this. The year that precedes AD 1 is 1 BC; there is no year 0. SEE ALSO **Julian Day**.

Caliban One of two outer **satellites** of Uranus in a retrograde orbit discovered by Brett J. Gladman and Philip D. Nicholson in 1997.

Callisto The second-largest of Jupiter's **Galilean satellites**, with a diameter of 4800 km (2980 miles), and the outermost of them. It has the lowest average density, at 1.86 g/cm^3, indicating that it contains a high proportion of water ice. The surface of the satellite is dark and very heavily cratered – in fact it is the most densely cratered object known. As well as the dense craters, there are several large, multiringed **impact features**, the largest of which is Valhalla, with an overall diameter of 4000 km (2500 miles). The **Galileo** probe found evidence for a 10-km (6-mile) deep subsurface ocean 100 km (60 miles) down, and detected a weak magnetic field and a thin carbon dioxide atmosphere. SEE ALSO **satellite**.

Caloris Basin The largest **impact feature** on Mercury. About half of it was photographed by the Mariner 10 probe in 1974–5. It is 1300 km (800 miles) in diameter and, like a number of similar features elsewhere in the Solar System, has a multiringed structure.

Calypso A small **satellite** of Saturn, discovered by Bradford Smith and others in 1980 between the two Voyager encounters. Calypso is irregular in shape, measuring 30 × 16 km (19 × 10 miles). It is coorbital with Tethys and Telesto; Calypso and Telesto orbit near the L_5 and L_4 **Lagrangian points**, respectively, of Tethys's orbit around Saturn (SEE **coorbital satellite**).

Camelopardalis A barren northern constellation, representing a giraffe. Its brightest star, Beta, is of magnitude 4.0 and has a wide magnitude 8.6 companion. NGC 1502 is an open cluster.

canals A network of dark linear markings on the surface of Mars, reported to exist by observers from the 1870s until well into the 20th century. They were first described by Angelo **Secchi** and Giovanni **Schiaparelli** as *canali*, the Italian word for "channels." In some English reports *canali* was mistranslated as "canals" – with the connotation that they were artificial constructions. This idea was seized upon by, in particular, Percival **Lowell**, who was convinced that the canals were artificial waterways built by intelligent beings to irrigate a dying planet. Despite the demonstration by Eugène **Antoniadi** that the canals were illusory, this romantic notion of Mars persisted in many minds until spacecraft visited Mars and sent back pictures that showed no sign of a canal network.

Cancer A dim and largely barren zodiacal constellation, between Gemini and Leo. In legend it represents the crab sent by Juno, queen of the gods, and trodden on by Hercules during his fight with the multiheaded Hydra. Beta, magnitude 3.5, is its brightest star. Zeta is a multiple system; the main components are of magnitudes 5.0 and 6.2, and the brighter star is again a close double. R Cancri is a Mira variable, range 6.1 to 11.8, period 362 days. RS Cancri is a semiregular variable (5.1 to 7.0, period about 120 days) and so is X Cancri (5.6 to 7.5, period about 195 days). The large open cluster **Praesepe**, M44, is flanked by Gamma, 4.7, and Delta, 3.9, known as the Asses. Also in Cancer is M67, a very old open cluster visible in binoculars.

Canes Venatici A constellation introduced by **Hevelius** on his map of 1690, representing the dogs Asterion and Chara, held by the herdsman Boötes. The only star above fourth magnitude is Cor Caroli ("Charles's Heart"), magnitude 2.9. This name was given to it by Edmond Halley, in honor of King Charles I of England. The star is the prototype of the class of *magnetic variable* stars whose spectra contain lines that vary in intensity. It is a double, with a companion of magnitude 5.6, separation 20 arc seconds, excellent for small telescopes. Y is a semiregular variable, ranging in magnitude from 5.0 to 6.5 in a period of approximately 160 days. The famous **Whirlpool Galaxy**, M51, lies in Canes Venatici. Other spiral galaxies are M63, M94, and M106. There is also the 6th-magnitude globular cluster M3, which is distinct in binoculars.

Canis Major A constellation representing one of Orion's hunting dogs. It is distinguished by the presence of **Sirius**, the brightest star in the sky. Other bright stars are Epsilon (Adhara), magnitude 1.4; Delta (Wezen), 1.8; and Beta (Mirzam), 2.0. Beta is variable over a very small range and is a prototype of the class of variables known either as *Beta Canis Majoris stars* or *Beta Cephei stars*. There are three fairly bright variables: R (5.7 to 6.3, 1.14 days, Algol type), W (6.4 to 7.9, irregular), and UW (4.8 to 5.3, 4.39 days, Beta Lyrae type). There is one prominent open cluster, M41, of 5th magnitude, south of Sirius.

Canis Minor A constellation representing the smaller of Orion's two dogs. The first-magnitude star **Procyon** makes it easy to locate. Beta (Gomeisa), magnitude 2.9, is a slightly variable **shell star**. There are no other objects of note, although three Mira variables (V, R, and S) can rise to above magnitude 8 at maximum.

cannibalism SEE **cluster of galaxies**.

Cannon, Annie Jump (1863–1941) American astronomer. Working at the Harvard College Observatory, initially as an assistant to E. C. **Pickering**, she reorganized the classification of stars into spectral types, and classified the 225,000 stars in the *Henry Draper Catalogue* of stellar spectra.

Canopus The star Alpha Carinae, the second-brightest star in the sky, magnitude −0.62. It is a bright giant of type F0, more than 10,000 times as luminous as the Sun, and lies 313 light years away.

Capella The star Alpha Aurigae, magnitude 0.08 (the sixth-brightest in the sky), distance 42 light years. It is in fact a spectroscopic binary, consisting of two giant stars of types G6 and G2, with an orbital period of 104 days.

Capricornus One of the less prominent of the zodiacal constellations; it has been identified with the Greek god Pan. Its brightest star is Delta (Deneb Algedi), magnitude 2.9. Alpha (Algedi) is a wide optical double, magnitudes 3.6 and 4.2. The semiregular variable RT has a range from 6th to 9th magnitude and a period of 393 days. The most important cluster is M30, a globular of magnitude 7.5.

captured rotation SEE **synchronous rotation**.

carbonaceous chondrite A rare type of chondritic meteorite (SEE **chondrite**) with a higher-than-average content of carbon, in the form of organic compounds. Not counting the **volatiles**, the proportions of chemical elements in a carbonaceous chondrite are very similar to the composition of the Sun. These meteorites are therefore believed to have come from bodies that formed very early in the Solar System's history, which makes them important in the study of its origin.

carbon–nitrogen cycle A sequence of nuclear reactions which accounts for the production of energy inside main-sequence stars that are more massive and hence hotter than the Sun. It was first described in 1938 by Hans **Bethe**. The reactions involve the fusion of four hydrogen nuclei into one helium nucleus with the release of a huge amount of energy. Nitrogen and oxygen are formed as intermediate products. The presence of carbon is essential, but it behaves as a catalyst and remains unchanged at the end of the cycle. The process takes place at temperatures above 15 million degrees K. It is also known as the *carbon cycle* or the *carbon–nitrogen–oxygen cycle*.

carbon star (C star) A cool red giant star whose surface composition shows carbon-based molecules such as carbon monoxide (CO), cyanogen (CN), molecular carbon (C_2), and other compounds of carbon, and also the presence of much lithium. A star of about the Sun's mass in the red giant phase has carbon, produced by nuclear fusion at its core, brought to the surface by convection currents. The carbon reacts to form the molecules visible in its spectrum. Carbon stars are allocated spectral type C in the Morgan–Keenan system of **spectral classification**.

Carina Part of the dismembered constellation Argo Navis, the ship Argo. It is the brightest and richest part of Argo, representing the ship's keel, and contains **Canopus**, the second-brightest star in the sky. Apart from Canopus, its brightest stars are Beta (Miaplacidus), magnitude 1.7, and Epsilon (Avior), magnitude 1.9. Epsilon and Iota (2.2) form part of the **False Cross**. Upsilon is a double, 3.0 and 6.3. Carina's variable stars include the unique **Eta Carinae**, which lies within the nebula NGC 3372. Other variables with maximum above magnitude 6.5 are the

Cepheids U (range 5.7 to 7.0, period 38.77 days) and l ("ell," 3.3 to 4.2, 35.54 days), as well as the Mira-type stars R (3.9 to 10.5, 309 days) and S (4.5 to 9.9, 150 days). There are several open clusters visible to the naked eye: NGC 2516, NGC 3114, and IC 2581. The open cluster IC 2602 around Theta is spectacular when viewed with binoculars. There is also a 6th-magnitude globular cluster, NGC 2808.

Carme One of Jupiter's four small outermost **satellites**, discovered in 1938 by Seth Nicholson. It is in a retrograde – SEE **retrograde motion** (1) – orbit, and may well be a captured asteroid.

Cassegrain telescope A **reflecting telescope** in which the light received by the primary mirror is reflected back from the secondary mirror through a hole in the primary. The light is brought to a focus – the *Cassegrain focus* – on the axis of the telescope, where the image can be observed by the eye or by instruments. The combination of a concave primary (in the form of a paraboloid) and a convex secondary (usually a hyperboloid) tends to cancel **aberrations** (2), such as coma, of the sep-

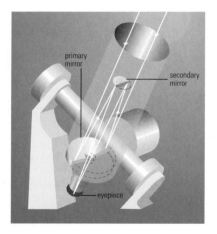

▲ **Cassegrain telescope** Light collected by the primary mirror is reflected from a hyperboloidal secondary mirror to the eyepiece through a hole in the center of the primary.

arate mirrors. The arrangement is used for both amateur and professional telescopes. It was devised in 1672 by the French priest Laurent Cassegrain (1629–93), whose identity was not established until the late 1990s. SEE ALSO **Schmidt–Cassegrain telescope**.

Cassini French family of Italian origin that produced four generations of astronomers and cartographers, all of whom ran the Paris Observatory.

Cassini, Giovanni Domenico, later also known as **Jean Dominique** (1625–1712), moved to Paris in 1669. He was the first to accurately measure the dimensions of the Solar System. He discovered the division in Saturn's rings that now bears his name, and also four of that planet's satellites. He measured Jupiter's rotation period, and also improved the tables of Jupiter's satellites, which helped Ole Römer in his determination of the velocity of light. The two most prominent were:

Cassini, Jacques (1677–1756), son of Giovanni. He accurately measured an arc of a meridian running through France. The results helped to show that the Earth is flattened at the poles, although he and his father had believed there to be a polar elongation. He determined the proper motion of the star Arcturus.

Cassini A joint space mission between NASA and the European Space Agency to explore Saturn and its moon Titan. Cassini was launched in November 1997 and is due to reach Saturn in November 2004. Cassini will go into orbit around Saturn, while the Huygens sub-probe will land on Titan.

Cassini Division The gap between Rings A and B in the ring system of **Saturn**, discovered by Giovanni Cassini in 1675. The Voyager spacecraft showed that it is not a true gap, but contains several narrow rings.

Cassiopeia A distinctive northern constellation, representing the mother of Andromeda. The five leading stars make up a W or M pattern, on the opposite side of the Pole Star from Ursa Major. Alpha (Schedar) is its brightest star, magnitude 2.2. Gamma, magnitude range 1.6 to 3.0, is the prototype of the variables known as **shell stars**. Rho is a highly luminous semi-regular variable, usually of around magnitude 5 but which on rare occasions falls to below 6. Other variables are R (4.7 to 13.5, 430 days, Mira type), RZ (6.2 to 7.7, 1.2 days, Algol type), and SU (5.7 to 6.2, 1.95 days, Cepheid). Eta is a fine colored binary, magnitudes 3.5 and 7.5; the orbital period is just under 500 years. Cassiopeia contains several open clusters resolvable in binoculars: M52, NGC 457, NGC 559, NGC 663 and M103.

Cassiopeia A The strongest radio source in the sky apart from the Sun, and the remnant of a supernova that exploded around 1660 but was not recorded on Earth. It lies about 10,000 light years away and appears optically as a faint nebula.

Castor The star Alpha Geminorum, magnitude 1.58, distance 52 light years. Castor is one of the finest visual binaries in the sky for small telescopes, consisting of main-sequence components of magnitudes 2.0 and 2.9, types A2 and A5, period 470 years. Each of the stars is itself a spectroscopic binary. A distant companion, Castor C, orbits the system. It is an eclipsing binary also known as YY Geminorum, made up of two faint red dwarfs, range 9.2 to 9.6.

cataclysmic variables (eruptive variables) A term used for a diverse group of stars that erupt in brightness, including **supernovae**, **novae**, **dwarf novae**, **flare stars**, **shell stars**, and some X-ray objects. With the exception of supernovae, these objects are very close binaries whose outbursts are caused by interaction between the two components. A typical system of this type has a low-mass secondary which fills its **Roche lobe** so that material is transferred on to the primary, which is usually a white dwarf. The transferred material has too much angular

▲ **Cassini** On its way to Saturn, the Cassini spacecraft passed through the Jupiter system. This image of Jupiter's Great Red Spot and its environs, together with Io and its shadow on the cloud-tops was obtained on December 12, 2000, from a distance of ¯9.5 million km (12.1 million miles).

momentum to fall directly on to the primary, and so forms an **accretion disk**. A hot spot arises on the outer edge of the accretion disk where the infalling material hits it. Outbursts occur when the flow rate varies, which can be at irregular intervals from weeks to many years.

catadioptric A type of telescope in which a lens–mirror combination forms the image. The **Schmidt–Cassegrain telescope** is the most common type of catadioptric design for amateur use; another is the **Maksutov telescope**. A catadioptric telescope is basically a reflector with a full-aperture lens at its top end. The lens corrects for **aberration** (2), and produces a divergent light-beam which is reflected on to the secondary mirror by a short-focus spherical or paraboloidal primary mirror. The design is compact and often used for portable instruments.

CCD Abbreviation for *charge-coupled device*, a solid-state electronic imaging device. A CCD is an integrated circuit whose upper layer contains a square array of closely spaced electrodes. A ray of light from a celestial object falling on an electrode is converted into an electronic charge, and the charges from all the elec-

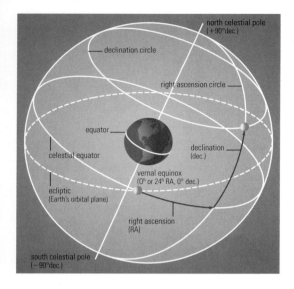

◀ **celestial sphere** A useful concept for describing positions of astronomical bodies is the celestial sphere. As shown, key points of reference – the celestial pole and celestial equator – are projections on to the sphere of their terrestrial equivalents. An object's position can be defined in terms of right ascension and declination (equivalent, respectively, to longitude and latitude).

trodes are "coupled" together and read off as a pattern – an instantaneous image of the object. A series of these images are fed to a computer and combined to create a picture, just as a photographic emulsion builds up a picture over the course of its exposure.

Pictures requiring hours of exposure on photographic emulsion can be recorded in minutes by using a CCD. Fainter objects can be detected, the effects of **light pollution** can be compensated for, and image processing by computer can be used to bring out detail. In professional astronomy CCDs have all but replaced traditional astrophotographic techniques, and they are becoming increasingly cheaper and readily accessible to amateur astronomers.

celestial equator The projection of the Earth's equator on to the celestial sphere, dividing the sky into the northern and southern hemispheres.

celestial latitude, longitude SEE **coordinates**.

celestial mechanics The branch of astronomy dealing with the motions of heavenly bodies. It uses the laws of physics to explain and predict the orbits of planets, satellites and other celestial bodies.

celestial meridian The great circle passing through the observer's zenith, the nadir, and the celestial poles, cutting the horizon at the north and south points.

celestial poles SEE **celestial sphere, pole**.

celestial sphere The imaginary sphere, of infinite radius and with the Earth at its center, upon which all the celestial bodies appear to be projected as though they were at a uniform distance. The most obvious behavior of the celestial sphere is its apparent daily east-to-west rotation, due to the axial spin of the Earth. The celestial sphere appears to rotate about the *celestial poles*, as though on an extension of the Earth's own axis, which points toward the celestial poles. Although the positions of the two celestial poles seem to be fixed, they are in fact slowly drifting because of the effect of **precession**.

The celestial sphere is divided into two halves by the *celestial equator*, which is equidistant from the poles. We can thus visualize a set of **great circles** similar to circles of latitude on Earth, called *declination circles*, each parallel to the equator. Declination circles allow us to specify the position of an object in terms of its angular distance north or south of the celestial equator.

Right ascension, the celestial equivalent of longitude, is also an extension of the terrestrial system. The zero of right ascension is chosen using the apparent motion of the Sun around the sky during the year, caused by the Earth's orbital motion around the Sun. The track traced by the Sun on the celestial sphere, called the **ecliptic**, indicates the plane of the Earth's orbit. The ecliptic is not the same as the celestial equator because the Earth's axis is tilted by about $23\frac{1}{2}°$.

For half of each year the Sun is in the northern hemisphere of the sky, and for the other half it is in the southern hemisphere. On two dates each year the Sun crosses the celestial equator; these points are known the **equinoxes**. It is the *spring equinox*, when the Sun crosses into the northern hemisphere, that is chosen as the zero of right ascension. Right ascension is measured eastward from this point but, unlike terrestrial longitude, it is measured in units of time rather than degrees, minutes and seconds of arc (SEE **sidereal time**).

Centaur One of a class of small objects orbiting the Sun between Jupiter and Neptune. The first to be discovered was **Chiron**. Roughly 50 were known by early 2003; they are typically 100 km (50 miles) in diameter. The Centaurs are probably icy **planetesimals** that have been perturbed inward from the Kuiper Belt and now occupy short-lived, unstable orbits.

Centaurus A brilliant southern constellation representing a centaur, with 13 stars above magnitude 3.5. The brightest is **Alpha Centauri**. Beta Centauri (Hadar or Agena) is magnitude 0.61, a giant of type B1. Alpha and Beta indicate the direction of the Southern Cross. Gamma is a close binary with an 85-year period; the components are equal, at magnitude 2.9, and their combined brightness is magnitude 2.2. R is a Mira-type variable, range

5.3 to 11.8, and the period is unusually long, 546 days on average. T is a semiregular variable (5.5 to 9.0, 90 days). The globular cluster **Omega Centauri** is the finest in the entire sky. There are also two bright open clusters, NGC 5460 and NGC 3766, and a fine planetary nebula, NGC 3918. The elliptical galaxy NGC 5128 is the radio source Centaurus A.

Centaurus A The galaxy NGC 5128, a strong radio source about 10 million light years away in the constellation Centaurus. Optically it appears as an elliptical galaxy crossed by a prominent dust lane. It is thought to result from the encounter of a massive elliptical galaxy with a smaller spiral with a high content of dust.

central meridian (c.m.) The imaginary line joining the north and south poles of a planet or the Sun, bisecting the disk. It is used by observers to estimate the longitude of features as the body rotates.

center of mass The point in a body, or a system of bodies, at which the total mass of the body or system may be regarded as concentrated. It is the point from which gravitational attraction of the system as a whole appears to act. SEE ALSO **barycenter**.

▲ **Centaurus A** Crossed by a prominent dark lane, this elliptical galaxy (also designated NGC 5128) is a powerful radio and X-ray source.

Cepheid variable One of an important class of variable stars that pulsate in a regular manner, accompanied by changes in luminosity. They take their name from the first of the type to be discovered, Delta Cephei, the variability of which was noted by John **Goodricke** in 1784. Cepheids periodically expand and contract, changing in size by as much as 30% in each cycle. The diagram shows the pulsations and light curve of Delta Cephei itself. A typical Cepheid variable has a surface temperature that varies between 6000 and 7500 K and a spectral type that ranges from F2 at maximum to G2 at minimum. The average luminosity is 10,000 times that of the Sun. Cepheids became important in cosmology in 1912, when Henrietta **Leavitt** discovered a simple relationship between the period of light variation and the absolute magnitude of a Cepheid. This relationship, the **period–luminosity** law, enables the distances of stars to be ascertained, not only in our Galaxy but also in other galaxies.

There are two classes of Cepheid: type I, or *classical Cepheids*, are younger and more massive than type II, or **W Virginis stars**. Both types follow a period–luminosity relationship, but their curves are different. The classical Cepheids, such as Delta Cephei itself, are yellow supergiants of **Population I**, while the second type are older **Population II** stars and are found in globular clusters and the center of the Galaxy. In using Cepheids to determine distances it is necessary to know which type is being observed. When Cepheids were first used to determine distances it was not known that there were two types. This resulted in applying the value for type II Cepheids to the classical Cepheids in error, which affected the measured distances to external galaxies. When the error was found, by Walter **Baade** in 1952, the result was to double the size of the Universe. Classical Cepheids are one to two magnitudes brighter than type II Cepheids of the same period. The periods of Cepheids are in the range 1–50 days; classical Cepheids have a somewhat shorter period on average than those of type II.

Cepheus A constellation in the far north, adjoining Ursa Major, but not very distinctive. In mythology, it represents the husband of Cassiopeia and the father of Andromeda. Alpha (Alderamin) is magnitude 2.4. Beta (Alfirk), magnitude 3.2, is variable over a small range, and is a prototype of the class known as either *Beta Cephei stars* or *Beta Canis Majoris stars*. Delta is the prototype **Cepheid variable**, with a range from magnitude 3.5 to 4.4 and a period of 5.37 days; it has a wide optical companion of magnitude 7.5. Mu, which William **Herschel** named the Garnet Star, is a semiregular red supergiant with a range of 3.4 to 5.1 and a period of about 2 years. T Cephei is a Mira variable (5.2 to 11.3, 388 days); VV is a giant eclipsing variable, 4.8 to 5.4, period 7430 days.

◀ **Cepheid variable**
Their extremely regular light variations make Cepheids valuable standard candles: the longer the period from one maximum to the next, the greater the star's intrinsic luminosity. As shown, the typical Cepheid light curve shows a rapid rise to peak brightness, followed by a slower decline to minimum.

Ceres The first **asteroid** to be discovered, by Giuseppe **Piazzi** on January 1, 1801. Its diameter of 934 km (580 miles) makes it the largest asteroid. It orbits in the main asteroid belt, at an average distance from the Sun of 414 million km (257 million miles), the distance of the "missing" planet predicted by **Bode's law**.

Cerro Tololo Inter-American Observatory An observatory on Cerro Tololo peak near La Serena, Chile, part of the US National Optical Astronomy Observatories. Its largest telescope is the Victor M. Blanco Telescope, a 4-m (158-inch) reflector, the twin of that at Kitt Peak. Other instruments include the 0.61-m (24-inch) Curtis Schmidt Telescope of the University of Michigan.

Cetus A large constellation, crossed by the celestial equator, but not very conspicuous; it represents the sea monster of the Andromeda legend. Its brightest star is Beta, magnitude 2.0. The most famous star in the constellation is the prototype long-period variable, **Mira**. Tau Ceti, magnitude 3.5, 11.9 light years away, is one of the closest Sun-like stars, being a main-sequence star of type G8. UV Ceti is a **flare star**; it is a pair of 12th-magnitude red dwarfs 8.7 light years away that exhibit large flares every few hours.

Chamaeleon A small constellation near the south pole, representing a chameleon. Its brightest stars are only of magnitude 4.1.

Chandrasekhar, Subrahmanyan (1910–95) Indian-American astrophysicist. He made significant contributions to the theories of stellar evolution, stellar atmospheres, and relativity. His theory of white dwarf stars suggests there is a limit to their mass (SEE **Chandrasekhar limit**).

Chandrasekhar limit The maximum possible mass for a **white dwarf** star, first calculated in 1931 by Subrahmanyan **Chandrasekhar**. The value is about 1.4 solar masses, although more recent computations suggest that a higher value is possible for a rapidly rotating white dwarf. If a star has a mass which exceeds the Chandrasekhar limit it will collapse to become a **neutron star** or possibly a **black hole**.

Chandra X-ray Observatory (CXO) A NASA satellite for studying the sky at X-ray wavelengths, formerly known as the Advanced X-ray Astrophysics Facility, launched in July 1999. It carries the highest-resolution **X-ray telescope** ever constructed. The main instruments are a CCD imaging spectrometer and a high-resolution camera. It is named in honor of Subrahmanyan **Chandrasekhar**.

charge-coupled device SEE **CCD**.

Charon The **satellite** of **Pluto**, discovered in 1978 by James Christy. Its diameter of 1186 km (737 miles) makes it the largest satellite in relation to its primary in the Solar System. Its mass is about 12% of Pluto's, and its density is about that of Pluto, 2.0 g/cm^3. It is unique among satellites in having a synchronous orbit with a period (6.4 days) matching the rotation period of its primary. From Charon's spectrum, it seems to have a grayish surface of water ice and some ammonia ice.

Chiron A small body in the outer Solar System, originally designated as an asteroid (no. 2060) following its discovery in 1977 by Charles Kowal. However, it has an eccentric, cometary orbit (period 50.7 years) which takes it from within Saturn's orbit out nearly as far as Uranus. In 1989, seven years before reaching perihelion, it was found to have a comet-like **coma** and was given a periodic comet designation (95P/Chiron). It has a diameter of roughly 200 km (125 miles), much larger than any comet. It is now classified as a **Centaur**.

chondrite A type of **stony meteorite**, consisting of an agglomeration of *chondrules* – millimeter-sized globules believed to have remained unchanged since they condensed out of the nebula from which the Sun and the Solar System formed. They are therefore more ancient than **achondrites**. Chondrites consist mainly of iron, calcium, aluminum, magnesium, and sodium silicates. They are classified into various types, including **carbonaceous chondrites**.

chromatic aberration An **aberration** (2) in lenses that results from the unequal refraction of the different colors that make up white light. Each color is brought to a focus at a different distance from the lens, causing the image of an object to be fringed with prismatic colors. Chromatic aberration is reduced by the use of an **achromatic lens**.

chromosphere ("sphere of color") The layer of the Sun's atmosphere between the photosphere and the corona. The chromosphere is normally invisible because of the glare of the photosphere shining through it, but it is briefly visible near the beginning and end of a total solar eclipse as a spiky red rim around the Moon's disk. At other times it can be studied by spectroscopy or with a **spectroheliograph**. The chromosphere is about 10,000 km (6000 miles) thick, with a temperature of 10,000–25,000 K. **Filaments**, **flocculi**, **plages**, and **prominences** originate in the chromosphere. SEE ALSO **supergranulation**.

Circinus A small constellation in the region of Alpha and Beta Centauri, representing a pair of compasses. Its only fairly bright star is Alpha, magnitude 3.2.

circumpolar stars Stars which from a given place of observation never dip below the horizon. Their polar distance (90° minus their declination) is less than the observer's latitude.

civil twilight SEE **twilight**.

Clementine A US Department of Defense space probe which in April and May 1994 imaged the entire surface of the Moon in several wavelength bands. This made possible not only the first complete mapping of the Moon at visual wavelengths, but also a detailed geological survey. Clementine returned the first tentative evidence that the floors of some south polar craters, permanently in shadow, may contain deposits of water ice.

closed universe In cosmology, a theoretical "model" universe in which there is enough matter to halt the expansion triggered in the **Big Bang**. It would then collapse back on itself, to end in an event the opposite of the Big Bang, sometimes called the *Big Crunch*. In a closed universe, the **curvature of space** is positive, the **deceleration parameter** is greater than 0.5, and the **density parameter** is greater than 1. COMPARE **open universe**.

cluster of galaxies A group of galaxies, which can range in size from a few dozen members, as in our own **Local Group,** to a rich cluster containing thousands of galaxies. Poor clusters are usually irregular in shape, but rich clusters tend to be spherical and condensed at the center, and with a giant elliptical galaxy often dominating the central region. Such giant galaxies have probably grown by *cannibalism*, their overwhelming gravity dragging in smaller galaxies. The nearest rich clusters of galaxies are the **Virgo Cluster**, 45 million light years away, and the Coma Cluster, 300 million light years away. Clusters are grouped into *superclusters* up to about 100 million light years across. The Local Supercluster, of which the Local Group is part, is centered on the Virgo Cluster.

CME ABBREVIATION FOR **coronal mass ejection**.

Coalsack A prominent dark nebula in the constellation Crux, the Southern Cross. It is seen in silhouette against the background stars of the Milky Way.

COBE ABBREVIATION FOR **Cosmic Background Explorer**.

coelostat Two mirrors arranged so that slowly rotating one of them counteracts the apparent rotation of the sky, enabling light from a celestial object to be directed along a fixed axis. The first mirror is mounted equatorially and is clock-driven at half the rate of the Earth's rotation. The beam of light it receives is reflected on to a second (fixed) mirror, which in turn reflects the beam into the instrument being used for the observation. A coelostat is used with heavy or fixed equipment, for example a spectro-heliograph. It is an improved form of **helio-stat** and **siderostat**. SEE ALSO **coudé**.

collimation The process of aligning the optical components of a telescope so as to bring the rays of light from the object being observed to a focus at the correct position. A refracting telescope requires only occasional adjustment of the object glass in its cell. Collimation is more of a problem with reflectors, as the primary and secondary mirrors both need to be adjusted, and the secondary can become misaligned quite easily. An incorrectly collimated telescope can suffer from **coma** (1).

color index The difference in brightness of a star at two different wavelengths, used as a measure of the star's color and hence its temperature. The brightness of the star is measured through colored filters, usually at blue (B) and yellow (V, for visual) wavelengths, and the difference between the two readings is the B −V color index. The B −V color index of cool red stars is positive (up to about +2.0), while that of hot blue stars is negative (about −0.5 maximum). A white star of spectral type A0 such as Vega has a B −V color index of exactly 0.0. A measurement at U (ultraviolet) wavelengths may also be taken, and a U −B color index derived. Other measurements in the red and infrared can also be used.

The color index of light from a star may change as the light passes through dust in interstellar space. Dust scatters blue light but allows red light to pass more freely, thereby reddening the star. This effect must be taken into account when determining the color index of a star. SEE ALSO **UBV system**.

▲ **Clementine** The Moon's south pole is seen in this image obtained by the orbiting Clementine spacecraft in 1994. This picture has been compiled from some 1500 individual images.

Columba A southern constellation adjoining Canis Major and Carina, representing a dove. Alpha (Phakt) is magnitude 2.6, but the constellation contains little of interest.

colure A great circle on the celestial sphere that passes through both celestial poles. SEE ALSO **equinoctial colure, solstitial colure**.

coma (1) A defect in an optical system which results in the image of a point appearing as a blurred pear-shaped patch with a flared appearance resembling a comet. It is caused by incident light striking a lens obliquely. The defect increases with distance from the center of the field of view.

coma (2) The spherical cloud of gas and dust surrounding the nucleus of an active **comet**.

Coma Berenices A northern constellation adjoining Boötes, representing the hair of Queen Berenice of Egypt. Its brightest stars, of 4th and 5th magnitudes, form a very wide star cluster. It is rich in faint galaxies; some are part of the **Virgo Clus-**

ter, but others are part of the separate and more distant *Coma Cluster*, 330 million light years away.

comes (plural **comites**) The fainter companion in a **binary star**.

comet A small, icy Solar System body in an independent, eccentric orbit around the Sun. The solid *nucleus* of a comet is small – that of **Halley's Comet**, for example, measures just 16×8 km (10×5 miles). The nucleus is made up of ices, largely water ice (plus some methane, carbon dioxide and carbon monoxide, with traces of organic substances), in which are embedded rock and dust particles, and is covered by a thin dark crust. As the comet approaches the Sun and gets warmer, evaporation begins, and jets of gas and dust escape through the crust to form the luminous, spherical *coma*. The coma can be huge, a million kilometers or so across, but extremely tenuous. Closer in, radiation pressure from the Sun and the solar wind may send dust and gas streaming away as a *tail*, tens of millions of kilometers long. The *gas tail* or *ion tail* is straight and long, up to 10 million km, while the *dust tail* is shorter, broader, and curved.

Comets are thought to originate in the **Oort Cloud**, where there may be as many as 6 trillion (6×10^{12}) of them, and in the **Kuiper Belt**. They are believed to be **planetesimals** left over from the formation of the Solar System. A comet may be perturbed (by a passing star, for example) and displaced toward the Sun. It may swing past the Sun on a parabolic or even a hyperbolic orbit and return to the outer Solar System, or enter an elliptical orbit as a *long-period comet*, with a period of thousands or even millions of years (SEE **parabola**). Further perturbation by one of the giant planets may drive it into a smaller orbit as a *short-period comet*, with a period of 200 years or less. Further such close encounters can subsequently modify the orbit.

With their loss of material at every approach to the Sun, periodic comets have a limited lifetime. The debris a comet leaves in its wake spreads around its orbit to form a **meteor stream**. Some comets, once they have expended all their gas and dust, may be strong enough to hold together and continue to orbit the Sun – the **Apollo asteroids**, **Amor asteroids**, and **Aten asteroids** may be extinct comets. Others, like Comets **Biela**, **West** and **Shoemaker–Levy 9**, have been observed to break up. Some objects first classed as asteroids have later behaved like comets – **Chiron**, for example, developed a coma.

Around a thousand comets have had their orbits calculated in detail, of which 150 are short-period comets. Several new comets are discovered every year. After its orbit has been confirmed, it is customary for a new comet to be given the name of its discoverer. Donati's comet (or, more for-

SOME PERIODIC COMETS AND ASSOCIATED METEOR SHOWERS						
Name	Discovery	Orbital period (years)	Perihelion distance (million km)	Eccentricity	Inclination (°)	Shower(s)
Encke	1786, 1818	3.28	50	0.85	12	Taurids
Pons–Winnecke	1819, 1858	6.37	188	0.64	22	Pons–Winneckids
Giacobini–Zinner	1900, 1913	6.61	155	0.71	32	Draconids (Giacobinids)
Tuttle	1790, 1858	13.5	149	0.82	55	Ursids
Tempel–Tuttle	1865	32.9	147	0.90	163 *R*	Leonids
Halley	240 BC	76.0	88	0.97	162 *R*	Eta Aquarids, Orionids
Swift–Tuttle	1737, 1862	135*	143	0.96	113 *R*	Perseids
Thatcher (1861 GI)	1861	415*	138*	0.98	80	Lyrids

approximate values   R indicates retrograde orbit

mally, Comet Donati), for example, was discovered by Giovanni Donati in 1858. A few comets, such as Halley's Comet and Encke's Comet, are named for the people who first calculated their orbit. In the case of codiscoveries, up to three names are allowed (from 1995, only two), as in Comet IRAS–Araki–Alcock (discovered independently by two amateur astronomers and the IRAS satellite). Formally, periodic comets are also given the prefix "P" plus a permanent number once the orbit is established, as in 1P/Halley. Non-periodic comets are similarly assigned the prefix "C."

commensurable Describing two numbers with a common factor. If the orbital periods of two planets around the Sun, or of two satellites round a planet, can be expressed as a ratio of small whole numbers, then the periods are called commensurable. Because of tidal interactions (SEE **tides**) between bodies of similar size, stable orbits tend to form at (nearly) commensurable distances. Examples are the orbital periods of Saturn and Jupiter, which are in a ratio of nearly 5 : 2, and the periods of Jupiter's **Galilean satellites**. With bodies which are very different in size, orbital **resonances** can have the effect of preventing bodies from having commensurable orbital periods, as with the **Kirkwood gaps** in the main asteroid belt and the gaps in Saturn's ring system. **Bode's law** was partly (but only by chance) successful in predicting undiscovered planets because of the commensurability of orbits.

comparator A device with which the images of the same star field on two photographs taken at different times can be rapidly alternated in the field of view. If any object has changed its position or varied in brightness during the interval between the times when the two photographs were taken, the comparator will show it up. Hunters of novae and asteroids use comparators. Variable stars reveal themselves as a pulsation, because they give images of different sizes according to their brightness. The *blink microscope* or *blink comparator* is a professional instrument in which the two images are viewed alternately in an eyepiece. The *stereo comparator* is a simple device for viewing the two images simultaneously with binocular vision; another arrangement uses two slide projectors and a rotating shutter.

Compton Gamma Ray Observatory (GRO) A NASA satellite for studying the sky at gamma-ray wavelengths, launched in April 1991. In its nine years of operation to June 2000, GRO increased the number of known gamma-ray sources tenfold, and recorded 2500 **gamma-ray bursts**.

conjunction The alignment of the Earth with two other bodies in the Solar System so that from the Earth the other two appear in (nearly) the same position. The term is used mainly for alignments with the major planets (SEE the diagram on page 46). A **superior planet** is at conjunction when on the opposite side of the Sun from the Earth. An **inferior planet** is at *inferior conjunction* when it lies between the Earth and the Sun, and at *superior conjunction* when it is on the far side of the Sun. Because planetary orbits are inclined at various angles to the plane of the ecliptic, alignments at conjunction are rarely exact, and conjunction is defined to occur when the two bodies have the same celestial longitude as seen from the Earth.

Two or more planets – including in this case the Moon – are also said to be in conjunction when they are close together in the sky.

constellation A region of the sky containing an arbitrary grouping of stars, forming an imaginary figure. The stars in a constellation lie at very different distances from us, so the groupings have no physical significance. The original number of 48 constellations given in Ptolemy's *Almagest* has grown to 88 (SEE pages 48–9), which were assigned boundaries on the celestial sphere by the International Astronomical Union in 1930. Constellations are very unequal in size and prominence; they range from the vast Hydra down to the tiny but

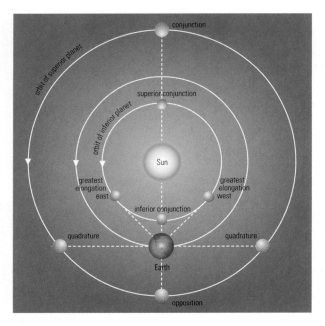

◀ conjunction

Superior planets come to conjunction with the Sun when on the far side of it, as seen from Earth (and are therefore lost from view). The inferior planets, Mercury and Venus, can undergo conjunction at two stages in their orbit. At superior conjunction, they lie on the far side of the Sun from Earth, while at inferior conjunction they are between the Sun and the Earth. Under certain circumstance, Mercury and Venus can transit across the Sun's disk at inferior conjunction.

brilliant Crux. Many of the names are drawn from ancient mythology, while more recent mapping of the southern-hemisphere skies introduced some modern-sounding names such as the Octant and the Telescope. Noteworthy stars within a constellation are designated by Greek letters followed by the name of the constellation in the genitive (possessive) case, as with Alpha Lyrae; these are often abbreviated in the form α Lyr.

continuous spectrum The unbroken sequence of colors, merging one into the other, produced when light is decomposed by refraction through a prism. An incandescent solid, liquid, or dense gas emits a continuous spectrum. The spectrum of a star consists of a continuous spectrum crossed by **absorption lines**. SEE **emission spectrum**.

continuum SEE **spacetime**.

coorbital satellite A satellite orbiting a planet at the same average distance as another satellite. Although the distances –

and hence the periods – are the same, the orbital inclinations and eccentricities may differ slightly. An example is **Helene**, a small satellite of Saturn with a very similar orbit to the larger satellite Dione.

Coordinated Universal Time SEE **Universal Time**.

coordinates Reference systems that can be used to define the position of a point or body on the **celestial sphere**. Several systems of coordinates are used in astronomy.

The most commonly used is the *equatorial system*, in which the reference plane is that of the celestial equator, and the coordinates are usually **right ascension** and **declination** (although **hour angle** and **polar distance** can be used alternatively).

In the *horizontal system* the reference plane is that of the observer's horizon, and the coordinates are **azimuth** and **altitude**.

In the *ecliptic system* the reference plane is that of the ecliptic, and the coordinates are celestial latitude and longitude. *Celestial latitude* is the angular distance between the object and the ecliptic, measured at right angles to the ecliptic. It is measured from 0° to 90°, positively toward the north ecliptic pole and negatively toward the south ecliptic pole. *Celestial longitude* is given by the angle between the object and the vernal equinox, measured along the ecliptic from 0 to 360° eastward from the vernal equinox. Celestial latitude and longitude are not measured with reference to the celestial equator; they are sometimes more correctly called *ecliptic latitude* and *ecliptic longitude*.

When dealing with the Galaxy, the system of *galactic coordinates* is used, in which the plane of reference is the galactic plane (the plane of the Milky Way). Its coordinates are galactic latitude and longitude. *Galactic latitude* is measured from 0° to 90° from the galactic equator, positively toward the north galactic pole. *Galactic longitude* is measured from 0° to 360° eastward along the galactic equator. As defined by the IAU, the position of zero galactic longitude is at RA 17h 46m, dec. −28° 56′ (epoch 2000).

Copernicus, Nicholas (1473–1543)

Polish churchman and astronomer. A canon at Frauenburg Cathedral and a local administrator, he pursued astronomy as a side-interest. The accepted view in his day was the geocentric (Earth-centered) universe as explained by **Ptolemy** nearly 1500 years before, with its complicated system of orbits. Through his study of planetary motions, Copernicus developed a heliocentric (Sun-centered) theory of the universe. In this *Copernican system*, as it is called, the planets' motions in the sky were explained by having them orbit the Sun. The motion of the sky was simply a result of the Earth turning on its axis, and the stars remained fixed as the Earth orbited the Sun because they were so very far away.

An account of his work, *De revolutionibus orbium coelestium*, was published in 1543. Most astronomers (and possibly even Copernicus himself) considered the new system to be merely a means of calculating planetary positions, and continued to believe in **Aristotle**'s view of the world. It was not until the discoveries by **Galileo** and

Johann **Kepler** that the reality of it began to take hold. SEE ALSO **aberration** (1).

Copernicus A large lunar crater, 93 km (58 miles) in diameter, in the Mare Imbrium. It is one of the most conspicuous craters, and is surrounded by a system of rays stretching for well over 500 km (300 miles).

Cordelia The innermost small **satellite** of Uranus, discovered in 1986 during the Voyager 2 mission. Cordelia and Ophelia act as **shepherd moons** to the planet's Epsilon Ring.

Coriolis force A force that appears to act on a body (such as a projectile or a mass of air) moving freely across the Earth so as to deflect it from a straight path. The deflection is to the right in the northern hemisphere, to the left in the southern. In fact the effect is a consequence of us observing the motion from within a rotating frame of reference – the spinning Earth. In 1735 the English meteorologist George Hadley (1685–1768) recognized the effect of the Earth's rotation on the movement of air currents; general deduction of the displacing force of the Earth's rotation was

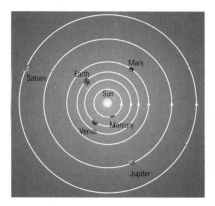

▲ **Copernicus, Nicolas** Copernicus's model of the Solar System put the Sun at the center, with Earth and the other planets in orbit about it. Replacing the earlier geocentric system, the Copernican system removed Earth from any special position in the Solar System.

THE CONSTELLATIONS

Name	Genitive	Abbreviation	Order of size
Andromeda	Andromedae	And	19
Antlia	Antliae	Ant	62
Apus	Apodis	Aps	67
Aquarius	Aquarii	Aqr	10
Aquila	Aquilae	Aql	22
Ara	Arae	Ara	63
Aries	Arietis	Ari	39
Auriga	Aurigae	Aur	21
Boötes	Boötis	Boo	13
Caelum	Caeli	Cae	81
Camelopardalis	Camelopardalis	Cam	18
Cancer	Cancri	Cnc	31
Canes Venatici	Canum Venaticorum	CVn	38
Canis Major	Canis Majoris	CMa	43
Canis Minor	Canis Minoris	CMi	71
Capricornus	Capricorni	Cap	40
Carina	Carinae	Car	34
Cassiopeia	Cassiopeiae	Cas	25
Centaurus	Centauri	Cen	9
Cepheus	Cephei	Cep	27
Cetus	Ceti	Cet	4
Chamaeleon	Chamaeleontis	Cha	79
Circinus	Circini	Cir	85
Columba	Columbae	Col	54
Coma Berenices	Comae Berenices	Com	42
Corona Australis	Coronae Australis	CrA	80
Corona Borealis	Coronae Borealis	CrB	73
Corvus	Corvi	Crv	70
Crater	Crateris	Crt	53
Crux	Crucis	Cru	88
Cygnus	Cygni	Cyg	16
Delphinus	Delphini	Del	69
Dorado	Doradus	Dor	72
Draco	Draconis	Dra	8
Equuleus	Equulei	Equ	87
Eridanus	Eridani	Eri	6
Fornax	Fornacis	For	41
Gemini	Geminorum	Gem	30
Grus	Gruis	Gru	45
Hercules	Herculis	Her	5
Horologium	Horologii	Hor	58
Hydra	Hydrae	Hya	1
Hydrus	Hydri	Hyi	61
Indus	Indi	Ind	49

THE CONSTELLATIONS (continued)

Name	Genitive	Abbreviation	Order of size
Lacerta	Lacertae	Lac	68
Leo	Leonis	Leo	12
Leo Minor	Leonis Minoris	LMi	64
Lepus	Leporis	Lep	51
Libra	Librae	Lib	29
Lupus	Lupi	Lup	46
Lynx	Lyncis	Lyn	28
Lyra	Lyrae	Lyr	52
Mensa	Mensae	Men	75
Microscopium	Microscopii	Mic	66
Monoceros	Monocerotis	Mon	35
Musca	Muscae	Mus	77
Norma	Normae	Nor	74
Octans	Octantis	Oct	50
Ophiuchus	Ophiuchi	Oph	11
Orion	Orionis	Ori	26
Pavo	Pavonis	Pav	44
Pegasus	Pegasi	Peg	7
Perseus	Persei	Per	24
Phoenix	Phoenicis	Phe	37
Pictor	Pictoris	Pic	59
Pisces	Piscium	Psc	14
Piscis Austrinus	Piscis Austrini	PsA	60
Puppis	Puppis	Pup	20
Pyxis	Pyxidis	Pyx	65
Reticulum	Reticuli	Ret	82
Sagitta	Sagittae	Sge	86
Sagittarius	Sagittarii	Sgr	15
Scorpius	Scorpii	Sco	33
Sculptor	Sculptoris	Scl	36
Scutum	Scuti	Sct	84
Serpens	Serpentis	Ser	23
Sextans	Sextantis	Sex	47
Taurus	Tauri	Tau	17
Telescopium	Telescopii	Tel	57
Triangulum	Trianguli	Tri	78
Triangulum Australe	Trianguli Australis	TrA	83
Tucana	Tucanae	Tuc	48
Ursa Major	Ursae Majoris	UMa	3
Ursa Minor	Ursae Minoris	UMi	56
Vela	Velorum	Vel	32
Virgo	Virginis	Vir	2
Volans	Volantis	Vol	76
Vulpecula	Vulpeculae	Vul	55

▲ **corona** Loops in the Sun's inner corona were imaged by the TRACE spacecraft in 1998. These loops are shaped by magnetic fields.

made by Gaspard de Coriolis (1792–1843). The Coriolis force accounts for the circulation of air around cyclones on the Earth, and is a driving force for the meteorology of the Earth and other planets with substantial atmospheres.

corona The outermost layer of the Sun's atmosphere, extending for many millions of kilometers into space. It springs into view as a white halo during a total solar eclipse; at other times it can be observed in visible light only by using a special instrument called a **coronagraph**. The corona emits strongly in the extreme ultraviolet and X-ray regions, and has been studied by spacecraft such as the **Transition Region and Coronal Explorer**. Its overall shape changes during the 11-year solar cycle, from regular and symmetrical at solar minimum to uneven with long streamers at solar maximum. The corona has a temperature of 1–2 million degrees K. When lines of the Sun's magnetic field connect, large currents flow and heat the corona to this high temperature.

The corona may be divided into the inner *K corona*, which shines by light from the photosphere scattered by high-energy electrons, and the outer *F corona*, which shines by light scattered by lower-energy dust particles. The K corona has a con-tinuous spectrum (the "K" standing for the German word *Kontinuum*), on which are superimposed the **Fraunhofer lines** of the F corona. There are also emission lines in the spectrum contributed by the *E corona*, consisting of lower-energy metal ions such as calcium present out to about two solar radii. SEE ALSO **coronal hole, coronal mass ejection, flare, Sun**.

Corona Australis A small southern constellation representing a crown or wreath. It contains no star above magnitude 4, but its arc of stars makes it easy to identify. It is sometimes called Corona Austrinus.

Corona Borealis A northern constellation, representing the wedding crown of Ariadne, adjoining Boötes. Alpha (Gemma or Alphecca), magnitude 2.2, is the only bright star, but the curve of five stars makes the constellation easy to find. In the "bowl" of the crown is the famous variable star **R Coronae Borealis**, prototype of a group of stars that fade suddenly and unpredictably. T Coronae Borealis is a recurrent nova, known as the Blaze Star, which erupted in 1866 and 1946.

coronagraph (sometimes spelled **coronograph**) An instrument, used in conjunction with a telescope, which allows the corona and prominences to be viewed by producing an "artificial eclipse," masking the light from the Sun's disk with a circular obstruction. Before the invention of the coronagraph, such observations were possible only during the brief duration of a total eclipse.

coronal hole A relatively cool region of the solar **corona**. Coronal holes are associated with weak regions of the Sun's magnetic field, and are sources of high-speed streams of the **solar wind**.

coronal mass ejection (CME) The eruption into interplanetary space of millions of tons of plasma from the Sun's corona, triggered by disturbances in the Sun's magnetic field. CMEs are more frequent at solar maximum (several a day) than at solar minimum (about one a week). They are often associated with eruptive **prominences** in the chromosphere and **flares** in the corona. CMEs travel at hundreds of kilometers per second, and have a marked effect on the **solar wind**. A strong burst of charged particles arriving at the Earth can cause *geomagnetic storms*, producing **aurorae** and interfering with radio communications.

Corvus A constellation adjoining Hydra and Crater, representing a crow. Its four brightest stars, of third magnitude, form a quadrilateral which is easy to find. Corvus contains the Antennae, a pair of 10th-magnitude interacting galaxies.

cosmic abundance The relative proportions of the various chemical elements (SEE **element, chemical**) in the Universe as a whole. It has been found from spectroscopic studies of the atmospheres of the Sun and stars that (with the exception of hydrogen and helium) the cosmic abundance is much the same as in the Sun. Stars are roughly two-thirds hydrogen and one-third helium, with other elements – chiefly

carbon and oxygen – accounting for only 2%. The chemical make-up of a particular star depends on its age (SEE **stellar evolution**). Some rare types of star, for example **Wolf–Rayet stars**, show unusual abundances of certain elements; others, such as *technetium stars*, contain elements it is difficult to explain by currently accepted theories of how stars evolve. SEE ALSO **interstellar matter**.

Cosmic Background Explorer

(COBE) A NASA satellite that studied the **cosmic microwave background** radiation. It determined that the microwave background has a **black body radiation** spectrum with a temperature of 2.73 K. In 1992 it detected slightly warmer and cooler spots in the background radiation, differing by about 30 millionths of a degree from the average; these *ripples* are attributed to density fluctuations in the early Universe that marked the first stage in the formation of galaxies. COBE was launched in November 1989 and operated until the end of 1993.

cosmic microwave background

Weak electromagnetic radiation from the

▲ **cosmic microwave background** This all-sky temperature map from COBE, from which the effects due to the motion of the Earth have been subtracted, shows "ripples" in the cosmic microwave background.

Although the extent of these variations is tiny – 30 millionths of a degree – they are believed to have been sufficient to trigger formation of the first galaxies after the Big Bang.

Universe, first detected in 1965 by Arno **Penzias** and Robert **Wilson** of the Bell Telephone Laboratories. The microwave background is black-body radiation at a temperature of 2.73 K and has an almost equal intensity in all directions in space. It is considered to be the remnant of the radiation from the hot, early Universe following the **Big Bang**. The background radiation increased its wavelength as the Universe expanded until today it peaks in the millimeter region of the spectrum. It is called the microwave background because it was first detected in the form of microwaves. Penzias and Wilson won the 1978 Nobel Prize for Physics for their discovery. SEE ALSO **Cosmic Background Explorer, Wilkinson Microwave Anisotropy Probe**.

cosmic rays High-energy atomic particles, moving at speeds approaching the speed of light, which enter the Earth's atmosphere from space. Cosmic rays (called "rays" because they were originally thought to be gamma-rays) consist mainly of the nuclei of atoms, mostly hydrogen, stripped of their electrons. While traveling through space they are called *primary cosmic rays*. When they enter the atmosphere they cause disintegration of the atoms they encounter and produce various atomic and subatomic particles, called *secondary cosmic rays*, which can be detected at ground level. The highest-energy cosmic rays come from X-ray sources such as the center of the Galaxy and the Virgo Cluster of galaxies. Cosmic rays of medium energy may originate in supernovae; low-energy cosmic rays are ejected from the Sun by solar **flares**.

cosmogony The study of how the Universe and objects within it were formed. The term is used most often for theories of the origin of the Universe (SEE **Big Bang, cosmology**) and for the origin of the Solar System. For the Solar System, the scenario now accepted by most astronomers is a development of the *nebular hypothesis* first proposed in the 18th century by Immanuel Kant and Pierre Simon de **Laplace**.

As a **molecular cloud** collapsed under its own gravity, a central condensation grew into a **protostar** which later became the Sun. Around the protostar, material from the cloud spread into a broad disk (SEE **proplyd**). Small grains of material stuck to one another by **accretion**, gradually forming **planetesimals**. (**Comets** and objects in the **Kuiper Belt** are believed to be a remnant population of these planetesimals.) The planetesimals collided with one another and grew into larger bodies, **protoplanets**, which swept up most of the remaining bits and pieces, as evidenced by **impact features** visible on many planetary bodies today. Gravitational perturbations between the resulting planets then moved them into their present orbits.

The different chemical compositions of the planets thus arose as a result of substances of different chemical composition condensing out at different distances from the Sun (SEE **volatile**). By this theory, planets are a normal consequence of star formation. Earlier "catastrophe" theories had explained the formation of planets by a chance event, such as material being torn from the Sun by the close approach of a passing star.

cosmological principle The principle that the Universe is both *homogeneous* and *isotropic*: in other words, that it is the same (on a sufficiently large scale) in all directions. The principle was assumed to be true by Einstein and others as it helped to solve the equations of general relativity. The discoveries that the **cosmic microwave background** is indeed isotropic, and that the large-scale distribution of galaxies in the Universe appears to be homogenous, show that the principle is real.

cosmological redshift SEE **redshift**.

cosmology The study of the structure of the Universe on the largest scale. The starting point for modern cosmology was the discovery in the 1920s by Edwin **Hubble** that the Universe is expanding. The rate of expansion is given by the **Hubble constant**. Other observational evidence shows that the Universe appears much the same in all directions (the so-called *isotropy*) and at all distances (*homogeneity*).

A *cosmological model* is a description of the Universe in terms of mathematical equations. These models aim to predict how the Universe behaves, and are tested by comparing them with observations. Cosmological models have included the **steady-state theory**, in which the Universe is not only the same in all places but also at all times. It therefore had no beginning, will have no end, and never changes at all when viewed on the large scale. This theory required matter to be created as the Universe expanded in order that the overall density of galaxies should not increase. For this reason it is also referred to as the *continuous creation* model. On the other hand, according to the **Big Bang** theory, the Universe was created in a single instant about 15,000 million years ago, and has been expanding ever since. In the future it may continue to expand or possibly collapse back on itself, depending on the total amount of matter and energy in the Universe – on whether it is above or below its **critical density**. An important consideration is the question of the **missing mass**: the amount of matter we see in the Universe is less than a tenth of the amount inferred from the motions of the galaxies.

Cosmologists have made several important advances in discriminating between cosmological models. The discovery of the **cosmic microwave background** radiation provided strong evidence against the steady-state theory. This background radiation is very uniform in all directions, which has led cosmologists to devise the so-called *inflationary* models (SEE **Big Bang**) in which there was a very rapid period of expansion shortly after the Big Bang. The Hubble constant is currently thought to have a value of 68 km/s/megaparsec. The **deceleration parameter** – which determines whether the Universe will expand for ever or collapse back on itself – has also been estimated, and seems to have a value close to 0.5, the value for a Universe which will just continue to expand for ever.

coudé An arrangement of auxiliary mirrors used with a telescope on an equatorial mount to direct the light path down the polar axis to a fixed focal point. The coudé system is used in conjunction with large, immovable instruments, such as a spectrograph, which are too massive to be mounted on the telescope itself. The word comes from the French *coudé*, meaning bent like an elbow, and is not the name of a person, as is sometimes thought. SEE ALSO **Nasmyth focus**.

Crab Nebula The nebula M1 (NGC 1952), about 6500 light years away in Taurus, the remnant of a supernova that was noted by Chinese astronomers in July 1054, when it shone as brightly as Venus, being visible even in daylight. It is now about 8th magnitude. The nebula was discovered in 1731 by the English astronomer John Bevis and independently by Charles **Messier** in 1758. It gained its popular name of the Crab Nebula after it was sketched by Lord **Rosse** to resemble that creature.

The Crab Nebula is an intense source of radio emission known as *Taurus A*, and is

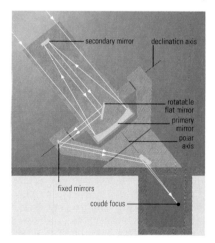

▲ **coudé** In this optical configuration, mirrors direct light from the telescope along the polar axis to a fixed observing position. This arrangement is advantageous for observations that require use of heavy or bulky detectors.

▲ **Crab Nebula** A supernova remnant, the Crab Nebula is an expanding gas cloud with filamentary structure and a pulsar at its heart.

also a source of X-rays (Taurus X-1). Gas ejected in the supernova explosion forms filaments that appear red on color photographs, surrounding an ionized gas that emits a yellow glow. The glow is **synchrotron radiation**, caused by electrons spiraling in an intense magnetic field. The synchrotron radiation extends from the radio domain into the visible, and the Crab Nebula is the only synchrotron nebula that can be seen in a small telescope. Inside the Crab Nebula lies the *Crab Pulsar*, the remaining core of the supernova of 1054. It spins 30 times a second, spraying out electrons to replenish the synchrotron radiation. The Crab Pulsar also flashes at visible wavelengths. SEE **pulsar**.

crater A circular formation, found on many bodies in the Solar System – planets, satellites, and asteroids. Nearly all craters are **impact features**, although on some bodies there are volcanic craters. A typical impact crater has a raised rim, or wall, and its floor is below the level of the surrounding terrain. Some have a central peak. Impact craters range in size from the microscopic (as found on lunar rocks returned by the Apollo missions) to features a third of the diameter of the body (as with Saturn's moon **Mimas**). Heavily cratered bodies such as **Callisto** are geo-

logically inactive and preserve a record of the bombardment that took place in the early Solar System, when there were still plenty of free **planetesimals**. Less cratered bodies are usually those on which geological activity has led to some degree of resurfacing, obliterating older impact sites.

Crater A small constellation adjoining Hydra and Corvus, representing a cup or chalice. Its brightest stars are of fourth magnitude and it contains no objects of note.

Crepe Ring A popular name for Ring C of **Saturn**. SEE ALSO **ring, planetary**.

Cressida An inner **satellite** of Uranus, discovered in 1986 during the Voyager 2 mission.

critical density The average density of matter in the Universe that would ensure it would just barely continue expanding for ever, without ever collapsing back on itself. The observed density of the Universe is less than a tenth of the critical value; the remainder is thought to be accounted for by dark matter. The ratio of the observed and critical densities is the *density parameter*, Ω. Hence, if the Universe has the critical density, then $\Omega = 1$. Ω is twice the value of $q0$, the **deceleration parameter**. SEE ALSO **missing mass**.

Crux The smallest constellation in the sky, popularly known as the Southern Cross. It is almost surrounded by Centaurus. Its brightest star, Alpha (Acrux), is of magnitude 0.77 and lies about 321 light years away. It is actually a visual double, magnitudes 1.3 and 1.7, both type B. With Beta (Mimosa, or Becrux), magnitude 1.2, Gamma (Gacrux, 1.6), and Delta (2.8), it makes a pattern that is more like a kite than a cross; the symmetry is disturbed further by Epsilon (3.6), the only other star above magnitude 4. Interesting objects include the **Jewel Box** cluster, and the **Coalsack**, a dark nebula.

C star SEE **carbon star**.

culmination The passage of a celestial body across the observer's meridian. At culmination, an object reaches the greatest altitude above, or least altitude below, the observer's horizon. In the case of circumpolar stars, both culminations are observable. *Lower culmination* occurs when a star transits between the pole and horizon; the star's **hour angle** is then exactly 12h. *Upper culmination* occurs when a star transits between the pole and zenith; the star's hour angle is then exactly 0h.

curvature of space The distortion of **spacetime** in the neighborhood of matter, one of the consequences of **Einstein**'s General Theory of Relativity. This curvature makes rays of light and particles of matter follow curved paths called *geodesics*. In **cosmology**, if the Universe contains enough mass it will have enough curvature to be a *closed universe*; if not, it will be an *open universe*.

cusp One of the points or horns of the crescent Moon, or of an inferior planet at crescent phase.

cusp caps The bright areas at the **cusps** of the crescent Venus that are sometimes reported by observers.

CXO ABBREVIATION FOR **Chandra X-ray Observatory**.

Cygnus One of the most distinctive constellations, often nicknamed the Northern Cross. Its brightest star is first-magnitude **Deneb**. Beta is Albireo, a beautiful colored double with a giant primary of type K3, magnitude 3.1, and a main-sequence companion of type B9, magnitude 5.1, separation 34 arc seconds. Chi has the largest visual range of any Mira variable, 3.3 to 14.2, period about 400 days. P Cygni is a B-type supergiant variable that has ranged between magnitudes 3 and 6 in the past, because of mass loss, and now appears around 5th magnitude. **61 Cygni** is a binary pair of orange dwarf stars, magnitudes 5.2 and 6.0. The constellation also contains the radio galaxy **Cygnus A**, the black hole candidate **Cygnus X-1**, the **North America Nebula**, and the **Cygnus Loop** supernova remnant.

Cygnus A The strongest source of radio emission outside our Galaxy, situated in the constellation Cygnus. It is thought to be caused by the collision of two galaxies. It is also a source of X-ray emission.

Cygnus Loop A supernova remnant consisting of a vast loop of gas ejected from a star that exploded about 30,000 years ago. It lies about 2500 light years away. The Loop is about 3° in diameter, and is expanding at about 100 km/s. Different parts of it bear the designations NGC 6960, 6979, 6992, and 6995. The brightest part, NGC 6992, is known as the Veil Nebula.

Cygnus Rift A prominent dark nebula that splits the Milky Way in the constellation Cygnus, also called the *Northern Coalsack*; its extension through Aquila to Ophiuchus is known as the *Great Rift*. The rift is a string of **molecular clouds** lying in the **Orion Arm**, about 2200 light years away.

Cygnus X-1 A strong source of X-ray emission in the constellation Cygnus, believed to contain a **black hole**. At the position of the source lies a spectroscopic binary star, HDE 226868, with an orbital period of 5.6 days. The visible star is a blue supergiant. Its invisible companion is calculated to be around 10 solar masses, far too great for a white dwarf or neutron star, so it may well be a black hole.

D

dark adaptation (dark adaption)
The process by which the human eye adjusts from vision under conditions of high illumination to night vision. It takes 20 minutes or more to develop fully, during which time the pupil widens and different receptors come into play. Dark adaptation is essential for observers leaving a brightly lit room to begin observing. It is instantly destroyed by any bright light, so observers use dim red lights when they need to consult star charts or record observations.

dark matter Unseen matter, inferred to exist in galactic halos and in the space between galaxies, that is thought to make up at least 90% of the mass of the Universe. It could exist in various forms, such as black holes, brown dwarfs, or unknown atomic particles, all of which are termed *cold dark matter*; or it could be in the form of a "sea" of fast-moving neutrinos, known as *hot dark matter*; or a mixture of both. SEE ALSO **MACHO, missing mass.**

dark nebula A cloud of gas and dust that is not illuminated, and can thus be seen only in silhouette against stars and bright nebulae beyond. They range in size from minute **Bok globules** to the naked-eye clouds of the **Coalsack** nebula in Crux and the gigantic Rho Ophiuchi Dark Cloud which covers about 1000 square degrees (2%) of the sky. Dust comprises only about 0.1% of the mass of a dark nebula, but is believed to play an important part in the formation of molecules in space, since the surface of the dust particles provides a site for atoms to adhere and combine into molecules. The interiors of these molecular clouds are very cold, typically only 10 K, which allows them to collapse under their own gravity into stars.

David Dunlap Observatory The observatory of the University of Toronto, Canada. Its principal instrument is a 1.88-m (74-inch) reflector, opened in 1935.

Dawes' limit SEE **resolving power.**

day The time taken by the Earth to rotate once on its axis. There are various definitions. An *apparent solar day* is the interval between two successive transits of the Sun across the meridian. This interval is variable because the Sun's apparent motion throughout the year is not uniform (SEE **apparent solar time**). For convenience, an imaginary *mean Sun* is defined which moves along the celestial equator at a constant rate, giving the *mean solar day*. A *sidereal day* is defined as the interval between two successive transits of the spring **equinox** across the meridian. This day is about four minutes shorter than the mean solar day because of the Sun's apparent eastward movement of about 1° each day. A mean solar day is 24h 03m 56.555s of mean sidereal time. A mean sidereal day is 23h 56m 04.091s of mean solar time. SEE ALSO **time.**

deceleration parameter (symbol q_0) A figure describing the rate at which the expansion of the Universe is slowing down. If q_0 is greater than 0.5, the expansion will eventually stop and reverse. If q_0 is less than 0.5, the Universe will expand for ever. Theoretical considerations suggest that the true value of q_0 should be exactly 0.5. SEE ALSO **critical density, missing mass.**

declination (dec., symbol δ) The angular distance of a celestial object north or south of the celestial equator. It is reckoned positively from 0 to 90° from the equator to the north celestial pole, and negatively from 0 to 90° from the equator to the south celestial pole. SEE ALSO **celestial sphere.**

decoupling era The time, about 300,000 years after the Big Bang, when the Universe had cooled sufficiently for the first atoms to form; also known as *recombi-*

nation. The **cosmic microwave background** radiation was released at the time of decoupling.

deep sky The Universe beyond the Solar System. Star clusters, nebulae, and galaxies are known as *deep-sky objects*; the term is not usually applied to individual stars.

Deep Space 1 The first space probe in NASA's New Millennium series, intended to test new technology; Deep Space 1 was powered by an ion drive propulsion system. It rendezvoused with asteroid 9969 Braille in July 1999 and with Comet Borrelly in September 2001.

deferent SEE **Ptolemy**.

degenerate matter A state of matter existing in stars in the final stage of their evolution (SEE **stellar evolution**), when they have ceased producing energy at their cores. Atomic nuclei and electrons are packed closely together at ultra-high densities, and the laws of classical physics no longer apply. Pressure ceases to be dependent on temperature and is a function only of density.

Deimos The smaller of Mars's two **satellites**, discovered in 1877 by Asaph Hall. It is a dark, irregular body, measuring about 15 × 12 × 10 km (9 × 7.5 × 6 miles), and may well be a captured asteroid.

Delphinus A small but distinctive constellation near Aquila, representing a dolphin. Its leading stars, of fourth magnitude, make up what looks like a widespread cluster.

Delta Aquarids A **meteor shower** which occurs at the end of July and the beginning of August. It has a double radiant in Aquarius, the more active radiant lying near the star Delta Aquarii. The shower is associated with Halley's Comet.

Deneb The star Alpha Cygni, magnitude 1.25, a supergiant of type A2, luminosity half a million times the Sun's, distance about 3230 light years. It is the brightest star in the constellation Cygnus.

density parameter (Ω) In cosmological models (SEE **cosmology**), a parameter that distinguishes between an open universe and a closed universe. The density parameter for our Universe therefore determines whether it will continue to expand for ever, or whether its expansion will halt and it will ultimately collapse back on itself. The parameter is the ratio of the actual density of the Universe to its **critical density** – the density it would have if it contained just enough mass to halt expansion but prevent collapse. If Ω is greater than 1 the Universe will collapse; if Ω is less than 1 the Universe will continue to expand. Ω = 1 corresponds to the critical density.

▶ **Deep Space 1** Test firing of the xenon ion engine of Deep Space 1. Launched in October 1998, the spacecraft had a highly successful encounter with Comet Borrelly in September 2001.

▲ **differential rotation** In a fluid body, such as a star or gas giant planet, the equatorial regions rotate more rapidly than the poles. As shown, a consequence of this is that a set of points lined up on the central meridian will become spread out in longitude over the course of a rotation. Points close to the equator will return to the central meridian earlier than those near the poles.

descending node The point at which an orbit crosses from north to south of a reference plane such as the celestial equator or ecliptic. SEE **node**.

Desdemona One of the small inner **satellites** of Uranus discovered in 1986 during the Voyager 2 mission.

de Sitter, Willem (1872–1934) Dutch cosmologist. His hypothetical *de Sitter Universe,* derived from Einstein's general theory of relativity, contained no mass, but provided the first theoretical indication of an expanding Universe.

Despina One of the small inner **satellites** of Neptune discovered in 1989 during the Voyager 2 mission. It appears to be the inner **shepherd moon** to the planet's Le Verrier Ring.

diagonal (1) An optical device, used in conjunction with a telescope's eyepiece, for deflecting the light path through 90° in order to make it easier for the observer to view through the eyepiece when the telescope is at an awkward angle. In this device, also called a *star diagonal,* either a mirror or a right-angled prism is set at 45° to the light path.

diagonal (2) Another name for the flat secondary mirror in a **Newtonian telescope**.

diamond-ring effect A short-lived phenomenon seen during a total solar eclipse, just before or after totality, when just the bright central part of the edge of the Sun's disk is visible as an intense point of light, together with the faint inner corona, giving the appearance of a diamond ring.

dichotomy The phase of Mercury or Venus when it appears exactly half-illuminated by the Sun, and the **terminator** is a straight line. The term is also applied, less commonly, to the half Moon.

differential rotation The rotation of different parts of a non-solid body at different rates. The deep atmospheres of the giant planets rotate slightly faster at the equator than at the poles. The Sun shows a similar differential rotation, but rather more pronounced. The stars and other objects in galaxies are in independent orbits and with periods that increase with distance from the galactic center, so galaxies may be regarded as showing differential rotation.

differentiation The process by which a planetary body evolves a layered structure. In a body which has grown to a sufficient size by **accretion**, energy supplied by gravitational compression and radioactive decay, supplemented by the kinetic energy of impacting bodies, will cause the interior to melt. Under the action of gravity, denser materials will then sink toward the center, where they will eventually form the *core,* and less dense materials will rise toward the surface, where they will form the *mantle* and the *crust.* Differentiation has given the terrestrial planets, for example, cores of nickel–iron and outer layers of a predominantly rocky composition. The distinct chemical compositions of some meteorites and asteroids indicate that they come from bodies which were large enough to have undergone differentiation before being broken up by impacts.

diffraction The slight sideways spreading of a beam of light as it passes by a sharp edge or through a narrow slit. Starlight passing through a telescope is diffracted by the edges of various components so that, instead of coming to a sharply focused point, it appears as a small disk of light surrounded by concentric *diffraction rings* of light, and radial *diffraction spikes*. The bright central disk is called the **Airy disk**.

diffraction grating A surface on which a very large number of equidistant parallel lines have been ruled very close together. Light striking a grating is dispersed by **diffraction** into a spectrum. There are metal gratings that function by reflection, and glass gratings that function by transmission. The lines are ruled at, typically, 100 to 1000 per millimeter. Diffraction gratings produce very high-quality spectra, and are used in astronomical **spectrographs**.

diffuse nebula A luminous cloud of gas in space. The term "diffuse" refers to the fact that diffuse nebulae cannot be resolved into individual stars, unlike star clusters and galaxies. They come in two varieties: **emission nebulae**, also known as **H II regions**, which shine by fluorescence; and **reflection nebulae**, in which the starlight reflects off dust particles.

Dione A medium-sized **satellite** of Saturn, diameter 1120 km (695 miles), discovered in 1684 by Giovanni Cassini. It has the second-highest density, at 1.44 g/cm³, of Saturn's main satellites. The surface is bright and icy, but one hemisphere is darker than the other. There are three main types of terrain: cratered terrain, cratered plains, and smooth plains. Evidence of past geological activity includes a number of **tectonic** features such as Palatine Chasma – a trough over 600 km (nearly 400 miles) long and up to 8 km (5 miles) wide. Dione shares its orbit with a small **coorbital satellite**, **Helene**.

direct motion (prograde motion)
(1) Orbital or rotational motion in the same direction as the Earth's motion: counterclockwise as seen from above the Sun's north pole. The orbital motion of all planets and most satellites is direct. Orbital

or rotational motion is direct if the orbital or axial inclination is less than 90°. COMPARE **retrograde motion** (1).

direct motion (prograde motion)
(2) The regular, west-to-east motion of Solar System bodies as seen from Earth relative to background stars. COMPARE **retrograde motion** (2).

Discovery
A series of NASA space probes built on a "faster, better, cheaper" principle: low-cost missions which can be designed and launched quickly. The first was the **Near Shoemaker** probe; another was **Mars Pathfinder**.

disk galaxy A name given to any type of **galaxy** in the form of a thin disk of stars in roughly circular orbits around the center, the brightness of which tails off with distance from the center. Disk galaxies include spiral galaxies and lenticular galaxies.

dispersion The separation of white light into its constituent colors by a lens, prism, or diffraction grating. This happens because the different wavelengths of light are refracted more at the red end than at the violet end of the spectrum. Dispersion has to be countered in telescope lenses to

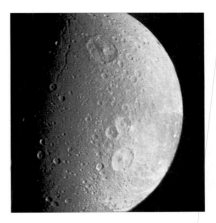

▲ **Dione** This Voyager image clearly shows Dione's uneven surface. Central peaks can be seen in the larger craters.

avoid **chromatic aberration**, but it is exploited in astronomical instruments such as the **spectrograph** to study the spectra of stars.

distance modulus The difference between the apparent magnitude (m) and the absolute magnitude (M) of a star. It is a measure of the star's distance (r) in parsecs:

$$m - M = 5 (\log r) - 5$$

diurnal motion The apparent daily motion of celestial bodies across the sky from east to west. It is a result of the Earth's axial rotation.

diurnal parallax SEE **parallax**.

Dobsonian telescope A Newtonian reflecting telescope on a simple form of **altazimuth** mount. The use of lightweight materials makes it possible to build a low-cost, large-aperture telescope that is portable, making the Dobsonian increasingly popular with amateur observers. A typical Dobsonian is mounted at its base, on a "rocker-box" assembly which rotates

▲ **Dobsonian telescope** This simple altazimuth mounting has enabled many amateur astronomers to construct large-aperture instruments comparatively cheaply. The Dobsonian offers considerable advantages of portability.

on a horizontal baseplate. It is named for its inventor, the American amateur John Dobson (1915–).

Dollond, John (1706–61) English optician. Formerly a silkweaver, in 1752 he joined the optician's business founded by his son, **Peter Dollond** (1730–1820). He found that the deviation of light rays could be achieved without **dispersion**, and combined crown and flint glass to make **achromatic lenses**. These he used in commercially successful refracting telescopes, which were further refined by Peter. John also invented the **heliometer**, a refractor modified so as measure small angular distances. Peter was the first to make an achromatic triplet lens.

Dominion Astrophysical Observatory An observatory near Victoria, BC, Canada, opened in 1917. It has 1.83-m (72-inch) and 1.22-m (48-inch) reflecting telescopes. The Dominion Radio Astrophysical Observatory, at Penticton, BC, has a 26-m (85-ft) radio dish.

Doppler effect The apparent increase in frequency (and decrease in wavelength) of radiation from a source moving toward the observer, or the similar decrease in frequency (increase in wavelength) of radiation from a source moving away. In everyday life, the Doppler effect is familiar as, for example, the change in pitch (i.e. frequency) of a siren as an ambulance speeds past.

In astronomy, its importance is that it explains the displacement of spectral lines. If a celestial object is moving toward the Earth, its light will be shifted to shorter wavelengths, toward the violet end of the spectrum. But if is moving away, then the shift will be toward longer wavelengths at the red end of the spectrum – a **redshift**. This *Doppler shift*, as it is called, makes it possible to detect and measure relative motion in the line of sight, in other words **radial velocity**, and rotation of celestial objects. The redshifts of distant galaxies provided the first evidence of the **expanding Universe**. The effect is named for Austrian physicist Christian Doppler (1803–53), who first described it.

Dorado A southern constellation, representing a goldfish. It contains the main part of the Large **Magellanic Cloud**, including the **Tarantula Nebula** and the highly luminous eruptive variable star S Doradus. Alpha, its brightest star, is magnitude 3.3.

Double Cluster Two large star clusters in the constellation Perseus, visible to the naked eye. They bear the designations NGC 869 and NGC 884, and are also known as h and χ (chi) Persei. They lie about 7200 and 7500 light years away.

double star Two stars which appear close together in the sky. There are two types of double star. If the stars are genuinely close together in space and are connected by gravity they are known as a **binary star**. Two stars which are quite distant from each other, but appear close together as a result of chance alignment, are known as an **optical double**.

doublet A lens with two components, which are either cemented together or separated by an air-gap. This construction is used to reduce **chromatic aberration**.

Draco A long, winding northern constellation representing the dragon slain by Hercules, extending between Ursa Major and Ursa Minor, with the dragon's head near the star Vega. Gamma (Eltanin), magnitude 2.2, is its brightest star. Alpha (Thuban), magnitude 3.7, was the north pole star in ancient times. On the whole Draco is a barren group but one of the stars in the head, Nu, is a double with a separation of 62 arc seconds; each component is of magnitude 4.9.

draconic month One revolution of the Moon around the Earth relative to its ascending **node**. It is equal to 27.21222 days of **mean solar time**.

Draper, Henry (1837–82) American pioneering spectroscopist. In 1872 he

▲ **Doppler effect** Electromagnetic waves emitted from a moving point source appear compressed ahead of, and more spread out away from, the direction of motion. The result is that, for light, the wavelength is shortened (blueshift) ahead of a source approaching the observer and lengthened (redshifted) behind a receding source such as a distant galaxy.

obtained the spectrum of Vega, the first time this had been done for a star other than the Sun. After his death, his widow made available the funds that made possible the *Henry Draper Catalogue*. His father, the English scientist **John William Draper** (1811–82), was a pioneering astrophotographer who took the first images of the Moon and the Sun's infrared spectrum.

Dreyer, Johan Ludvig Emil (1852–1926) Danish astronomer, who from 1874 lived and worked in Ireland, first at Lord **Rosse**'s observatory. He compiled the *New General Catalogue of Nebulae and Clusters of Stars* (SEE **NGC**) and its two supplementary *Index Catalogues*.

drive A means of moving a telescope to compensate exactly for the rotation of the Earth, so keeping the object being observed in the field of view. **Equatorial telescopes** are driven about their polar axis to track objects being observed. Computer control allows professional **altazimuth** mounted instruments to be driven simultaneously about their vertical and horizontal axes.

◀ **Dumbbell Nebula**
The brightest of the planetary nebulae, the Dumbbell takes its name from the two prominent lobes of material to either side of the central, illuminating star.

Dumbbell Nebula A large planetary nebula about 1250 light years away in the constellation Vulpecula, and designated M27. It was so named by Lord **Rosse** from its hourglass shape. The Dumbbell is visible in binoculars.

dwarf galaxy A galaxy that is much smaller than a normal galaxy and is of low luminosity. Such galaxies are usually elliptical or irregular. There are several dwarf ellipticals in the **Local Group** of galaxies.

dwarf nova A member of a class of irregular variable stars whose light curves resemble those of novae. Their luminosity stays the same for long periods, then rapidly increases, and finally slowly returns to normal. **U Geminorum stars** and **Z Camelopardalis stars** are the main types. Dwarf novae are close binary stars in which one component is a white dwarf and the other is a subgiant or dwarf.

dwarf star The most common type of star in the Galaxy, constituting 90% of its stars and 60% of its mass. Dwarfs are also known as **main-sequence** stars, from their position on the **Hertzsprung–Russell diagram**. The term "dwarf" refers to luminosity rather than size, so dwarf stars should be thought of as normal rather than diminutive. The Sun is a typical dwarf, roughly midway in the range of properties of dwarf stars.

The hottest and most massive dwarfs (spectral types O and B) have relatively short lifetimes, a few hundred million years or less, so they are rare in the Galaxy. The lifetimes of the lowest-mass dwarfs (type K and M) are so long that none has evolved from the main sequence since the Galaxy formed. Stars below 0.8 solar masses are known as **red dwarfs**, and those below 0.08 solar masses as **brown dwarfs**. SEE ALSO **stellar evolution, white dwarf**.

dynamical parallax The distance of a **binary star** determined from a relationship between the known masses of the components, the size of the orbit, and the period of revolution.

E

Eagle Nebula A bright nebula surrounding a cluster of new-born stars, collectively designated M16 or NGC 6611, in the constellation Serpens. In the nebula (known also as IC 4703) are dark columns of gas and dust, nicknamed the Pillars of Creation, in which stars are being born. The nebula and stars are about 6500 light years away.

early-type star A hot star of spectral type O, B, or A. The name was given when it was thought that the sequence of spectral types was an evolutionary sequence, hot stars being the youngest and cool stars the oldest. This is now recognized not to be the case, but the term has remained in use. COMPARE **late-type star**.

Earth The third major planet from the Sun, and the largest of the four inner, or terrestrial planets. From space the Earth is predominantly the blue of the oceans, plus the browns and greens of its land masses, the white polar caps, and a continually changing pattern of white cloud. Seventy per cent of the surface is covered by water, and it is this and the Earth's average surface temperature of 13°C (286 K) that make it suitable for life. The continental land masses make up the other 30%. Our planet has one natural satellite, the **Moon**. The main data for the Earth are given in the first table.

Monitoring the propagation of seismic waves from earthquakes has revealed the Earth's internal structure. Like all the terrestrial planets, there is a dense *core* rich in iron and nickel, surrounded by a *mantle* consisting of silicate rocks. The thin, outermost layer of lighter rock is called the *crust*. Continental crust can be as much as 50 km (30 miles) deep, but oceanic crust has an average depth of only 10 km (6 miles). The boundary between the crust and the mantle is called the *Mohorovičić discontinuity* (usually shortened to *Moho*) after its Croatian discoverer, Andrija

Mohorovičić (1857–1936). The mantle extends to a depth of 2890 km (1795 miles). The outer part of the core, down to 5150 km (3200 miles), is molten, but the inner core, 2460 km (1525 miles) in diameter, is solid because of the greater pressure there. The temperature and pressure at the Earth's center are estimated to be about 4000°C and 4 million bars (i.e. 4 million times the pressure at the surface). The solid inner core rotates two-thirds of a second faster than the layers outside it, and this, together with currents in the molten outer core, gives rise to the Earth's magnetic field.

The crust and the uppermost layer of the mantle together form the *lithosphere*, which consists of a number of tightly fitting slabs called *plates* – eight major ones and over twenty minor ones. The plates are all moving slowly with respect to one another, and are supported on a semimolten layer of mantle called the *asthenosphere*. Plates grow at *mid-oceanic ridges*, boundaries between adjoining plates where molten rock rises and solidifies into new oceanic crust, pushing the plates apart; an example is the Mid-Atlantic Ridge. Where two plates meet, one

▲ **Eagle Nebula** The "Pillars of Creation" are clearly visible in this European Southern Observatory image of the central part of the Eagle Nebula. These dark fingers of nebulosity are associated with star formation.

may be forced below the other, descending into the mantle and melting, causing earthquakes and volcanoes at the edge of the plate above. This is called *subduction*, and occurs principally in the so-called "ring of fire" around the Pacific Ocean's rim. Where plates collide and no subduction occurs, they crumple up to form folds that develop into mountain ranges like the Himalayas. All these interactions are known collectively as *plate tectonics*.

Because of tectonics, most of the Earth's surface is young compared with the planet's age of 4.6 billion years. The oldest rocks belong to the *Precambrian shields*, areas over 600 million years old, as in parts of Canada; some individual rocks have been dated to more than 4 billion years ago. Most impact craters in these areas have been weathered away, but some surface features are identifiable as impact structures.

The Earth's atmosphere consists of 78% (by volume) nitrogen, 21% oxygen, 0.9% argon, and 0.03% carbon dioxide, other gases such as neon making up the remaining 0.07%. In addition there can be up to 3% water vapor, depending on geographical location and weather conditions. There are higher concentrations of carbon dioxide and other gases, including sulfur dioxide, over industrial areas. The atmosphere produces a **greenhouse effect**, raising the surface temperature by 35 K.

The atmosphere is divided into a number of layers according to the way in which its temperature varies with altitude (SEE the table). The *troposphere* contains most of the atmosphere, and is where lifeforms are found and weather systems operate. The *stratosphere* contains the *ozone layer*, which absorbs the high-energy ultraviolet radiation from the Sun that is harmful to life. **Meteors** occur in the mesosphere, and **aurorae** in the thermosphere. The thermosphere is extremely rarefied, and its high temperature indicates the high kinetic energy of its molecules, rather than its heat content. Ionized atoms and molecules in the mesosphere and thermosphere constitute the **ionosphere**. Above the thermosphere is the exosphere, which contains the Earth's **magnetosphere** and **Van Allen Belts**, and merges into interplanetary space.

earthgrazer An alternative name for a **near-Earth asteroid**.

earthshine Sunlight reflected from the Earth on to the part of the Moon's nearside which is in the Sun's shadow. It is easiest to see when the Moon is a thin crescent, when it appears as a pale grayish light over the remainder of its disk. Earthshine is known popularly as "the old Moon in the new Moon's arms."

eccentric (1) Describing an orbit with a high **eccentricity**.

eccentric (2) SEE **Ptolemy**.

eccentric anomaly SEE **anomaly**.

eccentricity (symbol e) One of the elements of an **orbit**. It indicates how much an elliptical orbit departs from a circle. The eccentricity is found by dividing the distance between the two foci of the ellipse by the length of the major axis. A circle has an eccentricity of 0, and a parabola has an eccentricity of 1.

EARTH: DATA	
Globe	
Diameter (equatorial)	12,756 km
Diameter (polar)	12,714 km
Density	5.52 g/cm^3
Mass	5.974×10^{24} kg
Volume	1.083×10^{12} km^3
Sidereal period of axial rotation	23h 56m 04s
Escape velocity	11.2 km/s
Albedo	0.37
Inclination of equator to orbit	23° 26′
Surface temperature (average)	290 K
Orbit	
Semimajor axis	1 AU =
	149.6×10^6 km
Eccentricity	0.0167
Inclination to ecliptic	0° (by definition)
Sidereal period of revolution	365.256d
Mean orbital velocity	29.8 km/s
Satellites	1

EARTH: LAYERS OF THE EARTH'S ATMOSPHERE				
Layer	Upper boundary	Altitude (km)	Temperature (°C)	Pressure (mbar)
Troposphere	Tropopause	20*	−80*	250
Stratosphere	Stratopause	50	0	0.9
Mesosphere	Mesopause	80	−90	0.007
Thermosphere	—	500	1500	10^{-9}

The altitude, temperature and pressure are values at the upper boundary.
* Above the equator. At the poles the altitude of the tropopause is 10 km (6 miles).

echelle grating A **diffraction grating** designed to yield spectra of high **dispersion** and detail. Its lines are ruled farther apart than on an ordinary grating, and are shaped so as to produce high resolution over a narrow band of wavelengths at angles of illumination greater than 45°. Such a grating is used in an *echelle spectrograph*.

eclipse The partial or total obscuration of the light from a celestial body as it passes through the shadow cast by another body. A body may be eclipsed by the passage of another body between it and the observer, as in a **solar eclipse,** or by the intervention of another body between it and the source of the light it reflects, as in a **lunar eclipse.** The eclipse of a star by the Moon or by a planet or other Solar System body is called an **occultation.** SEE ALSO **annular eclipse, transit.**

eclipse year The interval between successive passages of the Sun through a given node of the Moon's orbit, equal to 346.62003 days of mean solar time. It differs significantly from other types of year because the nodes of the Moon's orbit move westward by over 19° per year (SEE **regression of the nodes**). Nineteen eclipse years are equal to 6585.78 days, almost exactly the same length of time as a **saros.**

eclipsing binary A type of **binary star** in which the orbital plane of the two stars is viewed almost edge-on, leading to mutual eclipses and consequent variations in the two stars' combined light output. Two eclipses may be expected during an orbital cycle, one usually causing a bigger drop in light than the other since the two stars are rarely equal in size or brightness. The deeper minimum is called the primary minimum, and the shallower minimum is the secondary minimum. If the stars are very unequal in brightness the secondary minimum may hardly be noticeable.

Eclipsing binaries are classified into three types on the basis of their light curves: **Algol** type (EA), **Beta Lyrae stars** (EB), and **W Ursae Majoris stars** (EW). The EA types have a constant light level out-

► **eclipsing binary** The light-curve (bottom) of an Algol-type binary can be accounted for by partial eclipses of each star by the other (top). At the deep primary minimum, the brighter, smaller star is largely covered by the fainter. At the secondary minimum, the brighter star passes in front of the fainter.

side eclipses, but EB and EW types show continuous variability. This is because one or both stars in the binary are elliptical in shape, having expanded to fill their **Roche lobe**. (If only one star has filled its lobe, the system is called *semidetached*; a system in which neither lobe is filled is known as *detached*.) The EW stars show more or less equal depths of primary and secondary minima. In such systems, both stars are thought to be filling their Roche lobes and are thus touching each other; they are known as *contact binaries*.

ecliptic The apparent yearly path of the Sun against the background stars. It intersects the celestial equator at the two equinoxes. The ecliptic is really the projection of the Earth's orbit around the Sun on to the celestial sphere. Because of the tilt of the Earth's axis the ecliptic is inclined to the celestial equator by an angle of about $23\frac{1}{2}°$, a figure known as the *obliquity of the ecliptic*. The ecliptic poles are the points 90° from the ecliptic, and lie in the constellations Draco and Dorado.

ecliptic coordinates SEE **coordinates**.

ecliptic limits The greatest angular distance the Sun or Moon can be from the Moon's nodes for which a **solar eclipse** or **lunar eclipse** can take place. The limits are 37° for the Sun at a solar eclipse, and 24° for the Moon at a lunar eclipse.

E corona SEE **corona**.

ecosphere The shell-shaped region of space surrounding the Sun or another star within which the temperature is such that any suitable planet would be capable of sustaining life. The dimensions of the ecosphere vary greatly with the type of star. The ecosphere of a small dwarf star would be narrow and lie close in, whereas that of a luminous giant would be broad and a considerable distance from the star.

Eddington, Arthur Stanley (1882–1944) English astronomer and physicist.

He pioneered the use of atomic theory to study the internal constitution of stars, and explained the role of radiation pressure in preventing stars from collapsing under gravity. Among his many discoveries were the mass–luminosity relationship, and the fact that white dwarfs contain degenerate matter. He supported and popularized Einstein's theory of relativity, and in 1919 he obtained experimental proof of the theory's prediction that gravity bends light by measuring star positions close to the Sun during a solar eclipse.

Eddington limit The maximum ratio of luminosity to mass that a star can have before radiation pressure overcomes the gravitational force holding it together, and the star consequently blows apart. It gives a theoretical limit to stellar mass of between about 50 and 100 solar masses.

effective temperature (symbol T_{eff}) The temperature a star would have if its output were **black-body radiation** at the same energy and at the same wavelengths as the star. As stellar spectra approximate closely to black-body spectra, a star's effective temperature is a good approximation to its actual surface temperature.

Effelsberg Radio Telescope A radio telescope near Bonn, Germany, operated by the Max-Planck-Institut für Radioastronomie, completed in 1970. With a diameter of 100 m (328 ft), it is the world's largest fully steerable radio dish. It is used either alone or as part of a **very long baseline interferometry** network.

Einstein, Albert (1879–1955) German–Swiss–American theoretical physicist. His work has had an enormous impact on 20th-century science. In astronomy, and in particular in cosmology, his theories of relativity have had profound effects. The Special Theory of Relativity (1905) gave the relation $E = mc^2$ between mass and energy; the General Theory of Relativity (1916) extended the theory to encompass gravitation. The General Theory has been borne out, for example, by its explanation of the advance of the perihelion of **Mercury**, the curvature of light by a gravitational field (SEE **Eddington, Arthur**), and the **gravitational redshift** of spectral lines. In the 1920s relativity made Einstein world-famous, but he never again matched the scientific work of his early years. He spent much time unsuccessfully seeking a unified field theory which would link relativity with electromagnetic forces. SEE ALSO **grand unified theory**.

Einstein ring A circular image of a distant source produced by a **gravitational lens**. Albert Einstein realized that if the lensing object has its mass concentrated in a single point and lies directly on our line of sight to a distant source, then in theory the source's image should be a perfect circle. In practice this is rarely seen, but incomplete circles – arcs – are often observed.

ejecta SEE **rays**.

Elara One of the four small **satellites** of Jupiter's intermediate group, discovered by Charles Perrine in 1905.

electromagnetic radiation (em radiation) The emission and propagation of energy in the form of periodic waves that can travel through a vacuum. It originates when charged atomic particles are accelerated, and it consists of oscillating electric and magnetic fields. There is a continuum of electromagnetic radiation – from long-wavelength radio waves of low frequency and low energy, through visible light waves, to short-wavelength X-rays and gamma-rays of high frequency and high energy – known as the *electromagnetic spectrum*. The speed at which em radiation travels in a vacuum (c) is given by its wavelength (λ) times its frequency (v):

$$c = \lambda v$$

This speed is constant, and is commonly known as the speed of light (SEE **light, speed of**). In the *quantum theory*, an em wave (particularly if its wavelength is that of visible light or shorter) can also be regarded as a stream of particles called *photons*, each having an energy (E) related to the frequency (or wavelength) of the radiation:

$$E = hv = hc/\lambda$$

With X-rays and gamma-rays, this energy, expressed in **electronvolts**, is a more important property than their wavelength.

Type of radiation	Gamma rays	X-rays	Ultra-violet	visible	Infrared	Radio waves
Wavelength	0.0001 0.001 0.01 0.1 1 10 100				1000 nanometers	
					10 100 1000 micrometers	
					1 10 millimeters	
					1 10 100 centimeters	
					1 10 100 1000 meters	
Radiated by objects at a temperature of ...	100,000,000	10,000,000				
		1,000,000				
			100,000			
			10,000 1000 100 10		1 K	

▲ **electromagnetic radiation** Spanning from short, gamma-ray wavelengths to long-wavelength radio emissions, the electromagnetic spectrum covers a wide range. Selection of appropriate wavelengths allows different processes and classes of objects to be observed. The visible region comprises only a tiny part of the electromagnetic spectrum.

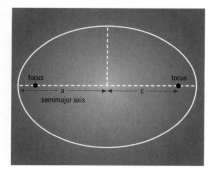

▲ **ellipse** The principal features of an ellipse are its two foci, its semimajor axis (a), its short (minor) axis and its eccentricity (e). In the case of an ellipse used to define a planetary orbit, the Sun occupies one of the foci while the other is empty. The planets' orbits are reasonably close to circular, but for most comets the orbital ellipse is extremely eccentric, with a very long major axis.

Photons also possess momentum, which means they can exert a pressure, called **radiation pressure**.

Almost all of our information about the Universe has been learned from studies of the electromagnetic radiation that reaches us from celestial objects. Much of the radiation that arrives at the Earth is absorbed by its atmosphere, but there are some bands of the electromagnetic spectrum that can reach the surface of the Earth and can be studied by Earthbound instruments. The two main ones are known as the *optical window* and the *radio window*. Other wavelength bands of cosmic radiation are detected and analyzed by instruments carried above the atmosphere by satellites or rockets.

electronvolt (symbol eV) A unit of energy used mainly in atomic physics for the energies of atomic particles. One electronvolt is the kinetic energy acquired by a particle carrying a charge equal to that on one electron when it passes through a potential difference of 1 volt. 1 eV is equivalent to 1.6021×10^{-19} joules, and, from the equation $E = h\nu$ (SEE **electromagnetic radiation**), corresponds to a frequency of

2.42×10^{14} hertz, or a wavelength of 1240 nm. An electron with a kinetic energy of 1 eV has a velocity of about 580 km/s.

element, chemical One of the basic materials of which everything in the Universe is composed. Elements are chemically homogeneous substances – that is, they cannot be decomposed by chemical means. The atoms of a particular element all have the same number of protons in their nucleus, but the number of neutrons can vary, giving rise to the different *isotopes* of the same element. The number of known elements is now over 100. Of these, 90 occur in nature, and the remainder have been prepared artificially, in laboratories. SEE ALSO **cosmic abundance**, **nucleosynthesis**.

element, orbital SEE **orbit**.

ellipse A closed curve, such that the sum of the distances of any point on it from two points within it – the *foci* (singular *focus*) of the ellipse – is a constant. In mathematics the ellipse is a type of conic section, so called because it is one of the intersections of a plane with a cone. The ellipse is important in astronomy because the closed **orbits** of planets, satellites, companion stars, comets, and so on are all close approximations to ellipses. (**Perturbations** make orbits depart from true ellipses.) It is also basic to **Kepler's laws**. Important parameters of the ellipse, and therefore of orbits, are the **semimajor axis** (*a*) and the **eccentricity** (*e*). The shape of the ellipse is governed by its eccentricity, which is given by $e = c/a$, where c is the distance from the center of the ellipse to one of its foci. For an ellipse, e is always less than 1.

elliptical galaxy A **galaxy** that is ellipsoidal in shape. Elliptical galaxies are composed of old stars and contain little gas or dust. They are the central dominant galaxies in very rich clusters, and many are powerful radio sources. Ellipticals are classified according to their degree of ellipticity (flattening), and given a designation from E0 to E7. They range widely in size, from 100 million stars for *dwarf ellipticals* to 10 million million stars for *supergiant ellipticals*.

ellipticity Another word for **oblateness**.

elongation The angular distance between the Sun and a planet or other Solar System body orbiting the Sun (or the angular distance between the Sun and the Moon). More accurately, it is the difference in the two bodies' celestial longitudes measured in degrees. A planet with an elongation of 0° is at **conjunction**; at 90° or 270° it is at **quadrature**; and at 180° it is at **opposition**. The inferior planets, Mercury and Venus, when at their maximum angular distance from the Sun, are said to be at *greatest elongation east* when following the Sun (setting at its latest time after it) and at *greatest elongation west* when preceding the Sun (rising at its earliest time before it). SEE the diagram at **conjunction**

emersion The reappearance of a celestial body from the shadow of another after it has been eclipsed or occulted.

emission nebula A cloud of gas in space that emits light. Ultraviolet radiation from nearby stars heats the gas of the nebula, causing its atoms to become ionized (lose their electrons). These electrons gain energy from the ultraviolet radiation. When they eventually recombine with the atoms they re-emit the energy they gained, some of it at visible wavelengths. The free electrons in the gas also lose energy in the form of radio waves, so emission nebulae can also be detected by radio telescopes.

Interstellar gas is mostly hydrogen, which is easily ionized by ultraviolet light. The symbol for ionized hydrogen is H II, and such emission nebulae are therefore also called **H II regions**. Hydrogen gives off its strongest light in the red, so emission nebulae appear red on photographs. However, to the eye they often look green because the eye responds more readily to a prominent pair of spectral lines of this color emitted by oxygen.

Hot stars that emit ultraviolet light are usually young, having recently formed from the gas and dust cloud around them. Thus the association between stars and gas clouds is not by chance. A famous example is the **Orion Nebula**, with the **Trapezium** stars inside it. Emission nebulae are very tenuous. Typically, every gram of material is spread through a volume of a million cubic km.

emission spectrum The **spectrum**, consisting of a series of bright lines, produced by a highly energetic source. Energy in the form of heat or electromagnetic radiation can be absorbed by an electron orbiting the atomic nucleus, causing it to jump to a higher orbit. When it returns to its old orbit, the electron emits radiation at

▶ **emission nebula** The Lagoon Nebula, M8, in Sagittarius is a classic example of an emission nebula. It shows strong H II emission, stimulated by ultraviolet radiation from young stars forming in its midst.

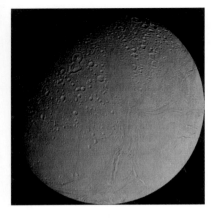

▲ **Enceladus** While heavily cratered in some regions, the surface of Enceladus has also been smoothed over by episodes of cryovolcanism. Various types of terrain can be seen in this composite image from Voyager 2.

a characteristic wavelength. The series of lines made up of emissions from different types of atom in the source is the emission spectrum. The hottest stars have many strong emission lines in their spectra. SEE ALSO **forbidden lines**.

Enceladus A **satellite** of Saturn, diameter 498 km (309 miles), discovered in 1789 by William Herschel. It is the most reflective body of any significant size in the Solar System, with an albedo approaching 1. Its surface consists of older, cratered terrain and more recently formed plains traversed by straight and curved grooves, and there are indications that **tectonic** and **volcanic** processes may still be operating. Tidal interaction with the larger satellite **Dione** probably explains why such a small body has been so geologically active. Material from Enceladus may be the source of Saturn's tenuous E Ring, in which the satellite is embedded.

Encke, Johann Franz (1791–1865) German mathematician and astronomer. In 1818 he computed the orbit of **Encke's Comet**, and predicted its return, and in 1837 he was the first to record the division in **Saturn**'s rings now named for him.

Encke also calculated the solar parallax from measurements of the 1769 transit of Venus.

Encke's Comet A periodic **comet** discovered by Pierre Méchain in 1786 and Jean Louis Pons in 1818. In 1819 **Johann Encke** computed its orbit and period, and proved that it was identical with three comets that had been previously observed. It was the second periodic comet to be discovered, and has the shortest-known period, at 3.3 years.

ephemeris (plural **ephemerides**) A table giving the predicted positions of a celestial object, such as a planet or comet, at given intervals.

epicycle SEE **Ptolemy**.

Epimetheus A small inner **satellite** of Saturn discovered during the Voyager missions. It is a **coorbital satellite** with **Janus**.

epoch (1) A point in time used as a fixed reference for comparison of astronomical data in, for example, star catalogs. **Precession** and the **proper motions** of stars cause their positions on the celestial sphere to change gradually, and observations made over any considerable period of time have to be reduced to a common epoch for them to be comparable. The starting point in the calculation of ephemerides is at a certain epoch. SEE ALSO **standard epoch**.

epoch (2) One of the elements of an **orbit**, defined as the time of perihelion passage.

Epsilon Aurigae An **eclipsing binary** star with a very long period, 27 years, magnitude 3.0 at maximum and a minimum of 4.1 during eclipse. The primary component is a supergiant of type F0. The eclipses last 700 days, the total phase taking 400 days; the last one was in 1983–4. No light is detected from the secondary component, which is thought to be surrounded by an obscuring disk of gas and dust which causes the eclipses; the secondary itself may actually be a close binary. At over 7000 light years away, Epsilon Aurigae is the most distant of the brighter naked-eye stars.

equant SEE **Ptolemy**.

equation of the center An irregularity in the motion of a body in an elliptical orbit. It is the difference between the true anomaly and the mean anomaly (SEE **anomaly**).

equation of time The difference between mean solar time, as shown on a clock, and apparent solar time, as shown on a sundial, caused by the combined effect of the eccentricity of the Earth's orbit and the **obliquity of the ecliptic**. The difference is zero four times a year: on April 15, June 14, September 1, and December 25. The maximum difference is 16 minutes and occurs in early November.

equatorial coordinates SEE **coordinates**.

equatorial telescope A telescope mounted such that it has one axis, the *polar axis*, parallel to the Earth's axis and therefore pointing to the celestial pole, and the other axis, the *declination axis*, at right angles to the polar axis and therefore parallel to the plane of the Earth's equator. Once an object is in the field of view the telescope needs only to be moved about the polar axis to follow the object. If the telescope is driven at the sidereal rate to counter the apparent rotation of the celestial sphere, the object will remain in the field of view.

There are various forms of equatorial mounting. The *German mounting*, so named because the first of the type was made by Joseph von **Fraunhofer**, is still used for small instruments. The telescope is attached at its center of mass to one end of the declination axis, which pivots about the top end of the polar axis; balance is achieved by attaching counterweights to the other end of the declination axis. Large, professional reflecting telescopes before

the age of computer-controlled **altazimuth** mountings required massive equatorial mounts in which the telescope could be supported on both sides of the declination axis and sometimes at the top and bottom of the polar axis, yet still have enough freedom of movement to access most of the sky. Such designs included the *fork mounting* and the *horseshoe mounting*.

A recent and radical departure from traditional equatorial mounts is the *equatorial table*. This is a flat surface on whose underside are curved casters which sit on motor-driven rollers. It allows amateur instruments like the traditionally altazimuth **Dobsonian telescope** to be given an equatorial capability.

equinoctial colure The **great circle** passing through the north and south celestial poles and the equinoxes (which are also called the equinoctial points).

equinox Either of the two points at which the Sun crosses the celestial equator. The *vernal equinox* (or *spring equinox*) occurs

▲ **equatorial telescope**. The equatorial mounting allows a telescope to be driven so that it can follow the apparent motion of the stars due to Earth's rotation. The equatorial is based around two main axes. The polar axis, around which the driving motor turns the telescope, is aligned parallel to Earth's axis of rotation. North and south movement is made along the declination axis.

when the Sun crosses the equator from south to north on or near March 21 each year. This point is also called the *First Point of Aries*. The other equinox is the *autumnal equinox*, which occurs on or near September 23 when the Sun crosses the equator from north to south. This point is also called the *First Point of Libra*. At the equinoxes, the Sun rises due east and sets due west.

Equuleus A very small and obscure constellation, representing a foal, next to Pegasus. Its brightest star, Alpha (Kitalpha), is only of magnitude 3.9. It has no features of particular interest.

Eratosthenes (*c.* 275–195 BC) Greek geographer, mathematician and astronomer. He is best known for making the first scientific calculation of the Earth's circumference, in which he measured the altitude of the Sun at Alexandria on a day when he knew the Sun to be overhead at a place a considerable distance away. How accurately he knew this distance, and hence how accurate his result was, are uncertain, but the method was sound. Eratosthenes also compiled a star catalog, and measured the **obliquity of the ecliptic**.

Erfle eyepiece A telescope eyepiece with low power and wide field of view, over 60°. It consists of three elements, at least

▲ **Eros** The rotation of Eros can be seen in this series of images obtained from the NEAR–Shoemaker spacecraft as it made its close approach to the asteroid in February 2000.

one of which is a doublet. (After German optician Heinrich Erfle, 1884–1923.)

Eridanus One of the largest constellations in the sky, representing a river; it extends from near Rigel in Orion to the south polar area. **Achernar** is its brightest star. Theta (Acamar) is a fine double, with components of magnitude 3.4 and 4.3. Epsilon, magnitude 3.7, is a nearby K2 main-sequence star, 10.5 light years away.

Eros Asteroid no. 433, discovered by Gustav Witt in 1898. It measures 33 × 13 km (21 × 8 miles), and rotates in 5.3 hours. Its orbit crosses that of Mars, and it is the largest of the **Amor asteroids**. The **Near Shoemaker** probe entered orbit around Eros in February 2000 and studied the asteroid for 12 months.

eruptive variables Another name for **cataclysmic variables**.

ESA ABBREVIATION FOR **European Space Agency**.

escape velocity The minimum velocity that a body, such as a rocket or a space probe, must attain to overcome the gravitational attraction of a larger body and leave on a trajectory that does not bring it back again. It depends on the mass (m) of the larger body and its radius (r), and is given by $\sqrt{(2GM/r)}$, where G is the gravitational constant. It is escape velocity rather than mass that determines the gases a body is able to retain as an atmosphere.

ESO ABBREVIATION FOR **European Southern Observatory**.

Eta Aquarids A **meteor shower** that occurs in May. Its radiant is in Aquarius, near the star Eta Aquarii. Like the Orionids, the shower is associated with Halley's Comet.

Eta Carinae A peculiar variable star lying within the nebula NGC 3372 in the constellation Carina, about 8000 light years away. It was originally cataloged at

ESCAPE VELOCITIES OF BODIES IN THE SOLAR SYSTEM	
Body	Escape velocity (km/s)
Sun	617.0
Mercury	4.4
Venus	10.4
Earth	11.2
Moon	2.4
Mars	5.0
Jupiter	59.5
Saturn	35.5
Uranus	23.5
Neptune	21.3
Pluto	1.1

third magnitude, but from 1833 began to vary irregularly, becoming at its brightest second only to Sirius. It then faded to just below naked-eye visibility. Material thrown off in the outburst forms the *Homunculus Nebula*, so called because its shape resembled a human figure to early observers. Studies of the light reflected from the dust cloud show that the central object varies in emission lines, especially those of hydrogen and neutral helium, with a period of 5.52 years. This suggests that the central object is in fact a binary system consisting of two stars of about 80 solar masses orbiting each other at a distance of about 20 AU.

Eudoxus of Cnidus (*c.* 408–355 BC) Greek mathematician and astronomer. His theory of 27 "homocentric" (centered on the Earth) spheres was the first ever attempt to account for the movements of the Sun, Moon, and planets on what we would now call a properly scientific basis. This picture of "heavenly spheres" would survive for two thousand years. According to the Roman historian Pliny, Eudoxus fixed the length of the year at 365.25 days, and according to the Roman architect Vitruvius he invented the sundial.

Euler, Leonhard (1707–83) Swiss mathematician. He made outstanding contributions to many areas of applied mathematics, especially theoretical astronomy. His analysis of the Moon's motion laid the foundations of lunar theory, and he made important advances in the study of the tides, planetary perturbations, and the orbits of comets. Euler's work on optical systems strongly influenced the technical development of telescopes and microscopes.

Europa The smallest of Jupiter's **Galilean satellites**, with a diameter of 3138 km (1950 miles), and the second-closest to the planet. Its density of 2.97 g/cm^3 indicates that it is a predominantly rocky body. Europa's smooth crust of water ice is crisscrossed by a network of light and dark linear markings. There are very few craters, so old ones must have been removed by some form of geological activity. A kind of ice **tectonics** might be operating on Europa. Images from the **Galileo** probe support this idea: they suggest that the dark lines may be deposits from geysers that erupted along long fissures, and that a few kilometers below the surface lies a layer of liquid water or "slush." SEE ALSO **satellite**.

European Southern Observatory (ESO) A consortium of ten European countries, including the UK. Its telescopes include a 3.6-m (142-inch) reflector, completed in 1975, and the **New Technology**

▲ **Europa** Rocky regions on Europa's surface are shown in brown. The pale areas toward the poles are icy plains. Long dark fractures can also be seen.

Telescope, both at La Silla Observatory; and the **Very Large Telescope** at the Paranal Observatory.

European Space Agency (ESA) An organization of European nations to promote space research and technology for peaceful purposes, founded in 1975. ESA has 15 member states: Austria, Belgium, Denmark, Finland, France, Germany, Ireland, Italy, the Netherlands, Norway, Portugal, Spain, Sweden, Switzerland, and the UK. Its headquarters are in Paris. ESA has designed and built the Ariane series of launch rockets, and operates a launch site at Kourou in French Guiana, on the Atlantic coast of South America.

EUVE ABBREVIATION FOR **Extreme Ultraviolet Explorer**.

evection An irregularity in the Moon's motion caused by perturbations by the Sun and planets. It amounts to a maximum of $1° 16'$ in a period of 32 days.

evening star Name sometimes given to the planet Venus when it is visible in the west after sunset.

event horizon The boundary surface of a **black hole** from within which nothing can escape. Observers outside the event horizon can therefore obtain no information about the black hole's interior. The radius of the event horizon is known as the **Schwarzschild radius**.

Evershed effect The radial outflow of gases within the penumbra of a **sunspot**. The effect was discovered in 1908 by English astronomer John Evershed (1864–1956), from the Doppler shift in the spectrum of a sunspot.

exit pupil For a telescope, the image of the primary mirror or lens as seen through the eyepiece. The point at which this image is formed is the optimum position for the observer's eye, so for comfort it should not be too close to the eyepiece itself (SEE **eye relief**).

exobiology The study of the possible existence of life beyond the Earth; also known as **astrobiology**. This includes looking for life forms on other planetary bodies in the Solar System, such as Mars and Europa; the study of organic molecules in nebulae, meteorites, and comets; the study of planetary surfaces and atmospheres; questions relating to the origin of life and the range of conditions under which it can survive; and the search for signs of life in space, such as incoming messages at radio or other wavelengths and investigating **extrasolar planets**. SEE ALSO **SETI**.

Exosat A European Space Agency satellite for X-ray astronomy, launched in 1983. It operated for three years, studying objects such as supernova remnants, X-ray binaries, and galactic nuclei.

expanding Universe The observation that each galaxy, or cluster of galaxies, is moving apart from the others, as revealed by the **redshift** in their light. The redshift is attributed to the Doppler effect and means that these objects are receding from us at velocities that are proportional to their distance (this is the **Hubble law**). The expansion was discovered in 1929 by Edwin Hubble, although the possibility had been suggested earlier on theoretical grounds by Georges **Lemaître**. SEE ALSO **cosmology**.

extinction SEE **atmospheric extinction**, **interstellar matter**.

extrasolar planet A planet orbiting a star other than the Sun. Stars with irregular **proper motions**, as often revealed by radial-velocity measurements, were long suspected of having planets orbiting them; **Barnard's Star** was a famous candidate. In 1992 the first confirmed extrasolar planets were detected, orbiting the pulsar PSR $1257 + 12$. In 1995 the first planet orbiting another main-sequence star, 51 Pegasi, was identified. By early 2003 around a hundred planetary systems had been discovered around a wide diversity of stars up to 180 light years away. Many of these *sub-*

stellar objects are in very close orbits and are many times the mass of Jupiter. The lower theoretical mass limit for a **brown dwarf** is about 0.013 solar masses (approximately 13 Jupiter masses), which by convention is taken as the upper mass limit for a giant planet.

Extreme Ultraviolet Explorer (EUVE) A NASA satellite launched in 1992 to survey the sky and study individual sources at very short ultraviolet wavelengths, a region of the spectrum not covered by previous satellites. It operated until December 2000.

extrinsic variable SEE **variable star**.

eyepiece (ocular) The system of lenses in an optical instrument nearest the observer's eye. Its function in a telescope used for visual observation is to magnify the image formed at the focus by the primary lens or mirror. Telescope eyepieces usually have two lenses, the *field lens*, farthest from the eye, and the *eye lens*. The magnification yielded by an eyepiece is given by the ratio of the focal length of the telescope's primary lens or mirror to the focal length of the eyepiece. There are numerous designs of eyepiece, varying in focal length, field of view, and magnification: SEE **Erfle eyepiece, Huygenian eyepiece, Kellner eyepiece, Nagler eyepiece, orthoscopic eyepiece, Plössl eyepiece, Ramsden eyepiece**. SEE ALSO **Barlow lens**.

NOTEWORTHY EXTRASOLAR PLANETS

Star	Mass of planet (Jupiter = 1)	Distance from star (AU)	Period of orbit (days)
HD 83443 b	0.35	0.04	2.99
HD 83443 c	0.17	0.17	29.83
HD 209458	0.63	0.05	3.52
51 Peg	0.46	0.05	4.23
υ And b	0.68	0.06	4.62
υ And c	2.05	0.83	241.3
υ And d	4.29	2.56	1308.5
GJ 876 b	1.89	0.21	61.02
GJ 876 c	0.56	0.13	30.12
HD 168443 b	7.64	0.29	58.10
HD 168443 c	16.96	2.87	1770
HD 114762	10.96	0.35	84.03
HD 82943 b	1.63	1.16	444.6
HD 82943 c	0.88	0.73	221.6
16 Cyg B	1.68	1.69	796.7
HD 10697	6.08	2.12	1074
47 UMa b	2.56	2.09	1090
47 UMa c	0.76	3.78	2640
HD 74156 b	1.55	0.28	51.60
HD 74156 c	7.46	3.47	2300
ε Eri	0.88	3.36	2518

eye relief In an optical instrument such as a telescope or binoculars, the distance between the eye lens of the **eyepiece** and the **exit pupil**. When the observer's eye is placed at the exit pupil, the whole of the eyepiece's field of view is visible. The eye relief must be sufficiently large for comfortable viewing.

F

Fabry–Pérot interferometer A type of **optical interferometer** for making accurate measurements of close spectral lines (SEE **spectrum**). The two parts of the incoming beam are recombined after passing through an arrangement of parallel plates called a *Fabry–Pérot etalon*. The beams under-go multiple reflections between the plates, which are highly reflective and accurately parallel. The instrument was developed by the French physicists Charles Fabry (1867–1945) and Alfred Pérot (1863–1925).

faculae (singular **facula**) Bright areas in the Sun's upper photosphere that herald the appearance of sunspots. They appear some hours before the spots, in the same place, but can remain for months after the sunspots have gone. Faculae are at a high-

▲ **faculae** Bright regions marking the location of clouds of hot hydrogen above active regions on the Sun, faculae are best seen, as here, close to the limb. Faculae are commonest around sunspot maximum.

er temperature than their surroundings, and so appear brighter. The word is Latin for "little torches."

falling star A popular (US) name for a **meteor**.

False Cross A pattern of stars made up of Iota and Epsilon Carinae, and Kappa and Delta Velorum. It is often mistaken for the Southern Cross, but it is larger, less bright, and more symmetrical.

Far Infrared and Submillimetre Space Telescope SEE **Herschel Space Observatory**

Far Ultraviolet Spectroscopic Explorer (FUSE) A NASA satellite launched in June 1999 to study the sky at wavelengths of 90–120 nm, in the ultraviolet. One objective is to determine the amounts of deuterium in various cosmic sources, which will refine our knowledge of conditions in the very early Universe.

F corona SEE **corona**.

field of view The angular diameter of the area of sky visible through an optical instrument. For a telescope or binoculars it depends on the eyepiece that is used: a higher-power eyepiece that increases the magnification will decrease the field of view. Wide-angle eyepieces such as the **Nagler eyepiece** have fields of 80° or more.

filament A solar **prominence** seen in silhouette against the Sun's bright disk as a dark, thread-like marking in photographs taken in the light of hydrogen alpha or the K line of calcium.

filar micrometer A device used in conjunction with the eyepiece of a telescope that incorporates cross-wires for measuring the angular size of an object or the separation between two objects. Superimposed on the field of view the observer sees two parallel wires and a third wire perpendicular to them. One of the parallel wires is movable, and the observer adjusts its position until the intersections of the two parallel wires with the third wire mark off the distance being measured. The angular distance may then be read from the micrometer scale.

filter A device used in conjunction with an instrument for receiving electromagnetic radiation which allows some wavelengths to pass, but blocks others. Filters for use with optical instruments such as telescopes are at their simplest a thin sheet of material placed over the full aperture of the instrument. Simple color filters made of gelatin bring out planetary features like Jupiter's Great Red Spot. More sophisticated filters that transmit only a narrow band of wavelengths (said to have a narrow *passband*) are used for observing the Sun, or objects such as nebulae. Multilayer or interference filters are used to combat **light pollution**. In **astrophotography**, color pictures can be created by combining separate exposures on suitable black-and-white emulsions taken through blue, green, and red filters. Filters are also used in **CCD** imaging. SEE ALSO **Sun**.

finder A small, low-power telescope mounted on a larger one and having a much larger field of view than the main telescope, enabling the observer to locate celestial objects more easily.

fireball An exceptionally bright **meteor**. Fireballs have been defined only loosely as meteors brighter than the planets; with the modern estimate of the maximum brightness of Venus, this would mean that all meteors brighter than magnitude -4.7 should be classified as fireballs. Some meteor showers, such as the Geminids, produce more fireballs than others.

first contact In a **solar eclipse**, the moment when the leading edge of the Moon's disk first touches the Sun's disk. In a **lunar eclipse**, it is the moment when

▲ **filament** The dark marking at top right in this image is a filament. It is seen in silhouette against the bright solar disk.

the Moon begins to enter the umbra of the Earth's shadow. The term is also used for the corresponding stage in eclipses involving other bodies.

first light The occasion on which a newly constructed telescope is first used to observe a celestial object.

First Point of Aries An alternative name for the vernal **equinox**, so called because it originally lay in the constellation Aries. Because of **precession**, it has now moved into neighboring Pisces.

first quarter One of the Moon's **phases**, when it is at quadrature on the way to being full and is half illuminated as seen from the Earth.

Flamsteed, John (1646–1719) English astronomer. He was appointed the first Astronomer Royal by King Charles II, with the task of obtaining accurate measurements of the Moon and stars for use in navigation. From Greenwich Observatory he used equipment fitted with sighting telescopes to measure with unprecedented precision the positions of thousands of stars. Flamsteed was a perfectionist and slow to release his results. Much to his anger, The Royal Society published his uncorrected observations. The catalog of

▲ **flare** Violent releases of magnetic energy in the inner solar atmosphere above active regions are seen in spectroheliograms as bright patches or ribbons, described as flares. Flares are most common at times of high sunspot activity.

his corrected observations was not published until 1725, after his death; the numbers assigned to stars in this catalog are still used (SEE **Flamsteed numbers**).

Flamsteed numbers A series of numbers allocated to the stars in each constellation in order of right ascension, for purposes of identification. The numbers were allocated by later astronomers to stars in John **Flamsteed**'s star catalog.

flare A sudden and violent explosion on the Sun's surface, usually from the region of an active group of **sunspots**. In the *flash stage* a flare builds to a maximum in a few minutes, after which it gradually fades within an hour or so. Flares emit radiation across the whole electromagnetic spectrum, from gamma-rays to kilometer-wavelength radio waves. Particles are emitted, mostly electrons (some at half the speed of light) and protons, and smaller numbers of neutrons and atomic nuclei. A flare can cause material to be ejected in bulk, most spectacularly in the form of **prominences**, at speeds that can exceed the Sun's escape velocity. When energetic particles from flares reach the Earth they cause radio interference, magnetic storms, and more intense **aurorae**. (It may be that **coronal mass ejections**, in which billions of tons of gas are ejected into space from the corona, are responsible for these phenomena, and not flares.) Although not well understood, the origin of flares is believed to be connected with local discontinuities in the Sun's magnetic field. SEE ALSO **flare star**.

flare star A variable star, usually a red dwarf, whose luminosity from time to time increases unpredictably and by as much as several magnitudes in a very short time (only a few seconds), and then decreases to its normal value in about one minute to one hour. Flare stars are also known as *UV Ceti stars*, after the best-known example, UV Ceti, which is a pair of 13th-magnitude red dwarfs. Outbursts of flare stars are sometimes accompanied by an increase in radio emission. Large flares may give out from 10 to 100 times as much light as the rest of the star. Flare stars are often binary, notably the *BY Draconis stars*, which show additional variations of a few tenths of a magnitude attributed to large "starspots" passing across their disk as they rotate. The low mass and relative youth of red dwarf flare stars means that steady hydrogen burning in their interiors may not yet have established itself. Flaring appears to be a consequence of instabilities in the young stars, allied with complex magnetic fields.

flash spectrum The **emission spectrum** of the solar chromosphere which is observable for a few seconds just before and after totality during an eclipse of the Sun.

flocculi (singular **flocculus**) Small features which give a mottled or granulated appearance to the Sun's chromosphere.

focal length The distance between the center of a lens or curved mirror and its focus.

focal ratio The ratio of the **focal length** of a lens or curved mirror to its diameter. A focal ratio of, say, 8 is written as *f*/8.

focus (1) For a lens or curved mirror, the point at which is formed the image of a distant source lying on the axis of the lens or mirror.

focus (2) SEE **ellipse**.

following (abbreviation *f*) Describing the trailing edge, feature, or member of an astronomical object or group of objects. For example, the *following limb* of the Moon is the edge facing away from its direction of motion; and the *following spot* (or *f*-spot) of a group of sunspots is the last of the group to be brought into view by the Sun's rotation.

Fomalhaut The star Alpha Piscis Austrini, magnitude 1.16, distance 25 light years, luminosity 17 times that of the Sun. It is a main-sequence star of type A3 surrounded by a **proplyd**.

forbidden lines Emission lines (SEE **emission spectrum**) in the spectra of some celestial objects which, when they were first observed, could not be identified with lines in spectra produced under laboratory conditions on Earth. It is now known that they are produced by atoms in what is called a metastable state. On Earth this state is very short-lived, but in highly rarefied nebulae, for example, it can last long enough for "forbidden" lines to be emitted.

fork mounting A type of mounting used for a large **equatorial telescope**.

Fornax A southern constellation originated by **Lacaille**, representing a chemical furnace. The only star above fourth magnitude is Alpha, magnitude 3.9, a double. Fornax includes a cluster of galaxies.

fourth contact (last contact) In a **solar eclipse**, the moment when the trailing edge of the Moon's disk last touches the Sun's disk. In a **lunar eclipse** it is the moment when the Moon's trailing edge leaves the umbra of the Earth's shadow. The term is also used for the corresponding stage in eclipses involving other bodies.

Fraunhofer, Joseph von (1787–1826) German physicist and optician, the founder of astronomical spectroscopy. He studied the diffraction of light through narrow slits and developed the earliest form of diffraction grating. In 1814 he observed and began to map the dark lines in the Sun's spectrum now called **Fraunhofer lines**, and was the first to appreciate their significance. He later found similar but differently distributed lines in the spectra of other stars. Fraunhofer solved many of the scientific and technical problems of astronomical telescopemaking. He manufactured achromatic lenses, was the first to build an **equatorial telescope** that realized the full potential of the design, and built the heliometer with which Friedrich **Bessel** measured the first stellar parallax.

Fraunhofer lines The dark **absorption lines** in the solar spectrum, caused by absorption at specific wavelengths in the upper, cooler, layers of the Sun's atmosphere. The most prominent ones at visible wavelengths are caused by the presence of neutral hydrogen (the **hydrogen alpha** or Hα line), singly ionized calcium (the calcium H and K lines), sodium, and magnesium. Although they were first observed

▲ **Fraunhofer lines** Dark lines interrupt the continuum of the solar spectrum. These result from absorption of light at specific wavelengths by atomic species in the Sun's atmosphere.

in 1802 by William Hyde Wollaston, they were first carefully studied from 1814 by Joseph von **Fraunhofer**.

Fred L. Whipple Observatory An

observatory located on Mount Hopkins, near Amado, Arizona, operated by the **Smithsonian Astrophysical Observatory** (SAO). It was opened in 1968, and later renamed to honor Fred **Whipple**, a former SAO Director. The main instruments are a 1.5-m (60-inch) and a 1.2-m (48-inch) reflecting telescope. The observatory runs the **MMT Observatory** jointly with the University of Arizona.

F star A star of spectral type F, with a surface temperature between 6100 and 7400 K on the main sequence; giants and supergiants of the same spectral type are about 300 K cooler. Main-sequence F stars range in mass from 1.2 to 1.7 solar masses, but most F-type giants and supergiants have evolved from considerably higher-mass stars. Bright examples are Canopus and Procyon.

full Moon The Moon's **phase** when it is at opposition, and its illuminated hemisphere is fully visible from the Earth.

fundamental stars Reference stars whose positions and **proper motions** have been determined with the greatest possible accuracy. Their coordinates are published in a *fundamental catalogue* and serve as the points to which positional measurements of other bodies can be related. The IAU has adopted 1535 stars as fundamental reference stars; relevant data are contained in the *Fifth Fundamental Catalogue*, published in Germany and known for short as the *FK5* from its German title.

FUSE ABBREVIATION FOR **Far Ultraviolet Spectroscopic Explorer**.

G

galactic coordinates SEE **coordinates**.

galactic halo The spheroidal distribution of old stars and globular clusters that surrounds the Galaxy. Its radius is about 50,000 light years. There appears also to be a much more extensive halo of **dark matter** extending out to a radius of at least 150,000 light years.

Galatea One of the small inner **satellites** of Neptune discovered in 1989 during the Voyager 2 mission.

galaxy A huge assembly of stars, dust and gas, an example of which is our own Galaxy. There are three main types, as originally classified by Edwin **Hubble** in 1925. *Elliptical galaxies* are round or elliptical systems, showing a gradual decrease in brightness from the center outward. They are given a designation from E0 to E7 in increasing degree of ellipticity. *Spiral galaxies* are flattened disk-shaped systems in which young stars, dust, and gas are concentrated in spiral arms coiling out from a central bulge, the *nucleus*; they are designated S, with lower case letters added to show how tightly the arms are wound, from tight (a) to loose (c). *Barred spiral galaxies* are distinguished by a bright central bar from which the spiral arms emerge; they are designated SB with letters from a to c appended, as for ordinary spirals. These three main classes are represented in the **tuning fork diagram**. In addition there are *lenticular galaxies*, systems intermediate between ellipticals and spirals, having a disk and nucleus, similar to spiral galaxies, but with no apparent spiral arms; they are classified S0. **Irregular galaxies** (designated Irr or Ir) are systems with no symmetry.

Current theories suggest that all galaxies were formed from immense clouds of gas at roughly the same time, soon after the Big Bang. In elliptical galaxies star formation took place rapidly over perhaps several hundred million years, using up all the interstellar gas. Spirals, however, are the result of a two-stage formation process. Gas was left over from the initial star formation which produced the bulge at their centers. The remaining gas rapidly settled into a disk, in which density waves formed spiral arms.

Galaxies can exist singly or in clusters that contain anywhere from just a few to thousands of members (SEE **cluster of galaxies**). Galaxies in clusters have a good chance of one near-collision in their life-

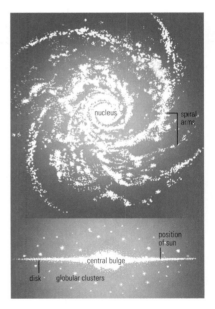

▲ **galaxy** The upper diagram shows the general outline for the structure of a spiral galaxy, of which our Galaxy is an example. Seen face-on (top), the galaxy has a fairly compact nucleus surrounded by an extensive disk containing star-rich spiral arms. Seen edge-on (below), the nucleus is revealed as a bulge, with the spiral arms confined to the flat plane of the disk. The galaxy as a whole is surrounded by a halo of globular clusters. As indicated, the Sun is located midway out in the disk of our Galaxy.

times. Interactions between gas-rich galaxies such as spirals can produce bursts of star formation, and such interactions are a possible trigger mechanism for generating **active galactic nuclei**, as in **Seyfert galaxies** and **quasars**. These objects are generally believed to be galaxies with nuclei in extreme states of activity. SEE ALSO **disk galaxy**, **low-surface-brightness galaxy**.

Galaxy, the The star system that contains the Solar System; the capital "G" distinguishes it from other **galaxies**. The Galaxy is spiral in shape and about 100,000 light years in diameter. Our Sun and Solar System are located at the edge of one of the spiral arms, 26,500 light years from the center. The spiral arms form a disk-shaped system, with a bulging core (or nucleus) in the direction of Sagittarius. Old stars, of Population II, are found in the Galaxy's core; younger, hotter Population I stars, together with interstellar dust and gas, make up the spiral arms (SEE **stellar populations**). The stars of the spiral arms form the Milky Way, running around the sky. It is uncertain how many spiral arms the Galaxy has, but there is good evidence, from infrared studies for example, that there is a modest bar at the center, and that

THE GALAXY: DATA

Diameter of disk	100,000 l.y.
Thickness of nucleus	6000 l.y.
Average density	7×10^{-24} g/cm^3
Total mass	2×10^{45} g = 10^{12} solar masses
North pole of galactic plane (epoch 2000)	RA 12h 51.4m, dec. +27° 7′
Point of zero longitude (epoch 2000)	RA 17h 45.6m, dec. −28° 55′
Galactic longitude of north celestial pole	123°
Distance of Sun from center	26,500 l.y.
Distance of Sun above galactic plane	14 l.y.
Rotational velocity of Sun	250 km/s
Period of Sun's revolution around center	200 million years
Absolute magnitude	−20.5

it should be classified as a barred spiral galaxy. Surrounding it is the spheroidal **galactic halo**, which is about 100,000 light years in diameter. In it are found very old Population II stars, including those in **globular clusters**.

Between us and the Galaxy's central bulge passes a prominent spiral arm, seen as the bright areas of the Milky Way in Centaurus, Crux, and Carina. This arm curls away from us in Carina where we see it virtually end-on, which accounts for the great accumulation of stars and nebulae in that constellation. Immediately outside this arm is a band of gas and dust, a region where stars have yet to form, which is visible as the dark rift in the Milky Way running from Cygnus to Crux. Beyond Carina we see a fainter arm that wraps itself round outside the Sun. It crosses the northern sky, gaining in brightness as it moves inward toward Cygnus.

The center of the Galaxy is marked by the radio source **Sagittarius A**, which has been adopted as the zero point of galactic longitude. The *galactic equator* is the plane of the Galaxy, running along the Milky Way. The whole Galaxy is rotating, but the rotational rate varies with the distance from the center. Our Sun circles the center at about 250 km/s, taking 200 million years per orbit. The Galaxy's age is thought to be about 10,000 million years, during which time the Sun would have completed approximately 50 revolutions.

The Galaxy is a member of a cluster known as the **Local Group**, which includes the Andromeda Galaxy and the Magellanic Clouds.

Galilean satellites The four chief satellites of Jupiter – **Io**, **Europa**, **Ganymede**, and **Callisto** – named collectively after **Galileo**, who observed them in 1610. All are in low-inclination, near-circular, **synchronous orbits**, and their orbital periods are in simple ratios to one another as a result of **resonance**. Apart from the Earth's Moon, they are the brightest satellites in the Solar System, and all would be naked-eye objects were it not for the glare from Jupiter itself. They pass

▲ **Galilean satellites** The four large satellites of Jupiter are arranged in order of distance from the planet: (left to right) Io, Europa, Ganymede and Callisto. The images were obtained by the Galileo orbiter.

through cycles of eclipses by Jupiter and its shadow, and transits across its disk. SEE ALSO **Jupiter, satellite**.

Galilean telescope A **refracting telescope** having a planoconvex objective and a planoconcave eyepiece. It forms an erect image and gives a small field of view. This type of instrument, first employed by **Galileo** to make astronomical observations, is no longer used in astronomy, but the same optical system is still used in opera glasses.

Galileo Galilei (1564–1642) Italian astronomer, physicist, and mathematician. He was one of the first to use the telescope for astronomical observations, and he improved its design. He discovered mountains and craters on the Moon, and found the four satellites of Jupiter now known as the **Galilean satellites**. He observed the phases of Venus, and studied sunspots, from whose motion he deduced that the Sun rotates. He found that some celestial objects, such as the Milky Way and the Pleiades, are resolvable into many more stars than can be distinguished by the naked eye.

Galileo concluded that **Aristotle**'s picture of the world, then still widely believed, was wrong, and instead championed **Copernicus**'s heliocentric theory. This brought him into conflict with the Catholic Church, and led to his trial and house arrest for the last eight years of his life. His many advances in physics included the law of the pendulum and how to create a vac-

uum. His chief books were *Sidereus nuncius* (on discoveries with the telescope), *Dialogue on the Two Chief World Systems* (heliocentric Universe), and *Discourses and Mathematical Demonstrations on Two New Sciences* (mathematical physics).

Galileo A space probe to Jupiter, launched in October 1989. *En route*, Galileo flew past the asteroids **Gaspra** and **Ida** in October 1991 and August 1993 respectively, photographing both. Galileo entered orbit around Jupiter in December

▲ **Galileo** Repeated passes close to the four large satellites of Jupiter by the Galileo orbiter since 1996 have revealed these in more detail than ever before. This image shows the volcano Zal Patera on Io.

◄ **gamma-ray burst** GRB 00131 was detected by several spacecraft on January 31, 2000. This series of images obtained with the Very Large Telescope (VLT) shows the fading visible-wavelength afterglow, just left of center.

1995, after dropping off a subprobe which entered the planet's atmosphere. It detected a hydrogen/helium ratio similar to the Sun's, and an intense radiation belt 50,000 km (30,000 miles) above the upper cloud deck. From 1995 to 2002 Galileo made over two dozen fly-bys of the Galilean satellites, returning highly detailed views. Its mission ended in 2003.

Galileo National Telescope (TNG)
A 3.5-m (138-inch) reflecting telescope at **Roque de los Muchachos Observatory**, opened in 1997 as a national facility for Italian astronomers. (The initials are those of the Italian version of the name.)

Galle, Johann Gottfried
(1812–1910) German astronomer. In 1846 he and Heinrich D'Arrest were the first to locate the planet Neptune, close to positions calculated independently by John Couch **Adams** and Urbain **Le Verrier**. Galle discovered Saturn's faint C Ring, and determined the solar parallax (used to fix the scale of the Solar System) by observing the asteroid Flora.

gamma-ray astronomy
The study of very high-energy **electromagnetic radiation** from space with wavelengths of about 0.01 nm and shorter. Gamma-rays are absorbed by the Earth's atmosphere, so all studies have to be made by high-altitude balloons and spacecraft. Early surveys of the sky at gamma-ray wavelengths were made by the satellites SAS 2, launched in 1972, and COS-B, launched in 1975. They found a general concentration of emission in the galactic plane, with a peak toward the galactic center, plus several strong individual sources such as the Crab Nebula, the Vela Pulsar, and the quasar 3C 273. In 1991 NASA launched a major satellite for gamma-ray studies called the **Compton Gamma Ray Observatory**.

Their very short wavelengths mean that gamma-rays have very high energies, and they are emitted from regions of very high temperature associated with some of the most violent events in the Universe. Most of the gamma radiation from the Galaxy is produced by cosmic rays interacting with gas in the interstellar medium. Certain gamma-ray lines at specific energies have been discovered, one attributed to the mutual annihilation of electrons and positrons in the galactic center, and another to the isotope aluminum-26, which is produced by supernovae.

Outside the Galaxy, the main sources of gamma-rays are those that produce **gamma-ray bursts**. There is also a gamma-ray background, a diffuse radiation from all over the sky, like the cosmic microwave background; its origin is unknown.

gamma-ray burst (GRB) An intense flash of gamma-rays, which may last from about a second to a few minutes. GRBs were first detected in 1967 by a satellite monitoring nuclear explosions on Earth. Since they occur randomly across the sky, at the rate of a few per day, they could not originate in our Galaxy (they would then be concentrated near the Milky Way). They had to be relatively near or very distant. In 1997 it was found that GRBs are followed by an X-ray "afterglow," which enabled their positions to be pinpointed with greater accuracy. Their optical counterparts were found to be young galaxies with high redshifts, which makes GRBs extremely energetic events. In 1999 the **Hubble Space Telescope** imaged the source of a GRB while it was in progress: a 24th-magnitude galaxy. Theorists have suggested exotic models such as colliding neutron stars or even "hypernovae" – enormously powerful supernova events – to account for the colossal energy output of GRBs.

Gamow, George (1904–68) Russian–American nuclear physicist and cosmologist. His most famous work, coauthored with Ralph Alpher and Hans **Bethe**, pictured a hot and dense early Universe in which the nuclei of elements were built up from protons, and was a cornerstone of **Big Bang** theory. Gamow's greatest prediction, borne out by the detection of the **cosmic microwave background** radiation, was that the Universe should by now have cooled to a few degrees above absolute zero.

Ganymede The largest of Jupiter's **Galilean satellites**, with a diameter of 5262 km (3270 miles), and the largest satellite in the Solar System – bigger, though less massive, than the planet Mercury. With a density of 1.94 g/cm^3, Ganymede contains a high proportion of water. Its surface is a mixture of dark, heavily cratered terrain and a brighter, more lightly cratered terrain covered with meandering, intersecting grooves that suggest recent geological activity. The most prominent feature is the dark Galileo Regio, some 4000 km (2500 miles) across. The **Galileo** probe detected a weak magnetic field, probably originating in a liquid layer deep below the surface. SEE ALSO **satellite**.

gas giant SEE **giant planet**.

Gaspra Asteroid no. 951, discovered by Grigorii Neujmin in 1916. In 1991 it became the first asteroid to be imaged from close quarters when the probe **Galileo** passed it *en route* to Jupiter. It is irregular,

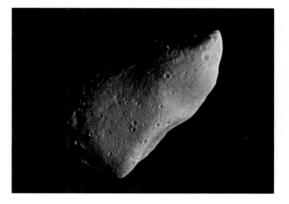

▶ **Gaspra** En route to Jupiter, the Galileo spacecraft made a fly-by of asteroid (951) Gaspra in 1991, revealing it to be an irregular-shaped, heavily cratered body. Gaspra lies in the main belt.

◄ Gemini Observatory
This infrared view of a star-forming region in the Milky Way was taken using the Gemini North telescope on Mauna Kea. Adaptive optics helped sharpen the image.

with a maximum dimension of about 20 km (12 miles), and is pitted with small craters.

Gauss, Carl Friedrich (1777–1855) German mathematician. His many achievements include valuable contributions to astronomy, geodesy, and physics. In 1801 he worked out how to calculate the orbit of the newly found asteroid Ceres from observations spanning just 9° of its orbit. This led him to the study of celestial mechanics, and on to a comprehensive study of the determination of cometary and planetary orbits, taking into account the effects of perturbations, and then to the theory of errors of observation, for which he developed the mathematical method of least squares. From 1807 Gauss was director of the observatory at Göttingen.

gegenschein A faint luminous patch on the ecliptic, visible usually only in the tropics on dark clear nights, directly opposite the Sun. It is caused by the scattering of sunlight back toward the Earth by particles in the zodiacal dust cloud, and is part of the **zodiacal light**. The word is German for "counterglow," by which name it was formerly also known.

Geminga One of the most powerful gamma-ray sources in the sky, in the constellation Gemini (hence the name). Geminga's energy is emitted as pulses of gamma-rays (and X-rays) with a period of 0.237 seconds, so it is most likely a spin-

ning neutron star: a short-wavelength version of a **pulsar**.

Gemini A large, prominent constellation of the zodiac. Its brightest stars are **Castor** and **Pollux**, named for the twins of Greek mythology. Eta is a semiregular variable, magnitude 3.2 to 3.9, with a period averaging 233 days. Zeta is a Cepheid variable, range 3.6 to 4.2, period 10.15 days. The open cluster M35 contains about 200 stars.

Geminids A **meteor shower** which occurs in December. Its radiant is in the constellation Gemini, near the star Castor. The shower, one of the most prolific, is associated not with a comet but with **Phaethon**, an asteroid.

Gemini Telescopes A pair of 8.1-m (320-inch) reflectors jointly funded by the USA, UK, Canada, Chile, Brazil, and Argentina. Gemini North is sited at Mauna Kea, Hawaii, and was dedicated in 1999. Gemini South is at Cerro Pachón, Chile, and opened in 2002. Together they provide complete sky coverage at optical and infrared wavelengths.

geocentric As viewed from or related to the Earth's center, as in *geocentric parallax* or *geocentric system*. COMPARE **heliocentric**.

geocentric parallax SEE **parallax**.

geodesic SEE **curvature of space**.

German mounting A type of mount used for an **equatorial telescope**.

Giacobini–Zinner, Comet A short period **comet**, orbiting the Sun in 6.59 years. It was discovered in 1900 by Michel Giacobini (1873–1938) and recovered in 1913 by Ernst Zinner (1886–1970). In 1985 it became the first comet to be studied at close hand when the International Cometary Explorer space probe passed through the comet's plasma tail.

giant molecular cloud (GMC) SEE **infrared astronomy**.

giant planet Any of the planets Jupiter, Saturn, Uranus, and Neptune, also known as the *gas giants*. They are so called because their masses are great compared with that of the Earth.

giant star A star that is placed well above the **main sequence** in the **Hertzsprung–Russell diagram**. Giant stars are larger than main-sequence (**dwarf**) stars of the same temperature, but giants of intermediate temperatures are smaller than the hottest dwarfs. Thus the term "giant" really refers to the relative luminosity of a star rather than its dimensions. Giant stars evolve from dwarf stars that have run out of hydrogen in their central regions, and so represent later stages in the lives of stars. Because of their high luminosity, giants are quite common among naked-eye stars, examples being Arcturus, Aldebaran, and Capella, but they are relatively rare overall. SEE ALSO **red giant**.

gibbous A **phase** of the Moon or a planet when more than half but less than all of its illuminated hemisphere is visible from the Earth.

Giotto A space probe built by the European Space Agency (ESA) to investigate **Halley's Comet**. It was launched on July

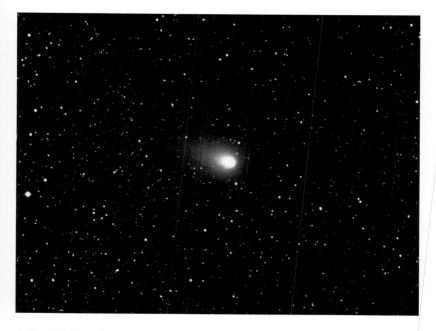

▲ **Giacobini–Zinner, Comet** This comet was imaged on November 1, 1998, by the Kitt Peak 0.9-m (36-inch) telescope. The teardrop-shaped coma shows strong bluish carbon monoxide emission.

2, 1985, from the ESA launch site at Kourou, French Guiana, and flew to within 600 km (375 miles) of the comet's nucleus on March 14, 1986. Giotto photographed the nucleus and analyzed the gas and dust given off. After the encounter Giotto was sent on a new trajectory to fly past Comet **Grigg–Skjellerup**, which it did on July 10, 1992.

glitch A sudden increase in the period of a pulsar. Pulsars slow down with age, but a few pulsars, particularly the Vela and Crab pulsars, occasionally show an abrupt speeding up. The largest of these glitches so far observed have been in the **Vela Pulsar**, and occur every few years. Typically, they make up for the slow-down of the previous few weeks. Glitches are believed to be caused by small, sudden adjustments in the internal structure of the neutron star.

Global Oscillation Network Group (GONG) A partnership between six observatories around the world engaged in closely monitoring solar oscillations. The Sun's surface oscillates, on both small and large scales, because of sound waves that propagate through it. These oscillations, which show up as small **Doppler shifts** in the solar spectrum, yield information on processes in the Sun's interior.

globular cluster A near-spherical cluster of very old (**Population II**) stars in the halo of our Galaxy. Globular clusters contain anything from 100,000 to a few million stars, concentrated tightly near the center, and with a diameter of a few tens to over 300 light years. Most globulars are at least 10,000 million years old, as deduced from the highly evolved state of their stars. This extreme age, combined with their distribution in the galactic halo, indicates that globulars formed while our Galaxy was condensing from a huge cloud of gas. Globulars can also be seen around other galaxies. They are more plentiful around elliptical galaxies than spirals – our Galaxy has about 140 known globular clusters, whereas the largest elliptical galaxies may have thousands. Although they occupy the spheroidal galactic halo, most globulars lie no further from the center of the Galaxy than does the Sun. As a result they congregate in the general area of sky containing the center of our Galaxy, in particular in the constellations Sagittarius, Ophiuchus, and Scorpius. Globular clusters orbit the galactic center, making periodic passages through the plane of our Galaxy. SEE ALSO **star cluster**.

globule SEE **Bok globule**.

GMC ABBREVIATION FOR giant molecular cloud (SEE **molecular cloud**).

GMT ABBREVIATION FOR **Greenwich Mean Time**.

Gold, Thomas (1920–) Austrian–American astronomer. With Hermann Bondi and Fred Hoyle he developed the **steady-state theory**, in which the continuous creation of matter drives the expansion of the Universe. In the 1960s, he explained that the signals from pulsars were beams of synchrotron radiation from rapidly rotating neutron stars.

GONG ABBREVIATION FOR **Global Oscillation Network Group**.

▲ **Giotto** The ESA Giotto probe recorded this close-up view of the nucleus of Halley's Comet on March 13, 1986. Shortly after this image was obtained, collision with a dust particle from the comet disabled the camera.

▶ **Global Oscillation Network Group** This is a computer representation of one of the almost 10 million modes of sound wave oscillation in the Sun, as monitored by helioseismologists working under the umbrella of GONG. Solar regions that are receding are shown in red, those that are approaching in blue.

Goodricke, John (1764–86) English astronomer, born in Holland. With the encouragement of his neighbor and collaborator, Edward Pigott, the deaf-mute Goodricke began at the age of 17 to study the brightest variable stars. In 1782 he announced the cause of Algol's variability (that it was what is now called an eclipsing binary), and the length of its period. Two years later he discovered the variability and periods of Beta Lyrae and Delta Cephei, the prototype Cepheid variable. At the age of 21 he died from pneumonia contracted while observing Delta Cephei.

Gould's Belt A region of bright stars and gas inclined at about 20° to the galactic plane. It contains the greatest concentration of naked-eye stars of spectral types O and B, and probably represents part of the local spiral arm of which the Sun is a member. It is named for the American astronomer Benjamin Gould (1824–96), who studied the region.

grand unified theory (GUT) A theory that attempts to "unify" three of the four forces of nature, and thus to demonstrate that they are different manifesta-tions of a single force. In descending order of strength, the four forces of nature are: the *strong nuclear force* (which binds atomic nuclei); the *electromagnetic force* (which holds atoms together); the *weak nuclear force* (which controls the radioactive decay of atomic nuclei); and **gravitation**. The

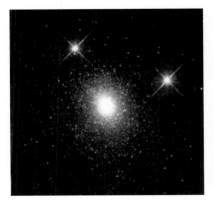

▲ **globular cluster** Containing at least 300,000 ancient stars, the globular cluster G1 orbits the Andromeda Galaxy. This Hubble Space Telescope image shows G1 to have a rich, compact core and less densely packed periphery.

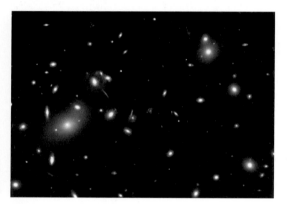

▲◀ gravitational lens (top) Light from a distant quasar may be bent in the gravitational field of a foreground galaxy to produce a distorted or multiple image. In this example, the quasar's light is split into four separate images, each arriving at the telescope along a slightly different path. (bottom) The massive galaxy cluster Abell 2218 in Draco acts as a gravitational lens, producing distorted, arc-shaped images of more distant objects in the same line of sight.

weak and strong nuclear forces operate over tiny distances within atomic nuclei, while the other two forces are infinite in range.

In the 1970s and 1980s, the unification of the electromagnetic and weak nuclear forces was first predicted and then demonstrated experimentally. GUTs are attempts to unify this *electroweak force* with the strong nuclear force. Many physicists believe that a more elaborate theory, a so-called *theory of everything* (TOE), will eventually show how all four forces are unified.

granulation The mottled appearance of the Sun's **photosphere**, caused by gases rising from the Sun's interior. When seeing conditions are very good the granulations can be resolved into small, nearly circular patches, called *granules*, surrounded by darker areas. Individual granules last only a few minutes.

graticule A system of reference marks or a measuring scale consisting of parallel vertical wires, or cross-wires, or a reticule (a grid of squares) placed in the focal plane of a telescope and covering the entire field of view. SEE ALSO **filar micrometer**.

grating SEE **diffraction grating**.

gravitation (gravity) The universal force of attraction between all particles of matter. *Newton's law of gravitation* states that the force of attraction F between two masses m_1 and m_2 is inversely proportional to the square of their separation r :

$$F = Gm_1m_2 / r^2$$

where G is the *gravitational constant*. This law is the basis of celestial mechanics.

Although at close range gravity is far, far weaker than the nuclear or electromagnetic force, it has a greater range and is the most important force on the large, cosmological scale. The whole concept of gravitation was reinterpreted by Albert **Einstein** in his General Theory of Relativity in terms of the curvature of **spacetime**.

gravitational lens A massive body, such as a star, black hole, or galaxy, which bends light or other radiation from a more distant object in whose line of sight it lies. The effect, which is predicted by Einstein's Theory of Relativity, was first detected during the total solar eclipse of 1919 by Arthur **Eddington**, who found that light from distant stars was bent as it passed the Sun. In 1979, the first example of gravitational lensing of light from a quasar was discovered. In such cases, the quasar's light may be split into two or four separate images by an intervening massive galaxy or cluster of galaxies. Other distant objects may have their light lensed into arcs, or even a complete **Einstein ring**. Gravitational lensing of quasars can be used to measure their distances. SEE ALSO **microlensing**.

gravitational redshift SEE **redshift**.

gravitational waves Waves emitted as the result of a violent disturbance in a gravitational field, such as a massive body being accelerated or distorted with a velocity that is an appreciable fraction of the velocity of light. Gravity waves were predicted by Einstein's General Theory of Relativity, but have proved elusive. Just as an oscillating charged particle emits electromagnetic waves, so a violently disturbed mass should emit gravity waves. Likely sources of gravity waves include close binary systems where at least one component is a neutron star or black hole. Because gravity is by far the weakest of the forces of nature, gravity waves are extremely feeble. Detectors based on **interferometry** are being developed which should be sensitive enough to pick up gravitational waves.

Great Attractor A hypothetical distant concentration of matter with a mass about 100,000 times that of the Milky Way, 150 million light years away in the direction of the constellations Hydra and Centaurus. It was suggested to account for the observed motions of galaxies, includ-

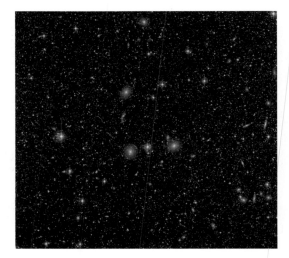

▶ **Great Attractor** The central region of ACO 3627, a huge galaxy cluster thought to form part of the Great Attractor, is seen here in an image obtained by the European Southern Observatory. The Great Attractor's gravitational pull is dragging all the galaxies in our local cluster toward it.

◀ **Great Red Spot** A 1996 Galileo orbiter image shows this huge storm, which has raged in Jupiter's atmosphere for hundreds of years.

ing those of the **Local Group**. Although there is a cluster of galaxies in the place in question, it now seems more likely that the gravitational pulls of several different clusters are responsible, and there is no single Great Attractor.

great circle A circle on a sphere whose plane passes through the center of the sphere. On the **celestial sphere** the celestial equator is a great circle, as are meridians (circles passing through both celestial poles).

greatest elongation SEE **elongation**.

Great Observatories Umbrella term for four major NASA spaceborne telescopes: the **Hubble Space Telescope**, the **Compton Gamma Ray Observatory**, the **Chandra X-ray Observatory**, and the **Space Infrared Telescope**.

Great Red Spot (GRS) A large oval area in the southern hemisphere of Jupiter. Since it was first observed by Heinrich Schwabe in 1831, it has varied in size and color, from about 11,000 to 14,000 km wide by 24,000 to 40,000 km long (7000 to 9000 miles by 15,000 to 25,000 miles), and from a deep red to a light pink, occasionally fading away altogether. In the late 1990s it was a pale tan color. It is an atmospheric phenomenon, a huge anticyclone, projecting above the surrounding cloud-tops. The red color may be produced by phosphorus.

Green Bank The location in West Virginia, USA, of the **National Radio Astronomy Observatory** and its Green Bank Telescope.

green flash An atmospheric phenomenon, observable under ideal conditions during the last seconds before sunset or the first seconds after sunrise, in which the uppermost segment of the Sun separates and flashes a green color. At these moments most of the sunlight has been absorbed in its long passage through the atmosphere, and only two narrow bands of wavelengths in the red and green are left. The atmospheric refraction of light is greater for short wavelengths (green rather than red), so the green flash appears.

greenhouse effect The warming of a planet's surface by heat trapped by its atmosphere. Incoming sunlight is absorbed at the surface and re-radiated at longer wavelengths, in the infrared – that is, as heat. If the atmosphere contains *greenhouse gases*, such as carbon dioxide, which are not transparent to the infrared, the heat is reflected back to the ground. The effect is the same as is produced by the glass enclosing a greenhouse. The result is to raise the planet's average surface temperature by an amount that depends on the atmospheric composition and density. On **Venus** the dense carbon dioxide atmosphere has raised the surface temperature to around 750 K. On Earth a smaller greenhouse effect has led to a surface temperature that suits many life forms, but some people believe that industrial emissions of carbon dioxide will increase the greenhouse effect, producing *global warming*, and should be halted.

Greenwich Mean Time (GMT)

Mean solar time for the meridian of Greenwich. In 1928 the International Astronomical Union recommended that GMT should be known as **Universal Time** for scientific purposes. Today, that usage has extended to legal and other purposes, although the term GMT is still used in many contexts. Greenwich was chosen as the world's prime meridian for timekeeping and navigation by the International Meridian Conference in 1884.

Gregorian calendar SEE calendar.

Gregorian telescope

A reflecting telescope described by Scottish mathematician James Gregory (1638–75) in 1663, five years before Isaac Newton constructed his first reflector. Light received by the parabolic primary mirror is reflected to a concave ellipsoidal secondary mirror, and then reflected back through a hole in the center of the primary to an eyepiece. It differs from the Cassegrain telescope in having a concave instead of a convex mirror for the secondary. For astronomical purposes this form of telescope has fallen into disuse because it is difficult to mount and handle.

Grigg–Skjellerup, Comet

The periodic **comet** with the second-shortest period known, 5.09 years. It was discovered in 1902 by New Zealander John Grigg (1838–1920), and recovered in 1922 by Australian John "Frank" Skjellerup (1875–1952). In 1992 it became the third comet to be encountered by a spacecraft when the Giotto probe passed within 200 km (125 miles) of it.

GRO ABBREVIATION FOR Compton Gamma Ray Observatory.

GRS ABBREVIATION FOR Great Red Spot.

Grubb, Howard (1844–1931)

Irish telescope-maker. In 1868 he took over the telescope-manufacturing business estab-

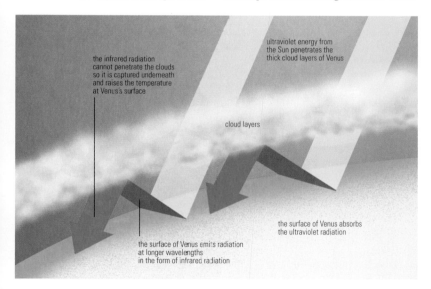

▲ **greenhouse effect** Short-wave radiation from the Sun readily penetrates the atmosphere to reach the ground. Longer wavelength radiation re-emitted from the ground is trapped by atmospheric constituents such as carbon dioxide, raising the planetary temperature. This greenhouse effect is significant in keeping Earth habitable, but has led to inhospitably high temperatures on Venus.

lished by his father, **Thomas Grubb** (1800–1878), and went on to make seven of the **astrograph** refractors used for the *Carte du Ciel* photographic sky survey, besides many other medium-sized and large telescopes for leading observatories. On his retirement in 1925 Grubb's business was acquired by Charles Parsons, son of the Third Earl of **Rosse**, who set up a new firm under the name of Sir Howard Grubb, Parsons & Co.

Grus A southern constellation, representing a crane. Its brightest star is Alpha (Alnair), magnitude 1.7. Mu and Delta are both optical pairs, wide enough to be separated with the naked eye.

G star A star of spectral type G, with a surface temperature of between about 5000 and 6000 K if it is on the **main sequence**, and between 4600 and 5600 K if it is among the giants. The coolest G-type supergiants are about 4200 K. G-type dwarfs range in mass from 0.9 to 1.1 solar masses. The giants and supergiants have generally evolved from more massive stars farther up the main sequence. Almost all G stars are dwarfs with compositions similar to the Sun. A few are subdwarfs with lower metal abundances. A small fraction of the G-type giants have peculiar chemical compositions that give them the name *barium stars*. Examples of G stars are the Sun, Alpha Centauri, and Capella.

Gum Nebula An immense emission nebula spanning 30 to 40° of sky in the constellations Puppis and Vela, containing the Vela Pulsar. The nearest part is 450 light years away, and its farthest part 1500 light years. It is named for the Australian astronomer Colin Gum (1924–1960).

GUT ABBREVIATION FOR **grand unified theory**.

H

H I region A cloud of neutral (i.e. not ionized) atomic hydrogen gas in interstellar space. H I (pronounced "h one") regions are typically of 10 to 100 solar masses and 10–20 light years across, with a temperature of around 100 K. They are about ten times as dense as interstellar matter, which is too low for molecular hydrogen to form in significant quantities.

H I regions do not emit visible radiation, but can be detected from their radio emission at a wavelength of 21 cm (SEE **twenty-one centimeter line**). Large radio telescopes can thus detect H I regions even in other galaxies. Different kinds of galaxy contain different proportions of H I. Our Galaxy and other spiral or irregular galaxies contain an appreciable fraction of their total mass in H I. Elliptical galaxies contain no (or extremely little) detectable H I.

In our Galaxy H I regions are heavily concentrated to a flattened disk in which a spiral pattern occurs. Spiral arms are made conspicuous, even in very distant galaxies, by hot, massive blue stars lying along the arms. These represent recent star-forming activity from the H I regions in the arms. Not all H I regions are capable of forming stars, only the densest ones that also contain molecular cores (SEE ALSO **giant molecular cloud**). Vast H I clouds, apparently devoid of luminous matter, have also been detected in the space between galaxies (SEE **Magellanic Clouds**).

H II region A cloud of ionized hydrogen in interstellar space. H II regions (pronounced "h two") are often embedded within **H I regions**, and have densities 10 to 100,000 times greater. Because of the high temperatures (8000–10,000 K) and densities of H II regions the pressure inside them is far higher than in the surrounding H I region or interstellar gas, so all H II regions are expanding, at around 10 km/s. Visible radiation from the H II region results from the *recombination* of protons and electrons to form hydrogen atoms; hot dust emits at infrared wavelengths; and *free–free radiation* at radio wavelengths is given out as electrons curve around protons. On images, H II regions often appear red due to the **hydrogen alpha line** at 656 nm, but they can look greenish when seen visually due to a prominent pair of **forbidden lines** at 496 and 501 nm emitted by oxygen.

Regions of ionized hydrogen can be produced by ultraviolet emission from old hot stars, from nova or supernova explosions, or from high-velocity **jets** impacting on interstellar gas. Conventionally, however these are not classified as H II regions but as, respectively, **planetary nebulae**, **supernova remnants**, and **Herbig–Haro objects**.

H II regions are stellar nurseries, the stars within them having recently formed through gravitational collapse within a **giant molecular cloud**. The life of H II regions is only a few million years, and their presence along with the hot stars marks out the arms in spiral galaxies. Complex H II regions are found in **starburst galaxies**. SEE ALSO **emission nebula**.

HA ABBREVIATION FOR **hour angle**.

Hα SEE **hydrogen-alpha line**.

Hadar Another name for **Beta Centauri**.

Hale, George Ellery (1868–1938) American astronomer. He planned and secured finance for the 1-m (40-inch) refracting telescope at **Yerkes Observatory**, completed in 1897, the 2.5-m (100-inch) Hooker reflector at **Mount Wilson Observatory**, completed in 1917, and the 5-m (200-inch) reflector at the **Palomar Observatory**, which entered regular service in 1949. Each was in turn the largest telescope of its day. The 5-m (200-inch) was named the Hale telescope in his honor. Hale was an accomplished solar observer, discovering among other things the magnetic fields of sunspots via the Zeeman effect in their spectra. He also invented the spectroheliograph.

Hale–Bopp, Comet

An exceptional long-period comet discovered in 1995 by American amateur astronomers Alan Hale (1958–) and Thomas Joel Bopp (1949–). It was exceptional for its size – the nucleus was about 40 km (25 miles) across – so that even though it approached the Earth no closer than 1.3 AU, it was still a brilliant object (mag. −0.7 maximum). Hale–Bopp emitted dust at a prodigious rate. As the nucleus rotated, with a period of 11.5 hours, the dust created a series of concentric shells, visible in the comet's inner coma. The gas and dust tails extended to 20° and 25° respectively, corresponding to actual distances of over 1 AU.

Hale Telescope

The 5-m (200-inch) reflecting telescope at **Palomar Observatory**.

Hall, Asaph

(1829–1907) American astronomer. His planetary studies covered the orbital elements of satellites, the rotation of Saturn, and the mass of Mars. On August 11 and 17, 1877, he discovered the two tiny satellites of Mars, which he named Deimos and Phobos. The largest craters on Phobos have been named Hall and Stickney (the maiden name of Hall's wife).

Halley, Edmond

(1656–1742) English scientist. His most famous achievement in astronomy was to realize that comets could be periodic, following his observation of **Halley's Comet** in 1682. He cataloged the stars of the southern hemisphere, discovered the proper motions of stars, suggested that the nebulae were clouds of gas, and speculated on the infinity of the Universe. He made the first complete observation of a transit of Mercury and showed how the results could be used to measure the Sun's distance. As the second Astronomer Royal, he embarked on a long series of lunar studies. It was Halley who persuaded Isaac **Newton** to write the *Principia*, and he also financed its publication. Halley's other achievements are numerous. He founded modern geophysics, charting variations in the Earth's magnetic field, and establishing the magnetic origin of the aurora borealis. In meteorology he showed that atmospheric pressure decreases with altitude, and studied monsoons and trade winds. In mathematics, he showed how to use mortality statistics to cost life assurance policies.

Halley's Comet

The brightest and best-known periodic **comet**. It takes 76 years to complete an orbit that takes it from within Venus's orbit to outside Neptune's. The orbit's inclination is 162°, making it retrograde. It was observed by Edmond **Halley** in 1682; later he deduced that it was the same comet as had been seen in 1531 and 1607, and predicted it would return in 1758. There are historical records of every return since 240 BC. Several space probes were sent to Halley on its 1986 return. The closest approach was by the probe **Giotto**, whose cameras showed the nucleus to be an irregular object measur-

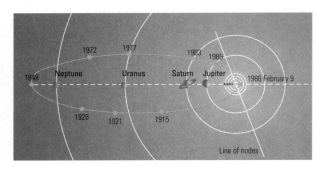

◀ **Halley's Comet** The elongated elliptical orbit of Halley's Comet takes it from aphelion beyond the orbit of Neptune to perihelion between the orbits of Venus and Mercury. The comet's orbit is retrograde, and it was last at perihelion on February 9, 1986.

▶ **halo** The most prominent components of spiral galaxies like the Milky Way are the stars, gas and dust populating their nuclear bulge and disk. Many galaxies are also surrounded by tenuous halos of hot plasma. This X-ray image of a Milky Way-type galaxy reveals how vast such a halo can be.

ing 15 × 8 km (9 × 5 miles). The main constituent was found to be water ice, but the surface was covered with a dark deposit, giving an albedo of only about 0.03. Some craters were imaged.

halo (1) A bright whitish ring, complete or partial, seen around a celestial body, usually the Sun or the Moon. It is caused by the refraction of light by ice crystals suspended in the atmosphere.

halo (2) The **galactic halo**, a spheroidal distribution of stars and globular clusters around our Galaxy.

halo population stars SEE **Population II stars**.

Harvard classification The system of spectral classification of stars developed in the late 19th century by E. C. **Pickering** at Harvard College Observatory. Stars were originally placed in a sequence from A to P, indicating decreasing strength of hydrogen absorption lines in their spectra. The rearrangement and pruning of these classes gave the sequence O, B, A, F, G, K, M, based on decreasing surface temperature, used in the *Henry Draper Catalogue*. This in turn gave way to the **Morgan–Keenan classification**.

harvest Moon The full Moon nearest the autumnal **equinox** (on September 23). The Moon's path is then almost parallel to the horizon, and so it rises only about 15 minutes later each evening instead of the usual half-hour or more, providing a succession of moonlit evenings formerly of great help to harvesters. In the southern hemisphere the harvest Moon is the full Moon nearest the vernal equinox (on March 21). SEE ALSO **hunter's Moon**.

Hawking, Stephen William (1942–) English theoretical physicist. He has used the General Theory of Relativity and quantum mechanics to study the **Big Bang** and **black holes**. He found that small black holes should lose energy – by emitting *Hawking radiation* – and eventually "evaporate."

Hayashi track The evolutionary path on the **Hertzsprung–Russell diagram** of a star beginning to form (SEE **protostar**) after it has collapsed from a gas cloud and before it reaches the main sequence. The Hayashi track lies above and to the right of the main sequence, and is nearly vertical. The star descends its Hayashi track, decreasing in luminosity but with little change in effective temperature. (After Japanese astrophysicist Chushiro Hayashi, 1920– .)

▲ **Helix Nebula** One of the closest planetary nebulae, the Helix Nebula (NGC 7293) is visually faint because its light is spread over a fairly large area of sky.

heavy elements SEE metals.

Hebe Asteroid no. 6, diameter 186 km (116 miles), discovered in 1847 by Karl Ludwig Hencke.

Hektor Asteroid no. 624, discovered in 1907 by August Kopff. It measures roughly 300 × 150 km (200 × 100 miles) and is the largest and brightest of the **Trojan asteroids** sharing Jupiter's orbit. Its elongated shape may indicate that it is a binary asteroid.

Helene A small **satellite** of Saturn, discovered in 1980 by Pierre Lacques and Jean Lecacheux. It is an irregular body, and is a **coorbital satellite** with Dione, orbiting near the L_4 **Lagrangian point** of Dione's orbit around Saturn.

heliacal rising In modern usage, the rising of a celestial body simultaneously with the rising of the Sun. Formerly it meant the moment when a bright star, such as Sirius, could just be seen rising before the Sun. Observation of this event was once of great importance for agricultural purposes, the heliacal rising of Sirius heralding to the ancient Egyptians the annual flooding of the Nile.

heliocentric As viewed from or in relation to the Sun's center, as in *heliocentric coordinates* and the *heliocentric system*. COMPARE **geocentric**.

heliocentric parallax (annual parallax) The **parallax** of a body determined from observations of its position made six months apart, using the diameter of the Earth's orbit as the baseline.

heliometer A refractor with the objective lens split across its diameter to give two images, formerly widely used for measuring very small angular distances.

heliosphere The region of space in which the Sun's magnetic field and the solar wind dominate the interstellar medium. It is similar in form to a planet's **magnetosphere**, with an outer *bow shock*, and a boundary called the *heliopause* where the energy of the solar wind particles has fallen to the level of galactic cosmic ray particles. Measurements continuing to be returned by the Voyager probes suggest that the heliosphere extends for 100 to 150 AU.

heliostat A mirror on an equatorial mount driven so that it follows the Sun's apparent motion across the sky and directs its light into a fixed instrument such as a solar telescope. A more sophisticated version is the **coelostat**.

helium flash The sudden commencement of helium fusion in the core of a red giant after the star's supply of hydrogen fuel has been exhausted. The helium flash is restricted to stars of less than about 2 solar masses; in more massive stars, helium fusion commences more gradually.

Helix Nebula A planetary nebula in Aquarius, NGC 7293. It has one of the largest angular diameters of any planetary nebula, a quarter of a degree, and at 300 light years away it is one of the nearest to the Sun.

Henry Draper Catalogue A catalog of stellar spectra, compiled by **Annie Cannon**

at Harvard College Observatory. It was named for Henry Draper (1837–1882), an American pioneer of astrophotography, whose widow supported the work financially. The catalog, completed in 1924, introduced the Harvard system of classifying stellar spectral types: O, B, A, F, G, K, M, in order of decreasing surface temperature. Stars are still widely known by their HD numbers from this catalog. SEE ALSO **spectral classification**.

Herbig–Haro object (HH object)
One of a class of small, faint nebulae independently discovered by George Herbig (1920–) and Guillamero Haro (1913–88). They are irregular in outline, contain bright knots, and are found in regions rich in interstellar material. HH objects are thought to be produced by very young stars hidden behind clouds of gas and dust. Powerful stellar winds from the young stars excite the gas, producing emission lines, while the light from the star is reflected by dust in other parts of the cloud.

Hercules
A constellation named for the great hero of Greek mythology, but not very conspicuous. Its brightest star is Beta (Kornephoros), magnitude 2.8. Alpha (Rasalgethi) is a semiregular red giant variable, range 2.7 to 4.0, and with a period that varies from 50 to 130 days; there are also longer-term variations with a period of about 6 years. It lies 382 light years away. It is also a double star, with a magnitude 5.4 companion, period about 3600 years. Zeta is a fine close binary, with a period of 34 years, magnitudes 2.9 and 5.5. The most interesting deep-sky objects are the two globular clusters M13 and M92.

Hermes
An unnumbered asteroid, about 1 km (0.6 mile) in diameter, discovered by Karl Reinmuth in 1937. On October 28 of that year it came within 800,000 km (500,000 miles) of the Earth, up to then the closest known approach of any asteroid. Although its period is less than 3 years, it has never been observed since.

Herschel, Caroline Lucretia
(1750–1848) German-born English astronomer, sister of William **Herschel**. In 1772 she joined her brother at Bath in England, becoming his housekeeper and astronomical assistant, editing and copying his papers, recording his observations, and preparing his catalogs. In the 1780s she became an observer in her own right, discovering many nebulae (as they were then called, including the galaxy NGC 253 in

► **Herbig–Haro object** High-speed jets from the young object HH 34, near right center, impact on the surrounding interstellar material to produce V-shaped shockwaves. HH 34 lies in Orion, in a region of ongoing star formation.

Sculptor) and eight comets. When her brother died in 1822, Caroline returned to her native Hanover, where she received many honors. She was the first woman to achieve real distinction in astronomy.

Herschel, John Frederick William

(1792–1871) English scientist and astronomer, son of William **Herschel**. At first an assistant to his father with observations and telescope-making, he went on to extend his father's work on double stars and nebulae, discovering over 500 more nebulae and clusters. In 1834 he took one of William's telescopes to the Cape of Good Hope and undertook a systematic survey of the southern sky, discovering over 1200 doubles and 1700 nebulae and clusters. He combined these and his father's observations into a *General Catalogue of Nebulae and Clusters*, which formed the basis for the later **NGC**. He made the first good direct measurement of solar radiation, and inferred the connection between solar and auroral activity. His *Outlines of Astronomy* (1849) was a standard textbook for many decades. He was also a pioneer of photography and its application in astronomy.

Herschel, (Frederick) William

(originally Friedrich Wilhelm Herschel) (1738–1822) German-born English astronomer. In 1757 he came to England as a musician and became a naturalized Englishman. He turned to astronomy in the 1770s, making telescopes and mirrors for his own use and later for sale. Herschel became famous in 1781 by discovering the planet Uranus, which he wanted to name after King George III; the next year he was appointed the king's private astronomer. He discovered two satellites of Uranus (1787), and two of Saturn (1789). He observed and cataloged many double stars, hoping to measure their distances by detecting the parallactic movement of the nearer against the further if they were optical doubles. He observed over 2000 nebulae and clusters, and published catalogs of them which later would be incorporated in the **NGC**. Herschel realized that the Milky Way is the plane of a disk-shaped stellar universe, whose form he calculated by counting the numbers of stars visible in different directions, and also noted the motion of the Sun toward a point in the constellation Hercules. In 1800 he discovered and investigated the properties of infrared radiation.

Herschelian telescope

A reflecting telescope in which the primary parabolic mirror is tilted so as to bring rays of light to a focus at the upper end of the tube, where an eyepiece is placed. It was devised by William Herschel to reduce the high loss of light from the secondary mirror of speculum metal used in the Cassegrain and Newtonian reflectors. This type of instrument is not now in use.

Herschel Space Observatory

A European Space Agency satellite, formerly known as the Far Infrared and Submillimetre Space Telescope, planned for launch in 2007. It will carry a cooled 3.5-m (138-inch) Cassegrain telescope for photometry and spectroscopy in the 57–670 μm wavelength range. It will orbit the Earth's L_2 Lagrangian point.

Hertzsprung, Ejnar

(1873–1967) Danish astronomer. He showed that the colors of the stars were related to their brightnesses, plotting one against the other on what is now called the **Hertzsprung–Russell diagram** and discovering the existence of giant and dwarf stars. All studies of stellar evolution stem from this diagram, but Hertzsprung's work remained largely overlooked until Henry Norris **Russell** independently developed the diagram in a slightly different form. Hertzsprung also did work on double and variable stars, and applied Henrietta **Leavitt**'s methods of stellar distance measurement.

Hertzsprung gap

The region on the **Hertzsprung–Russell diagram** between the main sequence and the giant branch where there are few stars. The gap exists because stars evolve rapidly through this region.

Hertzsprung–Russell diagram

(HR diagram) A plot of the **absolute magnitude** of stars against their **spectral type**; this is equivalent to plotting their **luminosity** against their surface temperature or **color index**. Brightness increases from bottom to top, and temperature increases from right to left. The diagram was devised by Henry Norris **Russell** in 1910, independently of Ejnar **Hertzsprung**, who had had the same idea some years before.

The HR diagram reveals a pattern in which most stars lie on a diagonal band, the **main sequence**, which includes our Sun. There is another region of **giant stars**, mainly K and M spectral types, which lies above the main sequence. There are further groupings above the main sequence called **supergiants** and **subgiants**. Below the main sequence are **subdwarfs** and **white dwarfs**.

The HR diagram is an important aid for interpreting astrophysical data. For example, the total range of stellar brightness is 27 magnitudes, corresponding to a factor of 10^{11} in luminosity, and stars' surface temperatures range from 2200 to 50,000 K. Luminosity and surface temperature are both related to the size of a star. Most main-sequence stars are about the size of the Sun. White dwarfs are about the size of the Earth, while red giants and supergiants can be as big as the Earth's orbit. About 90% of the stars in our neighborhood are main-sequence stars, about 10% are white dwarfs, and only about 1% are red giants or supergiants.

The existence of the three main groupings on the HR diagram means that a typical star passes through three very different stages during its life, and the progress of an evolving star can be followed on an HR diagram, from **protostar** to main sequence, then to giant star, and ending as a white dwarf (SEE **stellar evolution**). The diagram can also provide information about the ages of stars, and such dating is especially useful for establishing the ages of star clusters. Since the main sequence is a progression in stellar mass, the stars at the upper (most massive) end of the main sequence are the first to leave. As the clus-

ter ages, its main sequence burns down like a candle as stars of progressively decreasing mass and luminosity become red giants. The stage which this process has reached reveals the cluster's age. SEE ALSO **Hayashi track**, **Hertzsprung gap**, **instability strip**.

HET ABBREVIATION FOR **Hobby–Eberly Telescope**.

Hevelius, Johann (1611–87) German

astronomer. He was the last great astronomer to make observations with naked-eye sighting instruments for positional measurements, from which he produced a catalog of 1500 stars. He also made a map of the Moon, and discovered the Moon's **libration** in longitude.

Hewish, Antony (1924–) English

radio astronomer. He and Martin Ryle developed the technique of **aperture syn-**

▲ **Hertzsprung–Russell diagram** A plot of stars' luminosity against temperature. Most stars lie on the main sequence running from top left to bottom right. This diagram shows the principal regions of the Hertzsprung–Russell diagram, with giant stars above the main sequence to the top right, and dwarfs to the lower left.

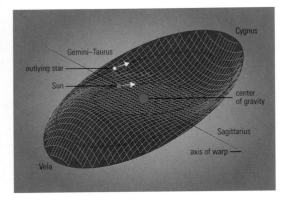

◀ **Hipparcos** Astrometric data from the Hipparcos satellite revealed warping of the disk of the Galaxy. The scale of the warping is exaggerated in this artist's rendition.

thesis in 1960. In 1967 his student Jocelyn **Bell Burnell** obtained the first signal from a pulsar, now known as CP 1919. For subsequent work on pulsars, Hewish shared (with Ryle) the 1974 Nobel Prize for Physics.

Hidalgo Asteroid no. 944, discovered by Walter Baade in 1920. Its diameter is about 50 km (30 miles), and its eccentric comet-like orbit takes it from just outside Mars's orbit out as far as Saturn. Although it has an asteroidal designation, it could well be a large extinct cometary nucleus.

high-velocity star A star in the galactic halo which does not share the rotation of the majority of stars in the Galaxy, including the Sun, around the galactic center. Consequently, it seems to have a large proper motion and a high velocity relative to the Sun. The orbits of high-velocity stars are usually eccentric, with a large inclination to the galactic plane. Most of them are **Population II stars**.

Himalia The largest of the four small **satellites** making up Jupiter's intermediate group. It was discovered by Charles Perrine in 1904.

Hipparchus of Nicea (*c.* 190–125 BC) Greek astronomer, geographer and mathematician. He made many accurate astronomical observations and compiled a star catalog (now lost but incorporated in Ptolemy's ***Almagest***) giving coordinates and magnitudes. He discovered the **precession** of the equinoxes. He found irregularities in the motion of the Sun and Moon, and from observations of eclipses made commendable estimates of their distances and sizes.

Hipparcos A satellite launched in 1989 by the European Space Agency, designed for **astrometry**. The name is a contraction of High Precision Parallax Collecting Satellite, and is also an allusion to **Hipparchus of Nicea**, who recorded accurate star positions more than two millennia previously.

Hipparcos operated until August 1993, measuring parallaxes and proper motions of 120,000 stars to a precision of 1 to 2 thousandths of an arc second, up to 100 times better than is possible from the ground. In addition, it performed two-color photometry of a million stars and measured their parallaxes and motions (though not to such a high accuracy). These two sets of observations were published in 1997 as the *Hipparcos Catalogue* and the *Tycho Catalogue*.

Hirayama family A group of **asteroids** having closely similar orbital elements (SEE **orbit**), in particular the **semimajor axis**. About a hundred such families are known, named usually after the largest asteroid in them. The *Flora*, *Eos*, and *Themis* families are three of the main ones. The fact that in many families the members have similar spectra suggests that they originate from one larger body which has been broken up by impact. These groupings were discovered by Japanese astronomer Kiyotsugu Hirayama (1874–1943) in 1928.

Hobby–Eberly Telescope (HET) A

reflecting telescope opened in 1997 at Mc-Donald Observatory, Texas, with an effective aperture of 9.2 m (360 inches). The main mirror consists of 91 hexagonal segments 1 m (40 inches) across, the images from which are directed to a common focus by a complex corrector system. The mounting is considerably simplified by having the telescope permanently aimed at an altitude of 55°, and moving only in azimuth. Even so, 70% of the sky is visible to it over the course of the year. It is named for William P. Hobby and Robert E. Eberly, American supporters of public education.

homogeneity SEE **cosmology**.

Homunculus Nebula SEE **Eta Carinae**.

horizon The great circle on the **celestial sphere** 90° from the observer's overhead point, the zenith.

horizontal coordinates SEE **coordinates**.

Horologium An obscure constellation of the southern sky introduced by **Lacaille**, representing a pendulum clock. Alpha, which is of magnitude 3.9, is the constellation's only star above magnitude 4.

Horsehead Nebula A dark nebula, with a very distinctive shape like the head and mane of a horse. It is silhouetted against the bright nebula IC 434 south of Zeta Orionis. IC 434 is colliding with an extensive dusty cloud on the east side of Orion, and the Horsehead is the most distinctive intrusion of that dark cloud. The Horsehead is 1600 light years away and measures some 3 light years across.

▶ **Horsehead Nebula**
Silhouetted against the emission nebula IC 434, the Horsehead Nebula in Orion is a very difficult object visually. Photographs such as this show the striking profile of the intruding dark nebula in the foreground.

horseshoe mounting A type of mounting used for a large **equatorial telescope**.

Horseshoe Nebula SEE **Omega Nebula**.

hot dark matter SEE **dark matter**.

hour angle (HA) The angle that the **hour circle** of an object makes with the **celestial meridian**. It is measured westward from the meridian, along the celestial equator to the point where the equator and hour angle first intersect.

hour circle A great circle on the **celestial sphere** passing through both celestial poles. **Declination** is measured along hour circles. On each half of an hour circle running from pole to pole, all points have the same **right ascension**.

Hoyle, Fred (1915–2001) English astrophysicist and cosmologist. With Hermann Bondi and Thomas Gold he developed the **steady-state theory**, in which the continuous creation of matter drives the expansion of the Universe. Although it was subsequently displaced by the Big Bang theory (which owes its name to a disparaging remark by Hoyle), it sparked research by Hoyle and others into **nucleosynthesis** in stars. Hoyle attracted controversy with unorthodox ideas, such as his suggestion that life was brought to Earth by comets.

HR diagram ABBREVIATION FOR Hertzsprung–Russell diagram.

HST ABBREVIATION FOR **Hubble Space Telescope**.

Hubble, Edwin Powell (1889–1953) American astronomer. At Mount Wilson Observatory he used the 2.5 m (100 inch) telescope to examine nebulae. He found that although some nebulae were clouds of gas lying within our Galaxy, many others – the so-called spiral nebulae – were resolvable as independent star systems. His **Hubble classification** of galaxies is still in use. In 1924, he measured the distance to the Andromeda Galaxy by identifying **Cepheid variable** stars in it. He attributed the **redshift** of the spectral lines of galaxies to the recession of galaxies, and hence to the expansion of the Universe. He established the velocity–distance relationship of recession (SEE **Hubble constant, Hubble law**).

Hubble classification A classification of **galaxies** according to their shape or structure, introduced by Edwin **Hubble** in 1925. The three major categories are elliptical (E), spiral (S), and barred spiral (SB); each category has several subdivisions depending on the observable shape or structure of the galaxy. Irregular galaxies (Ir) were not included in Hubble's original classification. SEE ALSO **tuning fork diagram**.

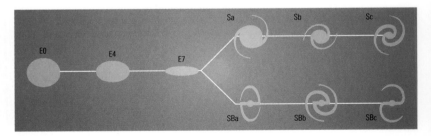

▲ **Hubble classification** A useful way of describing galaxy shapes, the Hubble classification splits these into three main classes. Elliptical galaxies (E) are subdivided on the basis of how far they are from spherical. The classification branches (and is hence sometimes known as the "tuning-fork diagram") to afford description of spiral (S) and barred spiral (SB) galaxies.

► **Hubble Space Telescope** Launched in 1990, the Hubble Space Telescope has provided astronomers with many results of fundamental importance. Here it is seen soon after release following the second servicing mission, carried out by Space Shuttle astronauts in 1997.

Hubble constant (symbol H_0) The rate at which the velocity of recession of galaxies increases with distance from us, according to the **Hubble law**. The zero subscript specifies that it is the expansion rate at the present time that is meant, for, according to the theory of the **Big Bang**, the rate changes with time (SEE **deceleration parameter**). Depending on what assumptions are made, modern values of H_0 range from about 50 to 100 km/s/megaparsec. Most recent researches suggest a value of 68 km/s/megaparsec.

The inverse of the Hubble constant is called the *Hubble time*, which gives a maximum age for the Universe on the assumption that there has been no slowing of the expansion. The Hubble times for the above values of H_0 are respectively 20×10^9 and 10×10^9 years, assuming that all the galaxies started to recede from a common point in the past, at the time of the Big Bang.

Hubble diagram A graph in which the apparent magnitude of galaxies is plotted against the redshift of their spectral lines. It is a straight line, demonstrating the linear relation between redshift and distance, as embodied in the **Hubble law**.

Hubble law The law proposed by Edwin **Hubble** in 1929 claiming a linear relation

between the distance (r) of galaxies from us and their velocity of recession (v), deduced from the redshift in their spectra. The figure linking velocity with distance is the **Hubble constant**. Hence $v = H_0 r$.

Hubble Space Telescope (HST) The largest astronomical telescope ever put into space, of Cassegrain design with a main mirror 2.4 m (94 inches) in diameter. It was launched by the space shuttle *Discovery* in April 1990 into an orbit about 600 km (375 miles) high. However, it was found to suffer from spherical **aberration** (2) as a result of a manufacturing error in its main mirror. Astronauts fitted corrective optics to the telescope during a servicing mission in December 1993.

The HST is designed to be repaired and serviced in orbit by space shuttle crews, and to allow its auxiliary instruments to be replaced with new ones. Its present instrumentation consists of the Advanced Camera for Surveys (ACS), a battery of three cameras capable of imaging from the mid-ultraviolet to the near-infrared regions of the spectrum; the Wide Field and Planetary Camera 2 (WF/PC-2, replacing an earlier version); the Space Telescope Imaging Spectrograph (STIS); and the Near-Infrared Camera and Multi-Object Spectrometer (NICMOS).

A mission scheduled for 2004 will see WF/PC-2 replaced with the Wide Field Camera (labeled WFC3) for near-ultraviolet imaging, and the Cosmic Origins Spectrograph (COS), also to operate in the ultraviolet. It is hoped that the HST will then continue to operate without further servicing until 2010.

Huggins, William (1824–1910) English amateur astronomer and spectroscopist. He used an inheritance to build an observatory and the first stellar spectroscope, and began a series of investigations, aided by neighbor and chemist William Miller (1817–70). In 1863 he demonstrated that the stars contain the same chemical **elements** that are present on Earth, and the next year confirmed the gaseous nature of a diffuse nebula. He investigated the motion of Sirius, and made the first discovery of a **moving cluster** (the Ursa Major cluster). In 1876 he took one of the first photographs of a star's (Vega's) spectrum. Huggins used spectroscopy to investigate many other objects, including comets, meteors, and the 1892 nova in Auriga.

Hulst, Hendrik Christoffel van de (1918–2000) Dutch astronomer. He carried out research on interstellar matter and the solar corona. In 1944 he predicted that hydrogen should emit radio waves with a wavelength of 21 cm (SEE **twenty-one centimeter line**), and in 1951 he and Jan **Oort** used Doppler shifts at this wavelength to map the Galaxy.

hunter's Moon The first full Moon after the **harvest Moon**, providing a succession of moonlit evenings in early October. In the southern hemisphere the hunter's Moon is in April or May.

Huygenian eyepiece A basic telescope lens commonly found on small refractors. It consists of two simple elements and is relatively free of **chromatic aberration**. (After Christiaan Huygens.)

Huygens, Christiaan (1629–95) Dutch mathematician, physicist, and astronomer. In 1655 he discovered Saturn's largest satellite, Titan, and the next year explained that the planet's telescopic appearance was due to a broad ring surrounding it. He was concerned with the development of the telescope and introduced the convergent eyepiece. Huygens' many contributions to physics include the idea that light is a wave motion, and the theory of the pendulum; he built the first pendulum clock.

Hyades A large open cluster several degrees across, 151 light years distant in the constellation Taurus. The brightest members are visible to the naked eye; in all it contains about 200 stars, most within a radius of 20 light years. Its estimated age is 650 million years. It is a **moving cluster** receding from us at 43 km/s. The star Aldebaran lies in the same field as the Hyades but is less than half as distant.

Hyakutake, Comet An exceptional comet, discovered by Japanese amateur astronomer Yuuji Hyakutake (1950–2002) in January 1996, which passed within 0.1 AU of the Earth. At its best, around March 25, it was of zero magnitude. The gas tail, which some observers reported to be over 90° long, underwent a disconnection event, and activity on the Sunward side of the nucleus was apparent in the form of jets and fans.

Hydra The largest constellation in the sky, representing the water snake killed by Hercules; it extends from Canis Minor to the south of Virgo. Its brightest star is Alpha (Alphard), magnitude 2.0. The Mira variable R Hydrae can reach magnitude 3.5 at its brightest, and fades to 10.9; its period is 389 days. There is one bright open cluster, M48.

hydrogen-alpha line (Hα line) The absorption or emission line of hydrogen in the red portion of the spectrum at a wavelength of 656.3 nm. The first line in the **Balmer series**, it is the third most conspicuous absorption line in the Sun's spectrum, and is responsible for the pink color of the solar chromosphere when viewed during a total eclipse. Light of this wavelength is often used for monochromatic study of the Sun with the spectrograph, and Hα filters are used by amateur observers for solar photography.

hydrogen spectrum The **spectrum** of atomic hydrogen. The hydrogen atom has the simplest structure of any atom (one proton as nucleus with one single orbiting electron), and the hydrogen spectrum is the simplest to interpret. As over 90% of the atoms in the Universe are hydrogen atoms, hydrogen lines feature prominently in the spectra of many celestial objects. The spectrum contains five principal series of lines associated with different levels of energy in the hydrogen atom, among them the **Balmer series** and the **Lyman series**. Each line corresponds to a quantum-mechanical process called a transition in which the electron "jumps" between orbits (SEE **emission spectrum**).

Hydrus A far southern constellation, representing a small water snake. There are few interesting telescopic objects. Beta Hydri, magnitude 2.8, is the brightest star in the constellation and the nearest fairly bright star to the south celestial pole.

Hygeia Asteroid no. 10, discovered by Annibale de Gasparis in 1849. It orbits in the main belt, and is the fourth-largest asteroid, with a diameter of 408 km (254 miles).

hyperbola An open curve. In mathematics the hyperbola is a type of conic section, so called because it is one of the intersections of a plane with a cone, and is formed when a plane cuts a cone at an angle to the cone's axis that is less than its slope. In astronomy a hyperbolic **orbit** is the path followed by a celestial body which passes another body but is not captured into an orbit around it. For a hyperbolic orbit the **eccentricity** is greater than 1.

hyperboloid The figure obtained by rotating a **hyperbola** about its axis. It is the shape used for the convex secondary mirror in a **Cassegrain telescope**.

Hyperion An outer **satellite** of Saturn, discovered independently by William Cranch Bond and his son George Philips Bond, and independently by William **Lassell** in 1848. It is irregular in shape, measuring 370 × 225 km (230 × 140 miles), and displays chaotic rotation – tumbling about no fixed axis as it follows its eccentric orbit. Although Hyperion is a low-density, icy body, its surface is dark. There are several large craters and a scarp, named Bond–Lassell, which extends for 300 km (200 miles). It seems most likely that Hyperion is a fragment of a larger body.

I

Iapetus A large outer **satellite** of Saturn, discovered by Giovanni Domenico Cassini in 1671. Its diameter is 1440 km (895 miles), and its density of 1.21 g/cm³ indicates that it is predominantly icy. However, although Iapetus's trailing hemisphere is bright, its leading hemisphere (the one facing the direction of orbital travel) is extremely dark, being completely coated by a deep-red material. (Like most of Saturn's satellites, Iapetus has **synchronous rotation**.) It seems more likely that the dark material was somehow deposited from outside rather than extruded from within, and it may be that it is dust ejected from **Phoebe**, the next satellite out.

IAU ABBREVIATION FOR **International Astronomical Union**.

IC ABBREVIATION FOR *Index Catalogue*, consisting of two supplements to the *New General Catalogue* (SEE **NGC**), produced in 1895 and 1908 by J. L. E. **Dreyer**. Objects in the *Index Catalogues* are still referred to by their IC numbers.

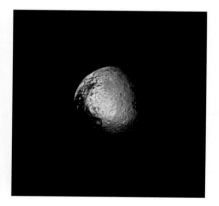

▲ **Iapetus** Features as small as 19 km (12 miles) across can be seen in this Voyager 2 view of Iapetus. It is remarkable in having a very dark leading hemisphere, which contrasts with the brighter trailing hemisphere.

Icarus Asteroid no. 1566, discovered by Walter **Baade** in 1949. It is an **Apollo asteroid** with a diameter of about 1.5 km (1 mile), and has a perihelion distance of 30 million km (19 million miles), within the orbit of Mercury.

Ida Asteroid no. 243, discovered by Johann Palisa in 1884. In 1993 it became the second asteroid to be examined from close quarters, by the probe Galileo *en route* to Jupiter. It is an elongated, cratered body, measuring 56 × 24 × 21 km (35 × 15 × 13 miles), and has a 1.6 × 1.3 × 1 km (1 × 0.8 × 0.6 mile) satellite, named Dactyl.

immersion The entry of a celestial body into the shadow of another at the beginning of an eclipse or occultation.

impact feature Any surface feature of a planet, satellite, asteroid, or other body with a solid surface, that results from the past impact of another body. The most obvious are impact **craters**. *Basins* are large shallow depressions marking the impact of a large body, which have often been filled in by upwelling fluid from below, like the lunar maria (SEE **mare**). *Multiringed basins*, like **Valhalla** on Callisto, are surrounded by concentric ringed structures representing "ripples" from the impact. Other impact features include **rays** radiating from craters, and *palimpsests* – the "ghosts" of craters flattened out or filled in since the impact that caused them. The term was also applied to the visible atmospheric disturbances that were caused by the collision of Comet **Shoemaker–Levy 9** with Jupiter.

inclination (1) (symbol *i*) The angle between the orbital plane of a body orbiting the Sun and the plane of the ecliptic.

inclination (2) (symbol *i*) The angle between the orbital plane of a satellite and the equatorial plane of the planet it orbits.

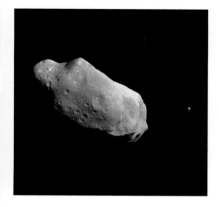

▲ **Ida** Asteroid (243) Ida was imaged by the Galileo spacecraft in August 1993. Its tiny satellite Dactyl is visible at right.

inclination (3) (symbol i) The angle between the orbital plane of a double star and the plane at right angles to the line of sight. The angle is zero if the orbit is seen in plan, and 90° if seen in profile.

inclination (4) (symbol i) The angle between the equatorial plane of a planet or satellite and its orbital plane; this is equivalent to the angle between the axis of rotation and the perpendicular to the orbital plane.

Indus A small southern constellation representing an American Indian. The only star above magnitude 3.5 is Alpha (3.1). Epsilon is the least luminous star visible to the naked eye: magnitude 4.7, distance 11.8 light years, type K4, absolute magnitude 6.9.

inequality The departure from uniform orbital motion of a body, due to **perturbations** by other bodies and the eccentricity of its orbit.

inferior conjunction SEE **conjunction**.

inferior planet Either of the planets Mercury and Venus, which orbit inside the Earth's orbit.

inflation A proposed early stage of the Universe, immediately after the **Big Bang**, in which for a brief fraction of a second the Universe expanded faster than the speed of light. Such a period of inflation explains the uniformity of the **cosmic microwave background**. SEE ALSO **cosmology**.

Infrared Astronomical Satellite (IRAS) A satellite, a collaboration between the USA, Netherlands, and UK, that surveyed the sky at infrared wavelengths. It was launched in January 1983 and operated for nine months, cataloguing nearly 250,000 sources. Its discoveries included several comets and asteroids, possible planetary systems forming around nearby stars, and **starburst galaxies**.

▶ **impact feature** The circular Manicouagan impact structure in Quebec, Canada, is revealed in this satellite image. The 70-km (44-miles) diameter feature was formed by an asteroid impact about 212 million years ago.

◀ infrared astronomy
This all-sky map of infrared sources was recorded by the Infrared Astronomical Satellite in 1983. The bright band running across the middle is the plane of the Milky Way.

infrared astronomy The study of **electromagnetic radiation** falling between the red end of the visible spectrum and radio wavelengths, between about 700 nm, or 0.7 μm, and 1 mm. The Earth's atmosphere is opaque to much of this wavelength range, although there are a number of atmospheric *windows* through which infrared observations can be made from high mountains. To avoid the Earth's atmosphere, observations are made from high-flying aircraft (SEE **Stratospheric Observatory for Infrared Astronomy**), from unmanned balloons, and from satellites such as the **Infrared Space Observatory**.

Astronomical sources of infrared radiation include the planets and other Solar System bodies, the stars and dusty regions of the Milky Way, and other galaxies (including **starburst galaxies**). Most of the interstellar dust and gas in our Galaxy is contained in *giant molecular clouds* (GMCs, SEE **molecular cloud**), such as the one in the Orion Nebula. Maps of GMCs at infrared wavelengths show the warm or hot regions where star formation is in progress or is about to begin. On a much smaller scale, highly compact and opaque **Bok globules** with just a few tens of solar masses are also potential star-forming regions. Infrared observations show the temperature of the globules to be very low, 10–20 K, indicating that they are at an early stage of gravitational collapse. Faint, wispy structures in interstellar space termed *infrared cirrus*, discovered by IRAS, are probably caused by infrared emissions from dust grains. These grains are composed of carbon in the form of graphite at a tempera-

ture of 30–40 K, ejected by the stellar winds of cool stars or supernova explosions.

Infrared wavelengths can penetrate the otherwise opaque clouds of dust in the plane of our Galaxy, so we can see sources toward the galactic center which are completely obscured at optical wavelengths. The observations show a high density of late-type stars and a large ring of molecular gas and dust, as well as many regions of active star formation. Beyond the Milky Way, infrared observations are a key to understanding the energy output of active galaxies and quasars and the physical processes operating within them.

In the Solar System, infrared observations allow us to work out the basic chemistry of the planets, their satellites, and the asteroids. Molecules in planetary atmospheres exhibit detailed infrared absorption spectra, which permit the composition and temperature of the atmospheres to be established. Spectral reflectance observations of asteroids, planetary satellites, and small planets without substantial atmospheres have been used to discover the mineralogical compositions of their surfaces. Infrared techniques have established the presence of water and carbon dioxide ices on Mars, water ice on many of the satellites of Jupiter, Saturn, and Uranus, and methane ice on Triton and on Pluto. Measurement of the black-body emission of small Solar System bodies has given us information on their sizes and surface albedos where no other technique would succeed.

Infrared Space Observatory (ISO) A European Space Agency satellite with a

0.6-m (24-inch) telescope for infrared astronomy, launched in November 1995. It operated until 1998.

inner planet Any of the planets Mercury, Venus, Earth, and Mars, which orbit inside the main asteroid belt.

instability strip The narrow region of the **Hertzsprung–Russell diagram** where pulsating stars are located. It extends from the brightest Cepheids down to pulsating white dwarf stars and also includes RR Lyrae stars, Delta Scuti stars, and dwarf Cepheids.

Integral A joint European, US and Russian satellite for gamma-ray astronomy launched in 2003; the name is short for International Gamma-Ray Astrophysics Laboratory.

interferometry The study of a celestial object by analyzing its interference pattern. When electromagnetic radiation emitted by a source is split into two and recombined, a pattern of interference fringes is produced that can be used to reveal more detailed information about the source than is possible from a single beam.

An *interferometer* is an instrument consisting of two or more detectors (telescopes or antennas) which collect signals from a single source. The signals are then combined to form the interference pattern. The longer the wavelength of the radiation being detected, the greater the separation between the detectors.

Various types of **optical interferometer** are used to measure the diameters of stars and resolve fine detail in their spectra. Interferometry is particularly important in radio astronomy, where the long wavelengths of the radiation mean that a single detector gives only a coarse resolution (SEE **radio interferometer**).

intergalactic matter Matter in the space between galaxies. There are no significant quantities of intergalactic dust, but there are clear signs of intergalactic gas. X-ray emission from clusters of galaxies comes from a hot, tenuous gas with a temperature of 10 to 100 million K between the galaxies in the cluster. Radio-emitting clouds of electrons ejected by radio galaxies are drawn out into a tadpole shape as if they were plowing through a resisting medium. How much intergalactic gas is primordial hydrogen and helium from the Big Bang falling into the clusters, and how much is matter expelled from the galaxies, is uncertain. Tidal interactions (SEE **tides**) between galaxies are common, and radio observations reveal streams of neutral hydrogen around interacting galaxies, for example M81 and M82. A similar gas stream, the Magellanic Stream, links the **Magellanic Clouds** with our Galaxy. It is possible that **dark matter** exists in the space between galaxies and galaxy clusters.

International Astronomical Union
(IAU) The controlling body of world astronomy, founded in 1919, with its headquarters in Paris. It is the authority for naming celestial bodies and their features.

International Atomic Time (TAI)
The international reference scale of atomic time, formed by the Bureau Interna-

▲ **Infrared Space Observatory** (ISO) The Eagle Nebula is shown here in a composite ISO image. Shorter infrared wavelengths are rendered in blue, longer wavelengths in red.

tional de l'Heure (BIH) in Paris from atomic clock data supplied by establishments worldwide. (The abbreviation TAI is derived from the name in French.) The fundamental unit of time is the atomic second, which is defined in terms of the resonance of the caesium atom. SEE ALSO **Universal Time**.

International Space Station (ISS)

A space station being built as a collaborative venture between NASA, the Russian Space Agency, ESA, Japan, and Canada. The first section was launched in 1998. More than 90 missions will be required to launch all the ISS's modules, which will total nearly 420 tons. The last is scheduled to be put in place in 2005.

International Ultraviolet Explorer (IUE)

A joint NASA, European Space Agency, and UK satellite, carrying a 0.45-m (18-inch) telescope. It was launched in 1978 and operated until 1996, taking ultraviolet spectra of stars and galaxies.

interplanetary matter The material in the space between the planets. It is made up of atomic particles (mainly protons and electrons) ejected from the Sun via the solar wind, and dust particles (mainly from **comets**, but some possibly of cosmic origin) and neutral gas in the plane of the ecliptic. The particles in the inner Solar System constitute the *zodiacal dust cloud*, scattered sunlight from which is responsible for the **zodiacal light**. An extremely tenuous ring of dust, over 50 million km (30 million miles) wide, orbits the Sun just outside the Earth's orbit. There is also the dust in **meteor streams**.

interstellar matter The matter contained in the regions between objects in the Galaxy. It includes clouds of *interstellar gas*, mainly hydrogen with some helium, and a small percentage (about 1% by mass) of interstellar dust grains. The hydrogen is mainly in a relatively cool neutral form at a temperature of 10–100 K (SEE **H I region**). Some very much hotter and more diffuse regions of ionized gas (**H II**

▲ **International Space Station** Photographed by a crew member on the Space Shuttle *Endeavour* in December 2001 is the International Space Station, with the Earth providing a dramatic background. The robot manipulator arm is clearly visible, as are several modules and solar arrays.

► **International Ultraviolet Explorer** The IUE, seen here in an artist's impression, studied comets, asteroids and the atmospheres of the giant planets. It also provided insights into stellar winds and hot, massive stars.

regions) also exist, together with dense clouds of molecular hydrogen and other molecules. The dust grains appear to be elongated in shape, about 100 nm long, and aligned by the galactic magnetic field. Most grains are thought to be composed of silicates or forms of carbon, and some have icy coatings.

The clouds of gas and dust are largely confined to the plane of the Galaxy and tend to be concentrated in the spiral arms. Our Galaxy contains about 10^{10} solar masses of material distributed between the stars, making up about 10% of its total mass. The density of these clouds is less than that of a laboratory vacuum on Earth, but they nevertheless have an obscuring effect on the light of stars behind them. The clouds scatter and absorb starlight passing through them, a phenomenon known as *interstellar extinction*. On average, starlight is dimmed by one magnitude for every kiloparsec it travels in the plane of the Galaxy. Red light is dimmed less than blue light, which causes a reddening of starlight.

The spectra of remote stars show **absorption lines** superimposed on the stellar spectra, produced by interstellar dust and gas. Sodium, calcium, potassium, iron, titanium, and other elements have been detected from these absorption lines. These elements are ejected from stars. Heavier elements, including lead, gallium, and krypton, are manufactured in the extreme conditions of a **supernova** explosion. SEE ALSO **nebula**.

interstellar molecules Molecules that exist in interstellar **molecular clouds**. Most have been identified by their emission lines at millimeter and centimeter wavelengths, although the first, including the cyanogen radical (CN) and the methylidyne radical and ion (CH, CH^+), were detected in the 1940s at optical frequencies. The hydroxyl radical (OH) was the first (in 1963) to be identified by radio astronomy. The list of interstellar molecules has been extended to more than 126 since, and includes hydrogen cyanide (HCN), ammonia (NH_3), methane (CH_4), methyl alcohol (CH_3OH), and more complex organic molecules such as the amino acid glycine.

intrinsic variable SEE **variable star**.

inverse-square law A reduction in the intensity of a physical quantity in proportion to the square of the distance from its source. For example, a planet three times the Earth's distance from the Sun would experience one-ninth the gravitational attraction of the Sun and receive one-ninth as much sunlight.

Io The innermost of Jupiter's four **Galilean satellites**, and the second-smallest, with a diameter of 3630 km (2980 miles). It has the highest density, at 3.53 g/cm³, indicating a lack of water ice in its make-up. Io's tidal interaction with Jupiter creates sufficient flexing and heating to make it by far the most volcanically active world in the Solar System. The surface, multicolored by the presence of sulfur, is being continually renewed. The active regions contain volcanic vents and calderas, lava flows, and volcanic plumes that send material to heights of up to 300 km (200 miles). Plumes prominent when the Voyager probes photographed Io included those named Pele, Loki and Prometheus. In the active regions there may be lakes of molten sulfur or even molten rock; elsewhere there are plains formed by volcanic deposits, and mountains. A thin atmosphere of sulfur dioxide surrounds Io, and a doughnut-shaped *plasma torus* of ionized (electrically charged) sulfur and oxygen surrounds its whole orbit and interacts with Jupiter's **magnetosphere** to create a *flux tube* carrying an electric current of about 5 million amperes (SEE the diagram). In 1996 the **Galileo** probe detected an iron core and magnetic field, and found the surface features to have changed significantly since the satellite was first imaged during the 1979 Voyager missions. SEE ALSO **satellite**.

ionosphere A region in the atmosphere of a planetary body in which there are free electrons and ions produced by ultraviolet radiation and X-rays from the Sun. The degree of ionization is greatly affected by solar activity.

The Earth's ionosphere extends from about 60–1000 km above the surface, and reflects radio waves of long wavelengths, although shorter wavelengths can pass through undisturbed. The reflecting power of these layers makes long-range broadcasting and telecommunications possible up to frequencies of about 30 MHz. The radio waves and microwaves from around 30 MHz to about 100 GHz that pass through the ionosphere are used in radio

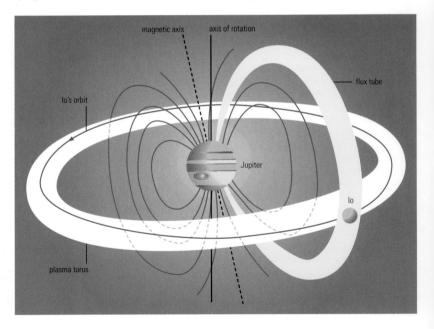

▲ **Io** Material ejected from Io's volcanoes forms a torus around Jupiter. A magnetic flux tube links Io and Jupiter: particles ejected during the satellite's volcanic eruptions can lead to enhancements of Jupiter's aurorae.

astronomy and satellite communications; they make up the *radio window*. The free electrons formed as a result of ionization do at times have serious adverse effects on the reception of radio signals.

IRAS ABBREVIATION FOR **Infrared Astronomical Satellite**.

Iris Asteroid no. 7, discovered by John Russell Hind in 1847. Although it ranks only about 25th in size, with a diameter of around 200 km (125 miles), it is the fourth-brightest.

iron meteorite A meteorite consisting mainly of iron (90–95%) and some nickel, with traces of other metals. Iron meteorites are classified into a number of groups according to the proportions of these other metals that they contain, and each group is thought to correspond to a different parent asteroid.

irradiation A physiological phenomenon in which a bright object viewed against a dark background appears larger than it really is. The effect can introduce errors into measurements made by an observer using a telescope.

irregular galaxy A galaxy which shows no symmetry. Some irregular galaxies are smaller than spirals and ellipticals, contain much gas, and are undergoing star formation. Some are classified as irregular for the only reason that they do not fit into other categories of galaxy.

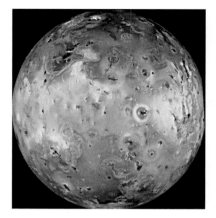

▲ **Io** A composite of images from the Galileo orbiter shows the volcanically active surface of Jupiter's moon Io. Intense volcanism driven by tidal stresses leads to rapid resurfacing of Io.

irregular variable A pulsating variable star whose variation in brightness does not follow any regular or predictable pattern. Irregular variables are red giants or supergiants.

ISO ABBREVIATION FOR **Infrared Space Observatory**.

isotropy SEE **cosmology**.

ISS ABBREVIATION FOR **International Space Station**.

IUE ABBREVIATION FOR **International Ultraviolet Explorer**.

J

Jansky, Karl Guthe (1905–50) American radio engineer. In 1931, while looking for sources of "noise" that interfered with radio communications, he discovered one located in the Milky Way, in Sagittarius. Jansky himself did not follow this up, but it led directly to the development of radio astronomy. The unit of flux density used in radio astronomy is named the *jansky* in his honor.

Janssen, Pierre César Jules (1824–1907) French astronomer. He specialized in solar spectroscopy and developed special instruments, including a spectrohelioscope for observing prominences without waiting for an eclipse. He discovered the helium line in the solar spectrum, and produced a photographic atlas of the Sun.

Janus A small inner **satellite** of Saturn, a **coorbital satellite** with Epimetheus. The two were discovered in 1978 when John Fountain and Stephen Larson re-examined photographs taken some years before.

▲ **Jodrell Bank** The Lovell Radio Telescope at Jodrell Bank was built in 1957 and originally known as the Mark I Telescope. It is still one of the world's largest fully steerable radio telescopes.

They are both irregular in shape, and are probably fragments of the same larger body. Either satellite may be the object reported in 1966 by Audouin Dollfus and given the name Janus, but not subsequently confirmed.

Jeans, James Hopwood (1877–1946) English mathematician and physicist. In astronomy he worked on stellar dynamics, the formation and evolution of stars, double-star systems, and spiral galaxies. In 1917 he proposed that the Solar System was formed from matter pulled out of the Sun by the gravitational attraction of a passing star, a theory later developed by Harold Jeffreys (1891–1989), but since abandoned. Jeans wrote many popular books on astronomy.

jet A stream of very-high energy particles emitted by objects such as **active galactic nuclei** and protostars. They are detectable by their radio emissions and may also be visible at optical wavelengths. In galaxies there are often two jets, emerging in opposite directions from the nucleus perpendicular to the galactic plane.

Jewel Box The open cluster NGC 4755 in the constellation Crux, around the star Kappa Crucis. The name derives from the varied colors of the stars in the cluster. It lies 7600 light years away.

Jodrell Bank The site at Macclesfield, Cheshire, UK, of the University of Manchester's Nuffield Radio Astronomy Laboratories, founded after World War II by Bernard **Lovell**. Its main instrument is the 76-m (250-ft) fully steerable dish opened in 1957 and named the Lovell Telescope in 1987; a major upgrade is due for completion in 2004. A second dish, elliptical in shape, 38 × 25 m (125 × 83 ft), was opened in 1964. These telescopes can be used individually or as part of the **MERLIN** array.

► **Jupiter** This view of Jupiter was assembled from four images obtained by the Cassini spacecraft on December 7, 2000. The Great Red Spot is seen at lower right. The shadow of the satellite Europa is the dark spot at lower left.

Jones, Harold Spencer (1890–1960) English astronomer. He was concerned with refining values for the distances and periods of Solar System bodies, and in 1941 obtained what was then the most accurate value of the Sun's distance. As the tenth Astronomer Royal, he planned the Royal Greenwich Observatory's move to Herstmonceux in Sussex.

Jovian The adjective for Jupiter. *Jovian planet* is another term for **giant planet**.

Julian calendar SEE **calendar**.

Julian Day A system of numbering days continuously, without division into years and months. It is used in astronomy when calculating events that recur over long periods, such as the solar cycle of sunspot variations. The Julian Day Number is the number of days that have elapsed since noon GMT on January 1, 4713 BC, which is defined as day zero. The idea was conceived in 1582 by the French chronologist Joseph Justus Scaliger. The name "Julian" was given by Scaliger in memory of his father, Julius Caesar Scaliger (1484–1558), and is not to be confused with the Julian calendar devised by the Roman Emperor Julius Caesar.

The Julian Date is given by the Julian Day Number for the preceding noon at Greenwich followed by the fraction of the mean solar day that has elapsed. Thus, the Julian Date of an observation made at 6 p.m. on January 1, 1995, is 2,449,719.25, the Julian Day Number being 2,449,719.

Note that in historical usage the year 1 BC is followed by the year AD 1, whereas in astronomical usage AD 1 is preceded by year 0, which in turn is preceded by year −1 (corresponding to the historical year 2 BC). Thus to convert years BC to astronomical years, subtract 1 and prefix a minus sign.

Juliet One of the small inner **satellites** of Uranus discovered in 1986 during the Voyager 2 mission.

Juno Asteroid no. 3, discovered by Karl Harding in 1804. It is the eighth-largest, with a diameter of 268 km (167 miles).

Jupiter The fifth major planet from the Sun, and the largest of the giant planets. Not counting the Sun, Jupiter contains around 70% of the mass of the Solar System. It is one of the brightest objects in the sky, with a magnitude at opposition of between −2.3 and −2.9. Its maximum apparent diameter is 50 arc seconds, and even at its most distant it is still 30 arc seconds across, so for most of the time it is the largest planetary disk visible. Through a telescope, Jupiter's yellowish elliptical disk is seen to be crossed by a number of brownish-red bands which show gradual, continual changes. Our knowledge of the planet owes much to visits by space probes: Pioneers 10 and 11, Voyagers 1 and 2, Ulysses, and **Galileo**. It became the focus of public attention in 1994 when Comet **Shoemaker–Levy 9** plunged into its atmosphere. The main data for Jupiter are given in the table.

The darker bands are called *belts*, and the lighter ones are known as *zones*. Their nomenclature is given in the second table. Changes occur in the pattern and colors of the belts and zones on a timescale of years, but in the long term the same fundamental structure persists. There is also much activity on timescales from days to weeks. Dark and light spots and streaks come and go. Most spots are short-lived. The white *ovals*, such as those that appeared on the

JUPITER: DATA	
Globe	
Diameter (equatorial)	142,800 km
Diameter (polar)	133,500 km
Density	1.33 g/cm³
Mass (Earth = 1)	317.8
Volume (Earth = 1)	1320
Sidereal period of axial rotation (equatorial)	9h 50m 30s
Escape velocity	59.5 km/s
Albedo	0.73
Inclination of equator to orbit	3° 07'
Temperature at cloud-tops	125 K
Surface gravity (Earth = 1)	2.69
Orbit	
Semimajor axis	5.203 AU = 778.3 × 10⁶ km
Eccentricity	0.048
Inclination to ecliptic	1° 18'
Sidereal period of revolution	11.86y
Mean orbital velocity	13.06 km/s
Satellites	60

STB in the 1940s, are rather more persistent. The most distinctive feature is the **Great Red Spot** (GRS) in the STZ, first observed by Heinrich Schwabe in 1831, since when it has disappeared and reappeared several times. Another prominent feature at the same latitude was the South Tropical Disturbance, first observed in 1901 as a gray bar across the STZ which grew to dominate the planet until its disappearance in 1940. A similar dark disturbance, the SEB Revival, erupts from time to time, as it did in 1993. Different types of streak are given names such as *projections*, *plumes*, and *festoons*. They are particularly common on the south edge of the NEB, as are small bright and dark spots.

All these cloud features indicate a dynamic and turbulent atmosphere. The clouds are drawn into horizontal bands by the rapid rotation and wind speeds of up to 400 km/h (250 mile/h). An equatorial jet-stream keeps the region between the equatorial belts rotating in about 9 hours 51 minutes (referred to as *System I* rotation), five minutes faster than the rest of the atmosphere (*System II*). Turbulence is often evident at the boundaries between Systems I and II. Individual features may drift backward and forward in longitude. The bright zones are high-altitude clouds supported by upwelling gases, while the dark belts are regions of descending currents. Eddies give rise to the spots, which are cyclones or (like the GRS) anticyclones. Aurorae and thunderstorms have been observed.

Jupiter's predominant atmospheric constituents are hydrogen and helium, hydrogen accounting for nearly 90%, and helium for most of the rest. But the upper atmosphere contains a considerable variety of chemical compounds, and they give the planet its multihued appearance. The highest features are red, due to phosphorus from lower levels carried aloft on rising currents. Ammonia crystals make up the high-altitude white clouds. The lower clouds that make up the belts are largely ammonium hydrosulfide; other substances present include methane, ethane, water, phosphine, hydrogen cyanide, carbon monoxide, and germanium tetrahydride. There is probably a complex cycle of chemical reactions, driven by ultraviolet radiation from the Sun, and also by high-energy lightning and auroral discharges.

The temperature at the cloud-tops is around 125 K, and the pressure is about 0.5 bar. Pressure and temperature increase quite rapidly with depth. Further down, where the temperature and pressure are similar to those at the Earth's surface, there are probably clouds of water ice crystals and water vapor. At 1000 km (600 miles) below the cloud-tops the atmosphere gives way to an ocean of liquid molecular hydrogen. The temperature here is about 1000 K. At a depth of 20,000 to 25,000 km (12,500 to 15,000 miles), under a pressure of 3 million bars, the hydrogen becomes so compressed that it behaves as a metal. At the center of the planet there is probably an iron–silicate core perhaps 15 to 20 times the Earth's mass, surrounded by an ice mantle of about 20 Earth masses. The core temperature is estimated to be 30,000 K. Jupiter emits about 1.7 times as much heat as it receives from the Sun.

The deep metallic hydrogen "mantle" gives Jupiter a powerful magnetic field and a very extensive **magnetosphere**. The magnetic axis is tilted by nearly 10° with respect to the rotational axis, and the magnetic field is nearly 20,000 times as strong as the Earth's. The magnetosphere is huge and has many unique features. It is several times the size of the Sun, and the magnetotail can stretch farther than the orbit of Saturn. The magnetic field traps a large quantity of plasma (charged particles). Low-energy plasma is contained in a vast *magnetodisk* in Jupiter's equatorial plane, while high-energy plasma is funneled into radiation belts which are ten thousand times as intense as the Earth's **Van Allen Belts**. The magnetosphere interacts with the innermost Galilean satellite, **Io**, to produce a *plasma torus* and a *flux tube*. Most of

JUPITER: NOMENCLATURE OF BELTS AND ZONES

Name	Abbreviation
North polar region	NPR
North north temperate zone	NNTZ
North north temperate belt	NNTB
North temperate zone	NTZ
North temperate belt	NTB
North tropical zone	NTropZ
North equatorial belt	NEB
Equatorial zone	EZ
South equatorial belt	SEB
South tropical zone	STropZ
South temperate belt	STB
South temperate zone	STZ
South south temperate belt	SSTB
South south temperate zone	SSTZ
South polar region	SPR

Jupiter's trapped plasma comes from Io's volcanoes. Some of the sodium from these volcanoes is ejected by Jupiter's magnetosphere and forms a vast "nebula" with a radius of roughly 35 million km (20 million miles). The magnetosphere is the source of the planet's powerful radio emissions. Some of this is **synchrotron radiation** from high-energy electrons in the radiation belts.

Jupiter has 60 known **satellites**, as of early 2003. The four major ones – Io, Europa, Ganymede and Callisto – are often referred to collectively as the **Galilean satellites**. Within the orbit of Io there are four small satellites. These eight together form a compact group orbiting close to Jupiter and in the planet's equatorial plane. The others are all minor objects, orbiting at distances from 7.5 to 28.5 million km (4.7 to 17.7 million miles) from the planet.

Jupiter also possesses a modest ring system (SEE **ring, planetary**), the brightest component of which is about 50,000 km (30,000 miles) above the cloud-tops.

K

Kapteyn's Star A red subdwarf, 12.7 light years away in the southern constellation Pictor, which has the second-largest proper motion known: 8.7 arc seconds per year. It is of magnitude 8.9, type M1, and is over 250 times less luminous than the Sun. It was discovered by the Dutch astronomer Jacobus Kapteyn (1851–1922).

K corona SEE **corona**.

Keck Telescopes A pair of identical telescopes at the W. M. Keck Observatory on Mauna Kea, Hawaii. Both have mirrors 10 m (396 inches) in diameter and consisting of 36 hexagonal segments. Keck I was completed in 1992, and its twin, Keck II, in 1996. They are run by the University of California and Caltech.

Kellner eyepiece An eyepiece with large **eye relief**, which makes it widely used for binoculars as well as telescopes. It is essentially a **Ramsden eyepiece** with one element replaced by an achromatic doublet (SEE **achromatic lens**). (Invented in 1849 by German optician Carl Kellner, 1826–55.)

Kepler, Johann (1571–1630) German mathematician and astronomer. He supported the Sun-centered Solar System put forward by **Copernicus**. In 1600 he went to Prague, becoming assistant to Tycho **Brahe**, on whose death in 1601 he succeeded as Imperial Mathematician to the Holy Roman Emperor, Rudolf II, with the task of completing tables of planetary motion begun by Tycho. From Tycho's accurate observations he concluded in *Astronomia nova* (1609) that Mars moved in an elliptical orbit, and went on to establish the first of his laws of planetary motion (SEE **Kepler's laws**). Erroneous reasoning led Kepler nevertheless to the second of these laws, and his desire to match celes-

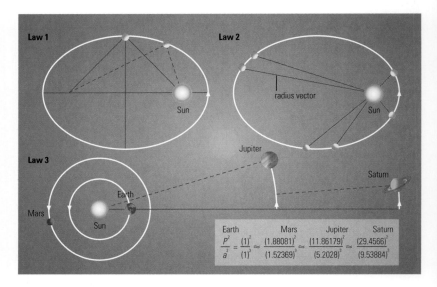

▲ **Kepler's laws** Planetary motions around the Sun are described by Kepler's three laws. The laws apply to all bodies in closed orbits around the Sun, and to satellites orbiting planets.

▶ **Kitt Peak National Observatory** This aerial view of Kitt Peak shows the many telescopes sited there. Largest, at the top right, is the 4-m (158-inch) Mayall Telescope. The triangular building in the foreground is the McMath–Pierce Solar Facility, the world's largest solar telescope.

tial and musical harmony led to the third law, published in *Harmonices mundi* (1619). The *Rudolphine Tables*, based on Tycho's observations and Kepler's laws, appeared in 1627 and remained the most accurate until the 18th century. Other works of his include *De stella nova*, on the supernova of 1604 (SEE **Kepler's Star**), and *Dioptrice* (1611), on optics and the theory of the telescope.

Kepler's laws The three fundamental laws governing the motions of the planets around the Sun (and other celestial bodies in closed orbits), first worked out by Johann **Kepler** between 1609 and 1619, and based on observations made by Tycho Brahe (SEE the diagram).

Law 1 deals with the shape of a planetary orbit: the orbit of each planet is an **ellipse** with the Sun at one of the foci.

Law 2 explains the varying speed of planetary motion such that the planet moves faster the nearer it is to the Sun: the line joining the planet to the Sun (the *radius vector*) sweeps out equal areas in equal times. This law is known as the *law of areas*.

Law 3 relates the size of the orbit to the period of revolution: the square of the period of revolution (P) is directly proportional to the cube of the mean distance of the planet from the Sun (a). In practice this means that P^2/a^3 is approximately constant for all bodies in orbit around the Sun.

Kepler's Star A **supernova** that appeared in October 1604 in the constellation Ophiuchus, reaching magnitude -3 and remaining visible to the naked eye until March 1606. It was studied by Johann **Kepler**. The smooth light curve is like that of a Type I supernova. The remnant is a radio source, and some faint nebulosity is visible optically. It was the last supernova seen in our Galaxy.

Kirkwood gaps Regions in the main **asteroid** belt where, because of **resonances** with Jupiter, no stable orbit is possible, and from which asteroids are therefore absent (or nearly absent). The gaps are at distances at which the revolution period of an orbiting body would be **commensurable** with Jupiter's, and Jupiter would perturb the body. There are pronounced gaps at $\frac{1}{3}$, $\frac{2}{5}$, and $\frac{1}{2}$ of Jupiter's period, for example. The gaps in the ring system of **Saturn** are kept clear by a similar perturbing effect of its satellites, in particular Mimas, whose period is twice that of a body at the distance of the Cassini Division. The American astronomer Daniel Kirkwood (1814–95) explained how the gaps came about.

Kitt Peak National Observatory An observatory near Tucson, Arizona, part of the US National Optical Astronomy Observatories, containing the world's largest collection of telescopes. Its largest instrument is the 4-m (158-inch) Mayall

Telescope, opened in 1973. It is also the site of the McMath–Pierce Solar Telescope, the world's largest solar telescope; the 3.5-m (138-inch) WIYN Telescope, opened in 1994; and the 2.3-m (90-inch) Steward Reflector of the University of Arizona.

Kohoutek, Comet

A **comet** discovered in 1973 by the Czech astronomer Luboš Kohoutek (1935–). When first detected it was nearly as far from the Sun as Jupiter, and was predicted to become as bright as the Moon at perihelion. Although it failed to live up to these expectations, it was studied extensively and advances in comet science were made as a result.

K star

A star of spectral type K, characterized by numerous absorption lines, especially of iron, titanium, and calcium, and the presence of molecular bands of CH and CN in its spectrum. Main-sequence K stars have surface temperatures ranging from 3550 to 4900 K. Giants are about 400 K cooler than this, and supergiants 300 K cooler still. At their lowest point on the main sequence, K stars are about half a solar mass, rising to 0.8 of a solar mass at their highest. As K-type giants and supergiants may be either evolved old stars of about one solar mass or evolved stars of higher masses, a wide range is represented. However, most of the K-type giants are stars of a few solar masses. Only a small fraction of K stars are variable: the coolest Cepheids have K-type spectra, as do the hottest of the Mira stars. A few per cent of K-type giants show spectral peculiarities with an overabundance of carbon and certain **heavy elements**. Bright examples of K stars are Pollux, Arcturus, one component of Alpha Centauri, and Epsilon Eridani.

Kuiper, Gerard Peter

(1905–73) Dutch–American astronomer. After World War II he embarked upon an observational program which revitalized lunar and planetary astronomy. He discovered the satellites Miranda (of Uranus) in 1948, and Nereid (of Neptune) in 1949, and found methane in the atmospheres of Uranus, Neptune and Titan, and carbon dioxide in the atmosphere of Mars. He suggested the existence of what is now called the **Kuiper Belt**. Kuiper was involved with many US missions to the Moon and planets, in particular with the Ranger and Mariner probes. He advocated carrying infrared telescopes on high-flying aircraft; the Kuiper Airborne Observatory was named in his honor.

Kuiper Belt (Edgeworth–Kuiper Belt)

A flattish zone stretching from the orbit of Neptune at 30 AU to about 1000 AU from the Sun, populated by **planetesimals** currently classed as asteroids. Its existence was proposed by Kenneth Edgeworth (1880–1972) in 1949 and by Gerard Kuiper in 1951 as a repository for comets or planetesimals left over from the formation of the Solar System. The concept was resurrected in the 1980s as a possible source of short-period comets, to account for the generally low inclinations of such comets, which the spherical **Oort Cloud** could not.

The first member of the Kuiper Belt was discovered in 1992, and by early 2003 over 600 had been found. Most are between 100 and 500 km (60–300 miles) in size, although larger ones exist: for example, Varuna may be as large as 1100 km (680 miles). There are almost certainly many smaller members. Members of the belt are often called *trans-Neptunian objects* (TNOs). Many, like Pluto, are in a 3 : 2 orbital resonance with Neptune, and are called *plutinos*.

Physically, Kuiper Belt objects are probably similar to comets. They show no comet-like activity because they are so far from the Sun. If they move inward into planet-crossing orbits they may become **Centaurs**, and then outgassing can start, as has been observed with **Chiron**.

L

Lacaille, Nicolas Louis de
(1713–62) French astronomer. He spent
the years 1751 to 1753 surveying the southern-hemisphere skies from the Cape of
Good Hope, and introduced 14 new southern constellations. In 1761 he made an
accurate measurement of the Moon's distance that took into account the Earth's
oblateness; this made possible a more
accurate method of determining terrestrial longitude.

Lacerta A small, obscure northern constellation between Andromeda and
Cygnus, representing a lizard, introduced
by **Hevelius**. Alpha, its brightest star, is
magnitude 3.8.

Lagoon Nebula A bright nebula in
Sagittarius, M8 or NGC 6523, surrounding the open star cluster NGC 6530. The
nebula is cut by a dark cloud which gives
it its name. The nebula and cluster lie at a
distance of 5200 light years.

Lagrange, Joseph Louis de
(1736–1813) French mathematician,
born in Italy. He made many contributions
to celestial mechanics, studying the
Moon's **libration**, the motion of Jupiter's
satellites, and perturbations and stability
in the Solar System. From his investiga-
tion in 1772 of the **three-body problem**
he found special solutions of the equations
which suggested the existence of the equilibrium positions now known as
Lagrangian points.

Lagrangian point One of the points at
which a celestial body can remain in a position of equilibrium with respect to two
much more massive bodies orbiting each
other. At these points the forces acting on
the smaller body cancel out. There are five
Lagrangian points, in the orbital plane of
the two large bodies (SEE the diagram).
Bodies at points L_1, L_2 and L_3, on the line
joining M_1 and M_2, are less stable than

bodies at points L_4 and L_5, which form
equilateral triangles with M_1 and M_2. The
Trojan asteroids lie in stable orbits near
the L_4 and L_5 Lagrangian points of
Jupiter's orbit around the Sun. (After
Joseph Louis de **Lagrange**.)

**Lalande, Joseph Jérôme (Le
Français) de** (1732–1807) French
astronomer. His main achievement was a
catalog of over 47,000 stars. He did much
to popularize astronomy in France. Lalande
observed Neptune in 1795, 51 years before
it was located by Johann Galle, without re-
alizing that it was a planet.

La Palma Location of the **Roque de los
Muchachos Observatory**.

Laplace, Pierre Simon de
(1749–1827) French mathematician and
astronomer. He used Isaac Newton's law
of **gravitation** to interpret the perturbations of planets and satellites, the shape
and rotation of Saturn's rings, and the stability of the Solar System. In 1796 he put

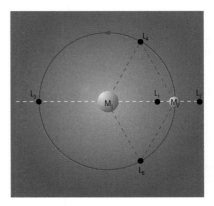

▲ **Lagrangian point** Many spacecraft are stationed
around the L_1 point between Earth and the Sun. The
L_4 and L_5 (60° ahead of and behind the planet)
Lagrangian points of Jupiter's orbit are occupied by
Trojan asteroids.

◀ **Larissa** Neptune's inner satellite Larissa was imaged during Voyager 2's fly-by in August 1989.

forward his *nebular hypothesis* of the origin of the Solar System, which, although wrong in detail, is broadly in agreement with the presently accepted view. His five-volume *Traité de mécanique céleste* published over the period 1799–1825 summarized 18th-century advances in celestial mechanics.

Large Binocular Telescope A twin reflecting telescope, sited at **Mount Graham International Observatory**, scheduled for completion in 2004. Each component has an 8.4-m (330-inch) primary mirror, giving a combined effective aperture of 11.8 m (465 inches). The centers of the mirrors are 14.4 m (567 inches) apart.

Large Magellanic Cloud (LMC) SEE **Magellanic Clouds**.

Larissa One of the small inner **satellites** of Neptune, first detected in 1981 by Harold Reitsema and colleagues when it occulted a star, and confirmed by the 1989 visit of Voyager 2. Larissa appears to be the inner **shepherd moon** to the planet's Adams Ring.

Las Campanas Observatory An observatory near La Serena, Chile, run by the Carnegie Institution of Washington. Its main telescopes are the twin **Magellan Telescopes** and the 2.5-m (100-inch) Du Pont Telescope, in operation since 1977.

Lassell, William (1799–1880) English brewer and amateur astronomer. From the observatory he built, he discovered several planetary satellites: Neptune's **Triton**, only 17 days after the discovery of Neptune itself, Saturn's **Hyperion**, and

Uranus's **Ariel** and **Umbriel**. He also discovered 600 nebulae. Lassell was the first to use the equatorial system of mounting for reflecting telescopes.

last contact SEE **fourth contact**.

last quarter One of the Moon's **phases** when it has almost reached western quadrature and is half illuminated as seen from the Earth.

late-type star A cool star of spectral type K, M, C, or S. The name was given when it was believed that the spectral types represented an evolutionary sequence, hot stars being the youngest and cool stars the oldest. COMPARE **early-type star**.

latitude A coordinate on the celestial sphere, or on the surface of a celestial body. *Celestial* (or *ecliptic*) *latitude* is the angular distance of a celestial body from the ecliptic; it is measured at right angles to the ecliptic, from 0° at the ecliptic to 90° at the ecliptic poles, positive to the north and negative to the south. *Galactic latitude* is the angular distance of a body from the galactic equator, measured from 0° to 90° at right angles to the galactic equator. *Heliocentric latitude* is the celestial latitude of a body as seen by an observer at the center of the Sun, rather than the center of the Earth. *Heliographic latitude* is the latitude of a point on the Sun's surface, north or south of the Sun's equator.

leap second The adjustment of one second in the radio time signal at midnight on June 30 or December 31. By international agreement the rate of radio time signals has been maintained since January 1972 in agreement with the International Atomic Time scale. However, the Earth's rotational period is slowing by about 0.003 of a second per day, so the radio time signals gradually diverge from Coordinated **Universal Time** (UTC), which is based on the Earth's daily rotation. To correct this divergence, the radio time seconds are retarded by the addition of one leap second as necessary.

Leavitt, Henrietta Swan (1868–1921) American astronomer. She specialized in variable stars, noticing that the brighter of the **Cepheid variables** tended to have the longer periods. From studies of Cepheids in the Small Magellanic Cloud she formulated the **period–luminosity law**, which was important in establishing the distance scale for other galaxies.

Leda The smallest of the four small **satellites** making up Jupiter's intermediate group; it was discovered by Charles Kowal in 1974.

Lemaître, Georges Edouard (1894–1966) Belgian priest and astrophysicist. In 1927 he argued that an expanding Universe would have originated in the radioactive explosion of a "primeval atom" or "cosmic egg" into which all mass and energy was concentrated. This was the first statement of what would become known as the **Big Bang** theory.

lens An optical component of an instrument, made of a transparent material (such as glass) which, by the material's property of refraction, modifies an incoming beam of light to help form an image of the light source. Single lenses have either two curved surfaces, or one curved and one plane surface. Lenses are used in various combinations in astronomical instruments. SEE **aberration, eyepiece, objective, telescope**.

lenticular galaxy A type of **galaxy** intermediate in form between spiral and elliptical types. Lenticular galaxies are designated S0, which indicates that they have the flattened form of spirals but not spiral arms. They appear lens-shaped, hence their name.

Leo A large and easily identified constellation of the zodiac, in mythology representing the lion killed by Hercules. Its brightest star is first-magnitude **Regulus**, part of the curved figure of stars, known as the Sickle, that forms the lion's head and chest. Gamma (Algieba) is a fine, wide binary, magnitudes 2.2 and 3.5, one K1-type giant and one G7-type giant, period 620 years. The Mira variable R Leonis can reach magnitude 4.4 (minimum 11.3, period 310 days). There are five galaxies in Leo on Messier's list: M65, M66, M95, M96, and M105.

Leo Minor A small constellation between Leo and Ursa Major. It has no star above magnitude 3.8.

Leonids A **meteor shower** that occurs in November. Its radiant is in the constellation Leo, in the "Sickle" asterism. The Leonids are associated with Comet Tem-

► **Lenticular galaxy** M85 (also designated NGC 4382) is a lenticular galaxy of type S0. It is the most northerly member of the Virgo Cluster, and is situated in the constellation Coma Berenices.

pel–Tuttle (SEE **meteor stream**). The shower is periodic, with modest displays in most years, but producing meteor storms at intervals of 33 years – many tens of thousands of meteors per hour were recorded in 1966 on the night of November 16–17.

Lepus A small constellation representing a hare hiding under the feet of Orion. Its brightest star is Alpha (Arneb), magnitude 2.6. It contains the very red Mira variable R Leporis (range 5.5 to 11.7, period 427 days) and the globular cluster M79.

Le Verrier, Urbain Jean Joseph

(1811–77) French astronomer. In 1846, independently of John Couch **Adams**, he predicted that an unknown planet (Neptune) was responsible for discrepancies between the calculated and observed orbital motions of Uranus. In 1859 he suggested that a planet which he prematurely named **Vulcan** lay within the orbit of Mercury, but despite some claimed sightings its existence was never confirmed.

Libra An obscure constellation of the zodiac, representing the scales of justice, but formerly seen as the claws of the scorpion, Scorpius, which it adjoins. The brightest star, Beta, is magnitude 2.6. Delta is an eclipsing binary of the Algol type, range 4.9 to 5.9, period 2.33 days.

libration A small oscillation of a celestial body about its mean position. The term is used most frequently in connection with the Moon. As a result of libration it is possible to see, at different times, 59% of the Moon's surface. However, because the areas that pass into and out of view are close to the limb and therefore extremely foreshortened, in practice libration has less of an effect on the features that can be clearly made out on the Moon's disk than this figure might suggest.

Physical libration results from slight irregularities in the Moon's motion produced by irregularities in its shape. Much more obvious is *geometrical libration*, which results from the Earth-based observer see-

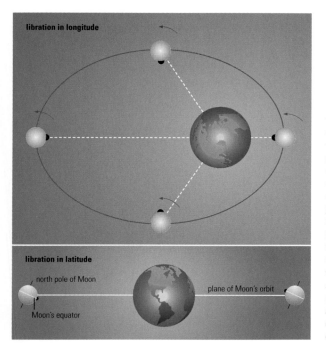

◀ **libration** Although the Moon is in synchronous rotation with respect to the Earth, ground-based observers can still, over the course of several lunations, see significantly more than just half of its surface. Libration in longitude (top) allows the observer to see slightly around the mean east or west limb, while the tilt of the lunar axis relative to the Moon's orbital plane allows regions beyond either pole to be seen on occasion.

ing the Moon from different directions at different times. There are three types of geometrical libration. *Libration in longitude* arises from a combination of the Moon's synchronous rotation and elliptical orbit (SEE the diagram). As a result, at times a little more of the lunar surface is visible at the eastern or western limb than when the Moon is at its mean position. *Libration in latitude* arises because the Moon's equator is tilted slightly from its orbital plane, so that the two poles tilt alternately toward and away from the Earth. A smaller effect is *diurnal libration*, by which the Earth's rotation allows us to see more of the Moon's surface at its western or eastern limb when it is rising or setting.

Lick Observatory The observatory of the University of California, sited on Mount Hamilton near San Jose, California. Its oldest telescope is a 0.91-m (36-inch) refractor, in operation since the observatory opened in 1888. The largest telescope is the 3-m (120-inch) C. Donald Shane Telescope, completed in 1959.

light, speed of (symbol *c*) The constant speed at which light and other **electromagnetic radiation** travels in a vacuum. Its value is 299,792,458.0 m/s. The first reasonably good estimate of *c* was made by Ole **Römer** in 1675. Albert **Einstein** showed with his Special Theory of Relativity that the speed of light in a vacuum is always the same, whatever the relative speed of the observer and the source of the light.

light curve A graph of the change in brightness with time of a variable star or other body showing brightness variations.

light pollution The detrimental effect of artificial lighting on astronomical observation and imaging. The problem has become much worse in recent years with the proliferation of such lighting and the emergence of new types of streetlamp. For amateurs, light pollution reduces the **limiting magnitude** (1) and makes deep-sky objects and comets, for example, difficult

or impossible to see, but the development of special **filters** and of **CCDs** and related technology is providing solutions. For professional astronomers it can impede or distort scientific observation, particularly where the area surrounding a once-isolated observatory has become built up.

light-time The time required for light to travel from any celestial object to the Earth. For example, light takes 8.3 minutes to reach the Earth from the Sun.

light year (symbol l.y.) A unit of distance measurement used in astronomy, equal to the distance traveled by light in a vacuum in one tropical year. 1 l.y. = 9.4607×10^{12} km = 0.3066 parsecs = 63,240 AU.

limb The edge of a celestial body which is visible as a disk, such as the Sun, the Moon, or a planet. SEE ALSO **following**, **preceding**.

limb darkening The decrease in the brightness at visible wavelengths of a celestial body toward its limb, giving it a vignetted appearance. The term is used most often in connection with the Sun, where, toward the limb, the observer is looking at the deeper, brighter levels of the Sun through a greater thickness of its atmosphere. The giant planets also show limb darkening.

limiting magnitude (1) The faintest magnitude of objects visible or recordable by photography or electronic means on a given occasion. The limiting magnitude depends on sky conditions, instrument aperture, and sensitivity of the eye or recording apparatus.

limiting magnitude (2) The faintest magnitude of objects recorded in a star atlas.

line of apsides SEE **apsides**.

line spectrum SEE **spectrum**.

LMC ABBREVIATION FOR **Large Magellanic Cloud**.

▲ **Local Group** Members of the Local Group are gravitationally bound to one another. As shown in this plot, which is centered on our Galaxy, the closest neighbors are the Magellanic Clouds and Sculptor Dwarf.

local arm SEE **Orion Arm**.

Local Group The small group of about 40 galaxies that includes our Galaxy, the two Magellanic Clouds, the Andromeda Galaxy, and the Triangulum spiral; most are dwarf galaxies. They are distributed over a roughly ellipsoidal space, about 5 million light years across. All the members are gravitationally bound so that, unlike more distant galaxies, they are not receding from us or from one another. The ten brightest (in absolute magnitude) are given in the table.

local time Time as measured at a given place on Earth, based on the mean solar day. Local time changes with longitude. A difference of 15° in longitude produces a difference of one hour in local time. To avoid the inconvenience of a multiplicity of local times, the Earth has been divided into 24 **time zones**, each of which has a standard time one hour different from that of the adjacent zone. In addition, *daylight saving time* (summer time) one hour in advance of standard time is kept in some countries for part of the year.

longitude A coordinate on the celestial sphere, or on the surface of a celestial body. *Celestial* (or *ecliptic*) *longitude* is measured eastward from the vernal equinox (the First Point of Aries) to the great circle passing through the pole of the ecliptic and the object to be measured. It is reckoned in degrees from 0 to 360°. *Galactic longitude* is the angle measured along the galactic equator from the direction of the galactic center to the point at which the galactic equator intersects the galactic circle of longitude of the object to be measured. *Heliocentric longitude* is the celestial longitude of a body as seen from the center of the Sun. *Heliographic longitude* is the longitude of a point on the surface of the Sun. It is measured from the solar meridian that passed through the ascending node of the Sun's equator on the ecliptic on January 1, 1854, at Greenwich mean noon, from 0 to 360° in the direction of the Sun's rotation.

long-period comet SEE **comet**.

long-period variables Pulsating red giants or supergiants with periods ranging from about 100 to 1000 days, usually of spectral type M or C. They are also known as **Mira** stars after the first of the type to be discovered. Each star has an average period which can vary by about 15%, so they are to some extent irregular. Their amplitudes range from 2.5 to 11 magnitudes, but the amplitude can also change appreciably from one cycle to the next. Long-period variables fall into three main divisions. Those that have a rise steeper than the fall tend to have wide minima and short, sharp maxima; as the asymmetry becomes greater, the period lengthens. Stars with symmetrical curves have the shortest periods. A third group show humps on their curves, or have double maxima and have long and short periods.

Lovell, (Alfred Charles) Bernard (1913–) English radio astronomer. From 1951 to 1981 he was director of the **Jodrell Bank** experimental station for radio astronomy, and oversaw the construction there of the world's first large steerable radio telescope. He used the telescope to pioneer the exploration of radio emission from space.

Lowell, Percival (1855–1916) American astronomer. In 1894 he built an observatory (now the **Lowell Observatory**) at Flagstaff, Arizona. From there he observed Mars, producing intricate charts of the so-called **canals**, which he ascribed to the activities of intelligent beings. He calculated that the orbital irregularities of Uranus and Neptune were caused by an undiscovered **Planet X**. His predictions led to the discovery of the planet Pluto in 1930 by Clyde **Tombaugh**. It was at Lowell's urging that Vesto Slipher carried out the observations that led to the discovery of galactic **redshifts** in 1916.

Lowell Observatory An observatory at Flagstaff, Arizona, founded by Percival Lowell in 1894. Its telescopes include a 0.61-m (24-inch) refractor installed by Lowell in 1896. At a nearby dark-sky site on Anderson Mesa is the 1.8-m (72-inch) Perkins Reflector of the Ohio State and Ohio Wesleyan universities. A Planetary Research Center funded by NASA is located at Lowell Observatory.

lower culmination SEE **culmination**.

low-surface-brightness galaxy (LSB galaxy) A galaxy that is very faint at optical wavelengths because it contains few stars or its stars are spread out. LSB

galaxies of many types are known, from dwarf irregulars to giant spirals. It is possible that, though difficult to detect, they are very common, and thus might account for some of the **missing mass** of the Universe.

L star A star of spectral type L, characterized by absorption lines of metal hydrides, such as FeH and CrH, and alkali metals such as sodium and potassium. With surface temperatures typically of 1500–2000 K, L stars are cooler than M stars, and the coolest are not true stars at all. The L type was introduced in 1998 for **brown dwarfs** and faint stars detectable at infrared wavelengths.

luminosity (symbol L) The absolute brightness of a star, given by the amount of energy radiated from its entire surface per second. Luminosity is expressed in watts (joules per second), or in terms of the Sun's luminosity. *Bolometric luminosity* is a measure of the star's total energy output, at all wavelengths. **Absolute magnitude** is an indication of luminosity at visual wavelengths.

luminosity class SEE **Morgan–Keenan classification**.

luminosity function A measure of the distribution of stars in space according to their **absolute magnitude**. It is usually expressed as the number of stars per cubic parsec for a given range of absolute mag-

THE LOCAL GROUP: THE MAIN MEMBERS			
Galaxy	Apparent mag.	Absolute mag.	Distance (thousand l.y.)
Andromeda Galaxy (M31)	3.4	−21.1	2365
Milky Way	—	−20.6	—
Triangulum Galaxy (M33)	5.7	−18.9	2590
Large Magellanic Cloud	0.1	−18.1	160
IC 10	10.3	−17.6	2675
M32	8.2	−16.4	2365
Barnard's Galaxy (NGC 6822)	9	−16.4	1760
M110 (NGC 205)	8.0	−16.3	2365
Small Magellanic Cloud	2.3	−16.2	190
NGC 185	9.2	−15.3	2020

nitude. The value of the luminosity function in the solar neighborhood rises to a maximum at absolute magnitudes of 14–15, then falls away for fainter magnitudes, implying that faint M-type dwarfs of about one ten-thousandth of the Sun's luminosity are the most abundant type of star locally. Luminosity functions can be plotted for other categories, such as particular types of star, star clusters, galaxies, or clusters of galaxies.

Luna A series of lunar probes launched by the former Soviet Union. The first three in the series were originally named *Lunik*. Luna 2 became the first man-made object to hit the Moon, on September 13, 1959. Luna 3, launched on October 4, 1959, obtained the first photographs of the farside of the Moon. Luna 9 achieved the first soft landing on the Moon on February 3, 1966. Luna 10, in March 1966, was the first craft to enter lunar orbit. In September 1970 Luna 16 made the first automatic return of a soil sample to Earth. Luna 17 carried the first unmanned lunar roving vehicle, **Lunokhod** 1.

Luna A A Japanese Moon probe, scheduled for launch in 2004. From lunar orbit it will release penetrators designed to impact the surface and send instruments 5 m (16 ft) down into the regolith.

lunar eclipse An **eclipse** of the Moon by the Earth. Since the Moon's orbital plane is inclined to the plane of the ecliptic, a lunar eclipse can happen only when the Moon is at opposition (i.e. at full Moon) and at the same time is at or near one of its nodes (SEE **ecliptic limits**). The Sun, Earth and Moon are then very nearly in a straight line. Lunar eclipses are visible from anywhere on Earth where the Moon is above the local horizon. The Moon does not become completely invisible during an eclipse because it is partially lit by sunlight refracted by the Earth's atmosphere, and it appears reddish because the blue component is removed from the sunlight by **scattering**.

A *total lunar eclipse* occurs when the Moon is entirely within the umbra, the dark central part of the Earth's shadow (SEE the diagram). The eclipse is *partial* when the Moon is partly within the umbra, and *penumbral* when the Moon passes through the penumbra but completely misses the umbra. The overall duration of a total lunar eclipse, from **first contact** to **fourth contact**, can be $3\frac{1}{2}$ hours; totality, the interval from **second contact** to **third contact**, can last for $1\frac{3}{4}$ hours. There are two or three partial or total lunar eclipses each year.

Lunar Orbiter A series of US probes put into orbit around the Moon to photo-

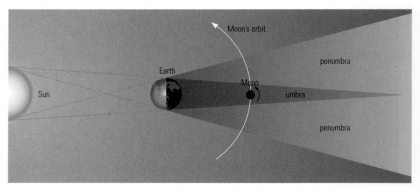

▲ **lunar eclipse** Earth casts a shadow to the right in this diagram. The shadow has two components – a dark central umbra, and a lighter area of partial shadow, the penumbra. If the full Moon – opposite the Sun in Earth's sky – passes through the central cone of the umbra, a total lunar eclipse will result.

► **Lunar Rover** Eugene A. Cernan carries out checks on the Lunar Rover on the Apollo 17 mission. He is at the Taurus–Littrow landing site.

graph its surface in detail far superior to anything obtainable from Earth. The photographs taken by the probes were processed on board and telemetered back to Earth. There were five probes in the series, all successful, from Lunar Orbiter 1 in August 1966 to Lunar Orbiter 5 in August 1967.

Lunar Rover (Lunar Roving Vehicle) One of the battery-powered cars used on the Moon by the crews of Apollos 15, 16, and 17 (SEE **Apollo program**). They enabled the astronauts to travel farther afield in their searches for rock samples.

Lunar Prospector A NASA probe to the Moon, launched in 1998. From polar orbit, its neutron spectrometer found strong evidence for the presence of water ice in the lunar regolith in both polar regions. The probe also obtained improved maps of surface composition and gravity variations, and found that the Moon has a small iron core.

lunar transient phenomenon (LTP) A short-lived, localized change in the appearance of a lunar surface feature, such as a glow or an obscuration. LTPs were reported by lunar observers for many years but treated with scepticism, but they

are now generally regarded as real events. In 1958 a carbon spectrum of activity in the crater Alphonsus was obtained, and instruments set up by the Apollo astronauts established a connection between LTPs and moonquakes. Since they are most common when the Moon is at perigee, LTPs are most likely to be releases of gas and dust induced by tidal forces (SEE **tides**).

lunation The time interval between two identical phases of the Moon, for example from one new Moon to the next new Moon. This corresponds to the Moon's **synodic period** or **synodic month** of 29.53059 days.

Lunik SEE **Luna**.

lunisolar precession The main component of **precession**, resulting from the gravitational effects of the Moon and Sun on the Earth. It amounts to about 50.4 arc seconds per year.

Lunokhod One of two lunar roving vehicles landed on the Moon by the former Soviet Union. All operations, such as traveling across the lunar surface, surveying the area, testing the mechanical properties of the surface and making chemical analyses of it, were controlled from Earth. Lunokhod 1 was placed on the Moon by Luna 17 in November 1970; Lunokhod 2 followed in January 1973.

Lupus A southern-hemisphere constellation, adjoining Centaurus, representing a wolf. Its brightest star is Alpha, magnitude 2.3, but it has no objects of particular interest.

Lyman series The series of lines in the ultraviolet region of the **hydrogen spectrum** analogous to the **Balmer series** in the visible part. The first Lyman line, Lα, is at 121.6 nm, and was first detected in the solar spectrum by photography from a rocket. The second, Lβ, is at 102.6 nm; Lγ is at 97.2 nm. In the spectra of **quasars**, Lyman lines are redshifted into the visible region. (After American physicist Theodore Lyman, 1874–1954.)

Lynx A dim northern constellation between Auriga and Ursa Major, introduced by **Hevelius**. Its leading star is Alpha, magnitude 3.1, but the only other star above magnitude 4 is 38 Lyncis, at magnitude 3.9.

Lyot, Bernard Ferdinand (1897–1952) French astrophysicist. He studied the Sun, and invented the **coronagraph** for observing the corona, and the **Lyot filter**. He also pioneered astronomical polarimetry, the measurement of the degree of polarization of light from celestial bodies, and invented a number of devices for the purpose.

Lyot filter A **filter** for observing the Sun at a particular wavelength, developed by Bernard **Lyot** in the 1930s. Sunlight is directed through a succession of alternating layers of quartz plates and polaroid sheets, and each separate passage reinforces some wavelengths and cancels others. The result is a number of very narrow (0.1 nm) and widely spaced wavebands which are easy to isolate.

Lyra A small but prominent constellation. In mythology, it represents the harp given to Orpheus by Apollo. Its brightest star is **Vega**. Lyra contains the famous "Double Double," the quadruple star Epsilon; the two main components, magnitudes 4.6 and 4.7, can be separated with the naked eye and each is again double. Beta (Sheliak) is an eclipsing variable, range 3.3 to 4.4, period 12.9 days; it is the prototype of the **Beta Lyrae stars**. Between Beta and Gamma (Sulafat), magnitude 3.2, lies the **Ring Nebula**, M57, a planetary nebula.

Lyrids A **meteor shower** that occurs in April, and has its radiant on the border of the constellations Lyra and Hercules, southwest of the star Vega. The Lyrids are associated with Comet Thatcher (1861 G1).

Lysithea One of the four small **satellites** making up Jupiter's intermediate group, discovered by Seth Nicholson in 1938.

M

McDonald Observatory The astronomical observatory of the University of Texas, on Mount Locke, Texas. It is home to the **Hobby–Eberly Telescope**, and also houses a 2.7-m (107-inch) reflector, opened in 1968.

MACHO Abbreviation for "massive compact halo object," an as-yet undetermined type of invisible body assumed to pervade in the Galaxy's halo (outer regions) to account for the **dark matter** it is supposed to contain. MACHOs could be brown dwarfs, extinct white dwarfs, or Jupiter-sized planets.

Maffei Galaxies Two nearby galaxies so close to the plane of the Milky Way that at optical wavelengths they are almost completely obscured by galactic dust. Maffei I is an S0 (lenticular) galaxy over 3 million light years away, on the edge of the Local Group. Maffei II is an S0 spiral but, at a distance of 15 million light years, it is unrelated to Maffei I and the Local Group. They were discovered as infrared sources in 1968 by the Italian astronomer Paolo Maffei (1926–).

Magellan A NASA space probe to Venus, launched in May 1989. It went into orbit around Venus in August 1990 and made a detailed radar map of 98% of the planet, revealing impact craters, volcanic mountains, lava flows and other features. In 1993 Magellan's highly elliptical orbit was circularized by repeated passes through the top of the planet's atmosphere, a technique known as *aerobraking*. Tracking of the probe in its new orbit allowed astronomers to map the gravitational field of Venus. Following a further lowering of the orbit in 1994, Magellan burnt up in the atmosphere.

Magellanic Clouds The two nearest galaxies to our own, visible to the naked eye like isolated patches of the Milky Way

in the southern sky. The *Large Magellanic Cloud* (LMC) lies in the constellations Dorado and Mensa and is about 6° across. The *Small Magellanic Cloud* (SMC) lies in the constellation Tucana and is about 3° across. They are named for the Portuguese explorer Ferdinand Magellan (1480–1521), who described them. Their distances are about 160,000 and 290,000 light years. Although usually classified as irregular galaxies, both Clouds are somewhat bar-shaped with faint outer structure that can be interpreted as rudimentary spiral arms. The mass of the LMC is about 10^{10} solar masses, and that of the SMC about 2×10^9 solar masses.

At the northeast end of the LMC lies the **Tarantula Nebula**, one of the largest and brightest groupings of hot stars and bright nebulosity known in any galaxy. Near here occurred **Supernova 1987A**, the first naked-eye supernova since 1604. The irregular distribution of star clusters and diffuse nebulae in the LMC contrasts with the more uniform stellar distribution in the SMC. Young stars in the LMC lie in a thin disk, which is seen nearly face-on (about

▲ **Magellan** During the first phases of NASA's Magellan mission, its orbit around Venus was highly elliptical (blue), and it used radar to map almost the whole planet. After three years, the orbiter used aerobraking techniques to reach a lower, circular orbit (red); it then mapped Venus's gravity field.

27° to the plane of the sky). There is no evidence for a spherical halo of the kind that surrounds our Galaxy.

The SMC is greatly extended in the line of sight, with a depth of about 60,000 light years, over five times the dimension it presents to us. It appears to have been distorted by tidal forces, possibly as a result of a close passage of the SMC and LMC some 200 million years ago.

The *Magellanic Stream* is a vast cloud of neutral hydrogen (SEE **H I region**) stretching between the Clouds and extending toward the Galaxy. It was probably torn out from the Clouds by gravitational interaction with the Galaxy.

Magellan Telescopes A pair of reflecting telescopes, both of aperture 6.5 m (255 inches) at **Las Campanas Observatory** in Chile. The telescopes were completed in 2000 and 2002, and can be used independently or together.

magnetohydrodynamics (MHD) The study of the behavior of plasma (a mixture of ions and electrons) in a magnetic field. A moving plasma is in effect an electric current, which interacts with the magnetic field to produce a force which modifies the motion of the plasma. The subject was pioneered by the Swedish physicist Hannes Alfvén, who predicted that waves (*Alfvén waves*) would propagate through a plasma, and he applied MHD to a variety of problems in astrophysics. Plasmas occur in many astronomical contexts, so MHD has proved fruitful in the study of solar phenomena such as sunspots, the interaction of the solar wind with planetary **magnetospheres**, star formation, and supernovae.

magnetosphere The region of space surrounding a planet in which the planet's magnetic field predominates over the **solar wind**, and controls the behavior of plasma (charged particles) trapped within it. The boundary of the magnetosphere is called the *magnetopause*, outside which is a turbulent magnetic region called the *magnetosheath*. Where the steady outward flow of the solar wind is first interrupted there is a **bow shock**. Downwind from the planet, the solar wind draws the magnetosphere out into a long, tapering *magnetotail*.

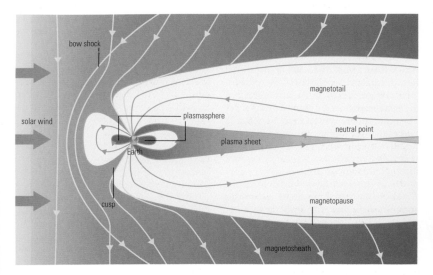

▲ **magnetosphere** The Earth's moving iron core gives it an active magnetic field that extends well beyond the planet. The arrows indicate the direction of the magnetic field.

Mercury, the Earth, and the giant planets have magnetospheres. The Earth's contains the **Van Allen Belts** of charged particles. Jupiter's magnetosphere is many times larger than the Sun, and contains a vast disk of plasma, and **Io**'s plasma torus.

magnification The factor by which an image produced by an optical system increases the angular size of an object. It is normally denoted by a multiplication sign, as in ×50 (SEE **binoculars**). The magnification afforded by a telescope is found by dividing its **focal length** by the focal length of the eyepiece being used with it. The higher the magnification, the dimmer the image and the more prone it is to the effects of bad seeing and defects of the instrument.

magnitude A measure of the brightness of celestial objects. In the 2nd century BC the Greek astronomer **Hipparchus of Nicea** classified the stars into six magnitudes, first magnitude being the brightest and sixth magnitude the faintest visible to the naked eye. In the 19th century it was found that a star of first magnitude was about 100 times as bright as one of sixth magnitude. It was decided to adopt a scale of magnitudes in which this ratio is exactly 100 : 1. Hence, 100 stars of magnitude 6 taken together are as bright as a single star of magnitude 1. The ratio of brightness between one magnitude and the next is thus the fifth root of 100, which is 2.512. Thus, differences of 2.5, 5, and 10 magnitudes correspond to brightness ratios of 10, 100, and 10,000 respectively. Very bright bodies, Sirius, Venus, and the Sun, have negative magnitudes. The faintest objects detectable from ground-based telescopes are about magnitude 30.

Apparent magnitude is the brightness of an object as seen from Earth. **Absolute magnitude** is the true brightness of an object, taking into account its distance from Earth. The difference between the apparent and absolute magnitudes is the *distance modulus*. SEE ALSO **bolometric magnitude**, **photoelectric magnitude**, **photographic magnitude**, **photovisual magnitude**, **visual magnitude**.

main sequence The region on the **Hertzsprung–Russell diagram** where most stars lie, including the Sun. It runs diagonally from hot, bright stars at the upper left down to cool, faint stars at the bottom right. The position of a star on the main sequence depends on its mass, the most massive stars being the brightest. The upper limit is about 60 solar masses at stars of spectral types O and B. The lower limit is at 0.08 of a solar mass with stars of spectral type M. Stars spend most of their lives on the main sequence, producing energy from the fusion of hydrogen to helium in their cores. The *zero-age main sequence* is where stars lie when they first start to burn hydrogen. As stars use up their internal hydrogen they move away from the main sequence. The more massive a star, the sooner it evolves off the main sequence. All stars on the main sequence are termed **dwarfs**, irrespective of whether they are larger or smaller than the Sun.

major planet Any of the nine largest planets of the Solar System – Mercury, Venus, Earth, Mars, Jupiter, Saturn, Uranus, Neptune, and Pluto – as opposed to the asteroids, which are also known as *minor planets*. Some astronomers feel that Pluto, because of its small size, does not merit being classified as a major planet.

Maksutov telescope A modification of the **Schmidt camera**, devised independently by Albert Bouwers (1940) and Dmitrii Maksutov (1944). The Schmidt's aspherical correcting plate is replaced by a concave lens which cancels out the chromatic and spherical aberrations – SEE **aberration** (2) – of the primary concave spherical mirror. A secondary mirror directs the light path through a hole in the primary to a Cassegrain focus, although various other configurations have been used. The Maksutov is a compact instrument that was popular with amateur observers, but it has now largely lost out to the **Schmidt–Cassegrain telescope**. (After Dmitrii Dmitrievich Maksutov, 1896–1964.)

mantle SEE **differentiation**.

many-body problem (*n*-body problem) The problem in celestial mechanics of finding how a number (*n*) of bodies move under the influence of their mutual gravitational attraction only. It is an extension of the **three-body problem**. The many-body problem has no general solutions. It requires a vast amount of computation to obtain good approximate solutions, but with modern computers this is now easily accomplished. Analyzing how the planets of the Solar System perturb one another, and how space probes such as the Voyager craft move between them, are examples of the many-body problem.

mare (plural **maria**) An extensive dark plain on the Moon, composed of basalt (solidified lava). Most of the maria were formed 3 to 4 billion years ago, after large meteorites struck the Moon and weakened the crust, and lava later erupted from below. They range from the vast Oceanus Procellarum, covering much of the north-western quadrant of the Moon, down to areas less than 100 km (60 miles) across. Nearly all maria are on the Moon's nearside, where the crust is thinner than on the farside. Maria is Latin for "seas," which is what early observers thought they were. Similar plains on Mars were also known as maria, but since detailed mapping by spacecraft these areas are now known as *planitiae*.

Mariner A series of US space probes to the planets. Mariner 2 flew past Venus in 1962, discovering that the planet was hot. Mariner 4 flew past Mars in July 1965, photographing craters on its surface. Mariner 5 passed Venus in October 1967, making measurements of the planet's atmosphere. Mariners 6 and 7 obtained further photographs of Mars in 1969. Mariner 9 became the first probe to enter planetary orbit when it went into orbit around Mars in November 1971. It made a year-long photographic reconnaissance of the planet's surface, and obtained close views of the two satellites, Phobos and Deimos. Mariner 10, the last of the series, was the first two-planet mission, passing Venus in February 1974 and then encountering Mercury three times, in March and September 1974 and March 1975.

Mariner Valley A series of enormous, interconnected canyons on Mars, stretching for 4500 km (2800 miles) just south of the equator. Individual canyons extend for up to 200 km (125 miles) and are 7 km ($4\frac{1}{2}$ miles) deep. Mariner Valley is a fault structure whose formation is probably linked to that of the volcanic Tharsis region at its western end. The name commemorates the **Mariner** probes to Mars; the official version is Valles Marineris.

Markarian galaxy Any of the galaxies cataloged by the Armenian astronomer Benjamin Eghishe Markarian (1913–85), distinguished by their strong ultraviolet emission. They include **Seyfert galaxies** and **quasars**.

Mars The fourth major planet from the Sun, and the second-smallest of the inner planets. Mars appears red to the naked eye, and is also known as the Red Planet. When viewed through a telescope it appears as a

▲ **mare** Mare Humorum is a lava-filled basin on the Moon, some 370 km (230 miles) in diameter. It has fewer impact craters than the areas around it, indicating its relative youth.

▲ **Mariner Valley** This composite Viking image shows the huge, sprawling canyon system of the Mariner Valley (Valles Marineris) on Mars. Ancient dry river channels run northward from the chaotic terrain at the east (right).

small orange-red disk with lighter and darker markings, and white patches often visible at one pole or the other. Because of its markedly eccentric orbit, its distance from Earth at opposition varies from 56 to 101 million km (35 to 63 million miles), with corresponding variations in apparent size and brightness of 14 to 25 arc seconds and magnitude −1.0 to −2.8. Favorable oppositions occur at intervals of 15 to 17 years.

Mars shows some variation in **phase**, appearing slightly gibbous at quadrature. Its axial tilt is similar to the Earth's, so it passes through a similar cycle of **seasons**, and seasonal variations in its appearance may be observed through a telescope. The polar caps vary in size; and the darker markings may vary in shape and extent, which was once thought to indicate seasonal changes in vegetation but is now known to be caused by seasonal dust storms covering and uncovering darker surface rocks. The planet was once imagined to be an abode of intelligent life, and some observers reported seeing linear surface markings which they believed were artificial **canals**. In 1996 it was announced that an **SNC meteorite** designated ALH 84001, recovered from Antarctica in 1984, appeared to contain the fossilized remains of primitive life forms. The main data for Mars are given in the table.

The surface of Mars reveals a long and complex history of geological activity. Although similar processes have operated on the Earth and Mars, the results on Mars have been dramatically different, mainly because of its smaller size and gravity, but also because of the absence of plate **tectonics**. The major difference in terrain is between the largely smooth, lowland volcanic plains of the northern hemisphere and the heavily cratered uplands of the south. There is also a prominent bulge, 5000 km (3000 miles) across, in the volcanic Tharsis region. The biggest volcanic structure on Mars – indeed in the whole Solar System –

MARS: DATA

Globe	
Diameter (equatorial)	6792 km
Diameter (polar)	6752 km
Density	3.91 g/cm^3
Mass (Earth = 1)	0.107
Volume (Earth = 1)	0.152
Sidereal period of axial rotation	24h 37m 23s
Escape velocity	5.0 km/s
Albedo	0.15
Inclination of equator to orbit	25° 11′
Surface temperature (average)	220 K
Surface gravity (Earth = 1)	0.38

Orbit	
Semimajor axis	1.524 AU = 227.9 × 10^6 km
Eccentricity	0.0934
Inclination to ecliptic	1° 51′
Sidereal period of revolution	686.980d
Mean orbital velocity	24.13 km/s
Number of satellites	2

▲ Mars The Viking spacecraft gave unprecedentedly sharp images of Mars. The prominent dark area left of center is Syrtis Major, and the bright area below it is Hellas Planitia.

is **Olympus Mons**. Nearby, on the Tharsis Ridge, are three other extinct volcanoes: Arsia Mons, Ascraeus Mons, and Pavonis Mons. On the Earth, plate tectonics can eventually move a volcano away from the underlying source of lava, forming a volcanic chain, but on Mars such volcanoes stayed in one place and grew to a great size. Mars has possibly been volcanically active in the very recent geological past.

Stretching east from Tharsis for 4500 km (2800 miles) is the network of giant canyons known collectively as the **Mariner Valley**. While some Martian canyons and similar features are geological faults, others, in the Mariner Valley and elsewhere, are clearly channels in which water once flowed across the Martian surface in the form of rivers. Some features have even been interpreted as dried-up seabeds. At some time in the past Mars must have had a very different climate, with a denser atmosphere. Perhaps all the water is now frozen into the subsurface permafrost layer; more likely is that the planet's low gravity has simply not been enough to hang on to all of its air and water, which have slowly been escaping.

Two large impact basins, Hellas (1800 km/1500 miles) and Argyre (900 km/550 miles) dominate the southern hemisphere. Here there are two terrain types: an ancient surface almost saturated with craters and crossed by many channels, and smoother, less heavily cratered *intercrater plains*. Many Martian craters show signs of erosion by the dust storms that blow across the surface. In places the dust forms dune fields that shift with the winds. Craters at higher latitudes appear smooth, probably because ground ice allows the crater walls to "creep."

The variable polar ice caps appear to be composed of solid carbon dioxide with underlying caps of water ice. The northern water cap is exposed in the summer when the carbon dioxide warms and becomes part of the atmosphere, but the southern cap always retains a carbon dioxide covering.

Before the space age the nomenclature for surface features was that introduced in the 19th century by Giovanni **Schiaparelli**, who produced a map bearing names from classical literature and of geographical locations on Earth. However, these *albedo features* visible from Earth represent differences in surface brightness, and few of them correspond to topographical features visible in images returned by probes. The modern nomenclature borrows the old names where appropriate; otherwise, large craters are named for astronomers and other people associated with Mars, small craters after towns and villages on Earth, and the various types of channel after the name for Mars in various languages or after rivers on Earth.

The atmosphere consists mainly of 95% carbon dioxide, $2\frac{1}{2}$% nitrogen, and $1\frac{1}{2}$% argon, with smaller quantities of oxygen, carbon monoxide, and water vapor. Although the atmospheric pressure at the surface is less than 1% of that on the Earth, a variety of meteorological phenomena do occur. Winds of up to 300 km/h (200 mile/h) drive seasonal dust storms that may last for weeks and lift dust particles to heights of 40 km (25 miles), sometimes obscuring the whole planet completely. Water vapor forms into a variety of cloud types, and produces morning mists and ground frosts. The surface temperature on Mars varies between 130 and 290 K.

The crustal rocks of Mars are iron-rich, giving the planet its red color; in composition they may resemble minerals like hematite and olivine. Much of the surface may be covered with a hard, ferrous clay-like deposit. Basalts cover the volcanic plains. From the planet's density, its interior structure probably consists of a mantle overlying a core. However, the extremely weak magnetic field suggests that if there is an iron or nickel–iron core it is not hot and fluid. As for its **magnetosphere**, Mars does possess an ionosphere, with which the solar wind interacts to produce a weak bow shock.

Mars has two tiny **satellites** in very close orbits: Phobos and Deimos.

Mars Exploration Rover

A NASA mission, scheduled for launch in 2003, to land two roving vehicles on the planet in 2004. The landings will be made in two locations, pinpointed from **Mars Global Surveyor** images, where there is clear evidence of past surface water.

Mars Express

A European Space Agency probe, with international participation, scheduled for launch in 2003. A polar orbiter will examine Martian geology, while the UK's *Beagle 2* lander carrying a suite of seven instruments will study the surface and atmosphere in detail.

Mars Global Surveyor

An American probe to Mars, launched in November 1996. It arrived in September 1997, but problems with aerobraking maneuvers delayed its achieving the desired polar orbit until 1999. The probe obtained over 300,000 high-resolution images of the surface, some of which showed conclusively that water had once flowed there.

Mars Pathfinder

A NASA probe to Mars, part of the **Discovery** program, launched in December 1996. It landed in July 1997, using a combination of parachute and airbags, which were deflated after touchdown. An 11.5 kg (25 lb) minia-ture rover vehicle, Sojourner, analyzed rocks in the vicinity of the lander, which was named the Carl Sagan Memorial Station. Lander and rover operated for nearly three months – well beyond expectations. The primary objective was to demonstrate a low-cost method of delivering an instrument package and a lander to the surface. Images of the terrain at the landing site, in Ares Valles, showed that water had once flowed across the surface.

mascon A high-density region below the Moon's surface; the word is derived from "mass concentration." Mascons cause localized increases in the Moon's gravitational field, which is how they were first detected by Lunar Orbiter 5 in 1967. The dozen or so mascons that are known have diameters of 50–200 km (30–125 miles) and lie about 50 km (30 miles) below the surface. They were probably formed when impacts from large meteorites blasted away the lighter crust, and the denser, underlying mantle material bulged upward.

▲ **Mars Global Surveyor** The existence of water on Mars, whether as subsurface liquid or as ice, is vital for the question of whether there was, or is, life on Mars. Several outflow channels in crater walls, including this example in Noachis Terra, were imaged by Mars Global Surveyor, and appear to resemble gullies on Earth.

◀ **Mathilde** Asteroid no. 253 was imaged by the NEAR Shoemaker probe. Its surface was seen to have several craters about 20 km (12 miles). It has a low albedo.

maser A cloud of molecular **interstellar matter** in which **electromagnetic radiation** emitted by high-energy molecules at a particular wavelength in the microwave region stimulates other molecules to emit radiation at the same wavelength. The word is an acronym of "microwave amplification by stimulated emission of radiation." A maser therefore amplifies natural microwave emissions, just as a laser does for light.

Maskelyne, Nevil (1732–1811) English astronomer. He was appointed the fifth Astronomer Royal in 1765. Building on a method of his for determining longitude at sea by observing the Moon, in 1767 he published the first *Nautical Almanac*, an annual book for the same purpose. In 1774 he measured the mass of the Scottish mountain of Schiehallion by seeing how far it caused a plumb-line to deviate, from which he calculated the average density of the Earth to be 4.7 g/cm^3 (the true value is 5.52 g/cm^3).

massive compact halo object SEE **MACHO**.

mass–luminosity relation A relationship between luminosity and mass for stars on the **main sequence**, derived theoretically by Arthur **Eddington** in 1924. It is expressed approximately as

$$L \propto M^x$$

where L is the luminosity, M is the mass, and the exponent x varies between 2 and 5 according to the mass. The relationship is not obeyed by white dwarfs, or by giants or supergiants. The relationship enables the mass of a star to be determined if its absolute magnitude is known. The theoretical upper limit (SEE **Eddington limit**) of stellar mass is about 100 solar masses, but very few stars exceed 30 solar masses. The lower limit of a star's mass is 0.08 of a solar mass.

Mathilde Asteroid no. 253, discovered in 1885 by Johann Palisa. It was imaged in 1997 by the Near Shoemaker probe *en route* to Eros, and was found to be an irregular object measuring 57 × 53 × 50 km (35 × 33 × 31 miles), with an unusually slow rotation period of 17.4 days. Mathilde is heavily cratered but with a very low density, about 1.3 g/cm^3, which suggests that it is a "rubble pile:" many fragments from a past impact held together by gravity.

Mauna Kea Observatory A major observatory in Hawaii, at an altitude of 4200 m (13,800 ft) – the world's highest and best observatory site for optical, infrared, and millimeter-wavelength astronomy. Its main telescopes include the twin 10-m (396-inch) **Keck Telescopes**; the 3.6-m (142-inch) Canada–France–Hawaii Telescope; the 3.8-m (150-inch) UK Infrared Telescope; the 15-m (49-ft) James Clerk Maxwell Telescope, owned by the UK and used for **millimeter-wave astronomy**; the Japanese 8.3-m (326-inch) **Subaru Telescope**; and one of the twin 8.1-m (319-inch) **Gemini Telescopes**.

Maunder minimum The period 1645–1715, when few sunspots were observed, as identified in the 1890s from historical records by British solar physicist Edward Walter Maunder (1851–1928). This was not simply the result of a lack of data from those years: recent studies, especially of tree rings from the period, reveal a genuine lull in solar activity. Also, the Maunder minimum correlates well with the long succession of below-average temperatures known as the *Little Ice Age*. The Maunder minimum was caused by a number of much longer-term fluctuations acting together to damp down the 11-year **solar cycle**.

mean anomaly SEE **anomaly**.

mean solar time Time based on the motion of the **mean Sun**, which travels at a uniform rate along the celestial equator.

mean Sun A fictitious Sun moving along the celestial equator at a constant rate equal to the average rate of the true Sun along the ecliptic. It completes its annual course with respect to the vernal equinox in exactly the same time as the true Sun. The true Sun travels at a variable rate along the ecliptic, and cannot therefore give a uniform measure of time.

Mensa A far southern constellation, introduced by **Lacaille**, representing the Table Mountain at the Cape of Good Hope where he worked. Mensa has no star above the 5th magnitude, but part of the Large Magellanic Cloud extends into it.

Mercury The innermost planet of the Solar System, and the smallest of the four inner planets. Mercury is difficult to observe because it is never more than 28° from the Sun, and its small angular diameter (8 arc seconds at greatest elongation) makes identification of surface detail very

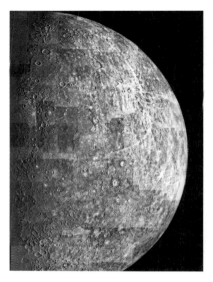

▲ **Mercury** Mariner 10 made three sets of observations of Mercury. This mosaic of images shows that the small planet is heavily cratered, indicating it has undergone no resurfacing for many millennia.

MERCURY: DATA	
Globe	
Diameter	4880 km
Density	5.43 g/cm³
Mass (Earth = 1)	0.0553
Volume (Earth = 1)	0.0562
Sidereal period of axial rotation	58d 15h 30m
Escape velocity	4.4 km/s
Albedo	0.06
Inclination of equator to orbit	0°
Surface temperature	100–800 K
Surface gravity (Earth = 1)	0.38
Orbit	
Semimajor axis	0.387 AU = 57.91 × 10⁶ km
Eccentricity	0.206
Inclination to ecliptic	7° 00′
Sidereal period of revolution	87.969d
Mean orbital velocity	47.89 km/s
Satellites	0

difficult, but through a telescope light and dark areas can be made out. As an inferior planet, Mercury passes through **phases**. It reaches magnitude −0.7 at brightest. A **transit** of Mercury across the Sun's disk occurs at intervals of either 7 or 13 years. The planet has no satellite. The main data for Mercury are given in the table.

Mercury rotates three times on its axis for every two revolutions of the Sun. As a result, the planet has a "hot pole" in each hemisphere – areas that spend longer facing the Sun and get heated more. Also, the **synodic period** of 116 days is very close to twice the rotation period. Thus, at greatest elongation west, say, Mercury presents to Earth almost the same face as at the previous greatest elongation west. This led earlier observers into thinking that Mercury had **synchronous rotation**. The true rotation period was established in 1965 by **radar astronomy**.

Very little was known about Mercury's surface until the Mariner 10 probe (the only mission to Mercury so far) made three

▲ **Mercury** Because Mercury is nearer to the Sun than the Earth, it displays phases. At (1) it is new, at (2) it is half-phase, at (3) full, and at (4) half-phase again.

close approaches to the planet in 1974 and 1975 and returned pictures of nearly half the surface. These showed a heavily cratered, lunar-like world. The largest impact feature, at 1300 km (800 miles) across, is the Caloris Basin, similar to the Moon's Imbrium Basin; the next-largest is the 625-km (390-mile) diameter crater Beethoven. Like the Moon, Mercury has **ray** craters. Unlike the Moon, it has *lobate scarps*: cliffs up to 3 km (2 miles) high and hundreds of kilometers long, and sometimes cutting across craters. They may be compression faults which formed when the planet's crust cooled and contracted. The *intercrater plains* appear to be volcanic, like the lunar maria, but unlike the maria they contain many small craters. There are also valleys and ridges.

Most craters on Mercury are named for writers, artists, and composers; plains are named for the word for Mercury in different languages; scarps take the names of famous explorers' ships; valleys are named for radar-astronomy observatories; and the names of ridges commemorate astronomers with a strong association with the planet.

The surface temperature at noon on the equator can reach over 800 K at perihelion, while on the night side it can fall to 90 K. Radar mapping of Mercury's polar

regions in 1991 and 1992 revealed what may be metallic deposits on the floors of craters permanently in shadow, where the temperature is estimated never to rise above 112 K. There is a very tenuous atmosphere, mainly of helium and sodium, deriving from the **solar wind**, but also from particles sputtered (knocked off) from the surface by solar radiation, and possibly from material slowly outgassing from the interior.

The presence of a weak magnetic field and the planet's high density for its size suggest that Mercury has a very large iron-rich core, proportionally far bigger than for any other Solar System body. This could be because a small body forming so close to the Sun would have trouble hanging on to lighter elements. Alternatively, Mercury could have suffered a collision with another body which blasted off most of its outer layers.

Mercury's orbital velocity is greater than that of any other planet. Consequently, the advance of its perihelion is that much more pronounced. (The advance of a planet's perihelion results from the slow rotation of the orbit's **semimajor axis** in the direction of the planet's revolution around the Sun.) The gravitational attraction of the other planets accounts for all but 43 arc seconds per century in Mercury's case. In the 19th century the discrepancy was attributed to an intra-Mercurial planet named **Vulcan**, but the answer was provided by **Einstein**'s General Theory of Relativity.

meridian SEE **celestial meridian**.

MERLIN (Multi-Element Radio-Linked Interferometer Network) An array of six radio telescopes linked to Jodrell Bank. Five of the telescopes are of 25 m (82 ft) in diameter at distances up to 127 km (79 miles) from Jodrell Bank; a sixth dish, of 32 m (105 ft) diameter at Cambridge, gives a maximum baseline of 217 km (135 miles).

Messier, Charles Joseph (1730–1817) French astronomer. He kept a record of various fuzzy extended objects that were in fixed positions in the northern sky so that he should not confuse them with comets, of which he was a successful hunter, discovering over twenty. But it is for the catalog of **Messier objects** – what we now know to be galaxies, clusters, and nebulae – that he is remembered today.

Messier object Any of the nebulae, star clusters, and galaxies, numbering over 100, which appear in the list begun by Charles **Messier**. Messier drew up the list so that he and other comet-hunters would not confuse these permanent, fuzzy-looking objects with comets. The first edition of the catalog was published in 1774, with supplements in 1780 and 1781. Not all the objects were discovered by Messier himself; several were found by Pierre Méchain. Objects in the catalog are given the prefix M, and are still widely known by their *Messier numbers*. Some of the best-known Messier objects are listed in the table.

metals (heavy elements) In astronomy, all elements heavier than helium. Population I stars (SEE **stellar populations**), which are young, have relatively high metal contents (up to 2%), while Population II stars, which formed first, have low metal contents (as little as 0.01%). The metal content of a star affects both its structure and its evolution.

meteor The brief streak of light in the night sky caused by a **meteoroid** entering the Earth's upper atmosphere at high speed from space. Meteors are popularly known as *shooting stars* or *falling stars*. They occur at altitudes of about 100 km (60 miles). The typical meteor lasts for a few tenths of a second to a second or two, depending on the meteoroid's impact speed, which can vary from about 11 to 70 km/s (7 to 45 mile/s).

Most meteor-producing meteoroids are low-density (0.3 g/cm^3) dust particles that originated in comets. They are rapidly eroded away by **ablation** and leave a trail of ionized and excited atmospheric atoms that is visible as the meteor. Most such meteors have a magnitude in the range 0–4.

Larger, denser meteoroids that originated in asteroids produce brighter meteors, the brightest of which may rival the full Moon or, in rare cases, even the Sun (SEE **fireball**, **bolide**). Ionization trails are detectable by radar (SEE **radar astronomy**), enabling meteors to be studied in the daytime.

A few meteors per hour may be seen on any clear, moonless night at any time of year. But at certain times of the year there are **meteor showers**, when meteors are more numerous than usual. Showers occur when the Earth passes through a **meteor stream** – dust particles spread around the

SOME OF THE BEST-KNOWN MESSIER OBJECTS		
Messier number	Popular name	Description
1	Crab Nebula	Remnant of supernova in Taurus
8	Lagoon Nebula	Emission nebula in Sagittarius
11	Wild Duck	Open star cluster in Scutum
13	—	Globular cluster in Hercules
17	Omega, Swan or Horseshoe Nebula	Emission nebula in Sagittarius
20	Trifid Nebula	Emission nebula in Sagittarius
27	Dumbbell Nebula	Planetary nebula in Vulpecula
31	Andromeda Galaxy	Spiral galaxy in Andromeda
33	Pinwheel Galaxy	Spiral galaxy in Triangulum
42	Orion Nebula	Emission nebula in Orion
44	Praesepe	Open star (or Beehive) cluster in Cancer
45	Pleiades	Open star cluster in Taurus
51	Whirlpool Galaxy	Spiral galaxy in Canes Venatici
57	Ring Nebula	Planetary nebula in Lyra
64	Black Eye Galaxy	Spiral galaxy in Coma Berenices
97	Owl Nebula	Planetary nebula in Ursa Major
104	Sombrero Galaxy	Edge-on spiral galaxy in Virgo

orbit of a comet. Most of the meteors appearing during the year are *sporadic meteors*, not associated with cometary orbits.

Meteor Crater (Arizona Meteor Crater, Barringer Crater)

A circular crater, about 1200 m (4000 ft) in diameter and 180 m (600 ft) deep, near Flagstaff in the Arizona desert. It is the best-known of the Earth's impact craters, and the first to be identified as such. The impacting body is estimated to have been a nickel–iron meteorite, about 40 m (130 ft) across and weighing a quarter of a million tons, which fell around 50,000 years ago. Quantities of meteoritic nickel–iron have been recovered from the plain around the crater, but most of the meteorite was probably vaporized on impact. The crater was discovered in 1891.

meteorite

That part of a large **meteoroid** that survives passage through the Earth's atmosphere and reaches the ground. Hundreds of tons of meteoroidal matter is swept up by the Earth each day. Most of it burns up in the atmosphere to produce **meteors**, but about 10% reaches the surface as meteorites and **micrometeorites**.

Meteorites whose fall is observed are known as *falls*. Their atmospheric passage produces a **fireball** or **bolide**, and analysis of the trajectories of some falls reveals them to have been in orbits like those of **near-Earth asteroids**. Over 99% of meteorite falls are unobserved. Meteorites discovered on the ground are known as *finds*. Many recent finds come from Antarctica, recovered from the ice surface.

There are three main classes of meteorite: **stony meteorites** are composed mainly of silicates of iron, calcium, aluminum, magnesium, and sodium; **iron meteorites** are composed of iron and nickel, and when polished and etched show characteristic **Widmanstätten patterns**; **stony-iron meteorites** have an intermediate composition. Stony meteorites, which make up 92% of all falls, are further subdivided into **chondrites** and **achondrites**. Iron meteorites account for just 7% of falls, but over half of all finds, iron being much easier than stony material to pick out on the ground. Just 1% of falls are stony-irons. The exterior of all meteorites is a *fusion crust* of material, melted by friction during passage through the atmosphere and then resolidified.

The largest known single meteorite weighs about 60 tons and lies where it fell at Hoba, in Namibia. Stony bodies weighing several tons often fragment before impact; the fall at Allende, Mexico, in 1969 deposited 2 tons of material over 400 km^2 (150 mile^2). The area over which fragments are scattered is called a *strewnfield*. Meteoroids weighing more than 100 tons that do not break up are not decelerated as much as lighter bodies, and produce impact craters such as **Meteor Crater**.

Most meteorites are fragments of asteroids, but a few are of lunar origin, and the **SNC meteorites** are believed to be of Martian origin. Analysis of meteorites can shed light on the geological histories of the parent bodies.

meteoroid

A particle or body less than 10 m (30 ft) across following an Earth-crossing orbit, with the potential to become a **meteor** or **meteorite**. Meteoroids are of cometary or asteroidal origin. Those of cometary origin are dust particles that are shed by **comets** and spread around comets' orbits, and are responsible for **meteor showers**. Those of asteroidal origin are fragments produced by collisions in the main **asteroid** belt (or they may be small asteroids in their own right), and may penetrate the Earth's atmosphere to land as **meteorites**.

meteor shower

The appearance of **meteors** from the same point in the sky, the **radiant**, around the same time each year. Nearly all showers are named for the constellation in which their radiant lies. There are a dozen or so major showers and many minor ones. Meteors which belong to a shower are called *shower meteors* to distinguish them from **sporadic meteors**, which do not. During major showers there is a build-up of activity to a *maximum*, when 10–100 shower meteors may be visible each hour from any one location (under ideal conditions; SEE **zenithal hourly rate**), and then a falling away. In a typical shower, activity lasts for around two weeks in all.

A shower occurs when the Earth passes through a **meteor stream**. In an *annual shower* the meteor rates vary little from one year to the next because the meteoroids in the stream are spread evenly around the orbit. With streams in which the meteoroids are bunched together in a swarm, meteor numbers are low except when the Earth intersects the swarm. Such *periodic showers* may produce a *meteor storm* (SEE **Leonids**).

MAJOR METEOR SHOWERS		
Shower	Dates	ZHR*
Quadrantids	Jan 1–6	100
Lyrids	Apr 19–25	10–15
Eta Aquarids	Apr 24–May 20	50
Delta Aquarids	Jul 15–Aug 20	20–25
Perseids	Jul 25–Aug 20	80
Orionids	Oct 15–Nov 2	30
Leonids	Nov 15–20	–
Geminids	Dec 7–15	100
Approximate zenithal hourly rate		

meteor stream A large quantity of **meteoroids** circling the Sun in a common orbit. Meteor streams are associated with **comets**, and consist of particles shed by comets at perihelion passage over the course of their active lifetimes. A **meteor shower** occurs when the Earth intersects a meteor stream. In a young meteor stream, cometary debris is bunched together in a *meteor swarm*, which produces a periodic meteor shower. As the stream evolves material is spread more evenly around the orbit. With time, the orbits of individual meteoroids move further away from the primary orbit to produce a torus-shaped stream; the showers such streams produce are low-activity showers lasting a number of weeks. Eventually, a stream will be indistinguishable from the "background" concentration of dust that pervades the inner Solar System (SEE **interplanetary matter**).

Metis One of the small inner **satellites** of Jupiter discovered in 1979 during the Voyager missions. It orbits within Jupiter's main ring.

Metonic cycle A 19-year cycle of **lunations**, after which the phases of the Moon begin to repeat themselves on the same days of the year. This is because 19 **tropical years** are equal to 235 **synodic months**, to within 2 hours. The discovery is often attributed to the 5th-century BC astronomer Meton of Athens. The cycle forms the basis of the Greek and Jewish calendars. SEE ALSO **saros**.

microlensing The action of a **gravitational lens** on a small scale. If **MACHOs** exist in the halo of our Galaxy they ought to make themselves known by microlensing the light of more distant stars. Such events would consist of temporary brightenings of the distant star lasting a few days. Stars of the Large Magellanic Cloud have been monitored since the mid-1990s, and a few hundred such brightenings have been observed.

micrometeorite A micrometer-sized **meteorite**. Micrometeorites are continually settling on the Earth's surface. The *micrometeoroids* that give rise to them are too small to be burnt up in the atmosphere; they are quickly decelerated and drift gently downward. The quantity that falls greatly exceeds the total of all other meteorites. There are three main types: spheres of high density, irregular compacted particles, and fluffy non-compacted particles of lower density.

Microscopium A small southern constellation representing a microscope, introduced by **Lacaille**. Its two brightest stars are only of magnitude 4.7.

microwave background radiation SEE **cosmic microwave background**.

midnight Sun The name given to the Sun when it is visible at midnight from inside the Arctic or Antarctic Circle. At the North Pole the Sun is continuously visible for the 6-month period for which it is north of the celestial equator, and continuously below the horizon for the 6 months it spends south of the celestial equator. A similar situation holds for the South Pole.

Milky Way The faint band of light visible on clear dark nights encircling the sky along the line of the galactic equator. It is the combined light of an enormous number of stars, in places obscured by clouds of interstellar gas and dust such as the **Coalsack**. It is in fact the disk of our **Galaxy**, viewed from our vantage point within it. It is brightest in Sagittarius, toward the galactic center.

millimeter-wave astronomy Observations of **electromagnetic radiation** at wavelengths from about 1 to 10 mm, at the shortest end of the radio spectrum. Millimeter waves are given out by molecules in dense clouds where stars are forming. Waves with a wavelength shorter than 1 mm are known as *submillimeter waves* and bridge the gap between radio and infrared astronomy (SEE **submillimeter-wave astronomy**).

millisecond pulsar A **pulsar** that flashes every few thousandths of a second, the first of which was discovered in 1982. Unlike normal pulsars, millisecond pulsars are not slowing down, and they have weak magnetic fields. Over forty are known, half of them in binary systems.

Mimas The smallest of Saturn's regular **satellites**, with a diameter of 398 km (247 miles), and the innermost, orbiting near the inner edge of the tenuous E Ring.

◀ **Mimas** This Voyager image of Saturn's tiny satellite Mimas was taken from 425,000 km (264,000 miles). It shows a heavily cratered world, with the giant crater Herschel prominent.

It was discovered by William Herschel in 1789. Mimas is heavily cratered; the largest crater, named Herschel, has a diameter of 130 km (80 miles).

minor planet An alternative term for an **asteroid** which tends to be preferred by professional astronomers.

minute of arc SEE **arc minute, second**.

Mir A Russian space station, the Core module of which was launched in 1986. Other modules were added: Kvant (in 1987), Kvant 2 (1990), Kristall (1990) and Priroda (1996). Mir served as a testbed for long-term human presence in space, and despite failures and accidents in the late 1990s was crewed almost continuously until 1999. With American participation since 1995, it was used to prepare the way for the **International Space Station**. It was destroyed on re-entry in 2001.

Mira The star Omicron Ceti, a red giant that is the prototype of the **long-period variables**, also known as *Mira stars*. It was the first star discovered to vary in a periodic manner. David Fabricius, a Dutch astronomer, first noticed Mira at 3rd magnitude on August 13, 1596. A few months later it vanished, but he saw it again in February 1609.

Mira's mean period is 332 days, but this is subject to irregularities. Its mean range is magnitude 3.5 to 9.1, but maxima as bright as 2.0 and minima as faint as 10.1 have been observed; its light curve is shown in the diagram. Its spectrum varies from M5 to M9. Mira has a faint companion, which is now known as VZ Ceti, a variable with a magnitude range of 9.5 to 12.0, possibly a white dwarf with an **accretion disk**. Mira lies 196 light years away.

Miranda The smallest and innermost of the five main **satellites** of Uranus, discovered in 1948 by Gerard **Kuiper**. Its diameter is 480 × 466 km (298 × 290 miles) and it has the lowest density of the five, at 1.35 g/cm^3. Miranda displays

► **Mira** The light-curve of Mira (o Ceti) shows a slow fall in brightness followed by a long minimum and then a steep climb up to a short, sharp maximum.

one of the most complex and baffling surfaces ever seen. A rolling, cratered terrain containing both old and new craters is interrupted by well-defined regions called *coronae* containing parallel or concentric light and dark bands. One of these, Inverness Corona, contains a bright, tick-shaped feature known as the Chevron, and faults running from it lead to cliffs 20 km (12 miles) high. These features, once thought to result from the breakup and reassembly of the satellite, are probably the result of extensive ice **volcanism** and geological faulting, or **differentiation** affecting only part of the surface.

mirror An optical component of an instrument (which may be plane, spherical, ellipsoidal, paraboloidal or hyperboloidal) which redirects an incoming beam of light and may also modify it so as to aid the formation of an image of the light source. Mirrors are used in many astronomical instruments, in particular the **reflecting telescope**. They are made of glass or quartz, ground and polished to achieve the required accuracy of curvature, and coated with a thin layer of a highly reflective material. Aluminum is the usual coating material as it is cheap and easily deposited on the glass. The primary mirrors of reflecting telescopes were once made of a copper–tin alloy called *speculum*, and this term was sometimes used to refer to the mirror itself. A variety of sophisticated and exotic techniques, such as honeycomb structures and the use of ceramic materials, are now being used to fashion primary mirrors with diameters of 10 m (33 ft) or more for professional reflecting telescopes.

missing mass Invisible or low-luminosity matter that surrounds individual galaxies and pervades clusters of galaxies. Analysis of the rotation of our Galaxy strongly suggests that the galactic halo is embedded in an enormous *corona* that is roughly 600,000 light years in diameter and contains at least 10^{12} solar masses of low-luminosity matter. Other galaxies have similar rotation curves, so they too are presumed to be surrounded by large, massive coronae. In addition, clusters of galaxies must contain a substantial amount of non-luminous matter otherwise there would not be enough gravity to hold the clusters together.

The issue of the missing mass also arises on the cosmological scale. By measuring the rate at which the cosmic expansion is slowing down (the so-called **decelera-**

▲ **Miranda** The surface of Uranus's satellite Miranda is unlike anything else in the Solar System. Cliffs and craters can be seen.

tion parameter), astronomers can deduce the average density of matter in the Universe. Estimates of the cosmic deceleration indicate that the average density of matter is near the **critical density** of 5×10^{-30} g/cm^3, which is equivalent to about three hydrogen atoms per cubic meter of space. But the average density of matter that astronomers actually see in space is about 3×10^{-31} g/cm^3. Thus the observed matter is less than one-tenth that needed to account for the behavior of the Universe. SEE ALSO **dark matter**.

Mizar A double star in Ursa Major, also known as Zeta Ursae Majoris, of magnitude 2.2. Its wide companion, Alcor, magnitude 4.0, can be seen with the naked eye. Mizar and Alcor are too far apart to be a genuine binary. However, with a small telescope a 4th-magnitude companion can be seen closer to Mizar, and this does form a very long-period binary with Mizar. In addition, Mizar, Alcor, and the companion to Mizar are all **spectroscopic binaries**.

MK system ABBREVIATION FOR **Morgan–Keenan classification**.

MMT Observatory An observatory on Mount Hopkins, near Amado, Arizona. It houses the 6.5-m (255-inch) MMT Telescope, which began operation in 2000. This instrument was originally opened in 1979 as the *Multiple-Mirror Telescope* (MMT), an innovative design in which six 1.8-m (72-inch) mirrors were mounted in a hexagonal array on an altazimuth mounting. The six images were collected at a common focus, giving the same light-gathering power as a 4.5-m (176-inch) telescope. The MMT closed for conversion in March 1998; the old initials have been retained.

TYPES OF MONTH	
anomalistic month	27.55455 mean solar days
draconic month	27.21222 mean solar days
sidereal month	27.32166 mean solar days
tropical month	27.32158 mean solar days

mock Sun SEE **parhelion**.

molecular cloud An interstellar cloud in which the gas is in the form of molecules. Over a hundred different **interstellar molecules** have been identified in these clouds. Their temperature, about 20 K, is enough to make them visible to infrared telescopes; optically they may show up as **dark nebulae**. They range in size from 1 solar mass to giant molecular clouds (GMCs) of 10^5 to 10^7 solar masses. GMCs are found in the spiral arms of galaxies, and are sites of star formation. The nearest is part of the **Orion Nebula** complex.

Monoceros A constellation, representing a unicorn, which adjoins Orion. Its brightest star, Alpha, is only of magnitude 3.8, but it is crossed by the Milky Way and contains some famous clusters and nebulae, including the **Rosette Nebula** and the open cluster M50. Also in Monoceros is the massive binary known as **Plaskett's Star**.

month The period of revolution of the Moon around the Earth. Various months, defined according to the choice of reference point, are listed in the table. The *calendar month* is one of the 12 divisions of the Gregorian calendar (SEE **calendar**), and is approximately equal in length to one synodic month. SEE ALSO the entries for each type of month listed in the table.

moon (with lower-case "m") Another word for **satellite**.

Moon (with capital "M") The Earth's only natural **satellite**. Apart from the Sun it is the brightest object in the sky because of its proximity, being at a mean distance of only 384,000 km (239,000 miles). Like the Sun, its apparent diameter is about half a degree. With an actual diameter of 3476 km (2160 miles) and a density of 3.34 g/cm^3, the Moon has 0.0123 of the Earth's mass and 0.0203 of its volume. The Earth and Moon revolve around their common center of gravity, the **barycenter**. Although it is so bright (the full Moon has a visual magnitude of -12.7) its surface rocks are dark, and the Moon's albedo is only 0.07. It is a cratered world, in many ways a typical Solar System satellite. It is the only other world whose surface features we can easily see from the Earth, and the only other world on which humans have walked.

As the Moon orbits the Earth, it is seen to go through a sequence of **phases** as the proportion of the illuminated hemisphere visible to us changes. One complete sequence, from one new Moon, say, to the next, is called a *lunation*. At the crescent phase, the dark side of the Moon is seen to be faintly illuminated by *earthshine* – sunlight reflected from the Earth. At new Moon or full Moon, **eclipses** can occur. An observer on Earth always sees the same side of the Moon because the Moon has **synchronous rotation**: its orbital period (the **sidereal month**) around the Earth is the same as its axial rotation period. The visible side is called the *nearside*, and the side invisible from Earth is the *farside*. In fact, the face the Moon presents to us does vary slightly because of a number of effects known collectively as **libration**.

The Moon has been studied by many space probes: the **Luna** and Zond series launched by the former Soviet Union, and the US series **Ranger**, **Surveyor**, and **Lunar Orbiter**. The manned **Apollo** missions and some of the later Luna probes returned samples of lunar material to the Earth for study. In the 1990s, the **Clementine** and **Lunar Prospector** probes gathered much valuable new information.

The surface features may be broadly divided into the darker maria (SEE **mare**), which are low-lying volcanic plains, and the brighter highland regions (sometimes called *terrae*), which are found predominantly in the southern part of the Moon's nearside and over the entire farside. There are **impact features** of all sizes. The largest are called *basins*, produced during the early history of the Moon when bombardment by impacting objects was at its heaviest. On the nearside, where the crust is thinner, some basins were subsequently filled with upwelling lava to produce the maria, which in turn became cratered. The greatest lunar

▲ **Moon** When it is at its brightest, at full Moon, the differences between the bright lunar highlands and the dark, lava-filled basins become very apparent.

depression, discovered by Clementine, is centered on the crater Aitken near the south pole, and is about 2500 km (1500 miles) in diameter – the largest such structure in the Solar System. The floor of Aitken is 13 km (8 miles) below its rim, the deepest point on the lunar surface. In the regions of craters at the north and south poles, whose interiors are in permanent shadow, water ice exists just below the surface. The smaller basins are similar to the largest craters, formerly called *walled plains*, which have flat floors and are surrounded by a ring of mountains, like the 298-km (185-mile) diameter crater Bailly. Other features are mountain peaks and ranges, valleys, elongated depressions known as **rilles**, **wrinkle ridges**, low hills called *domes*, and bright **rays** of ejecta radiating from the sites of recent cratering impacts.

Early lunar cartographers assumed that the darker lunar features were expanses of water, and named them after fanciful oceans (oceanus), seas (mare), bays (sinus), lakes (lacus), and swamps (palus). Other features are named for famous people, principally astronomers and other scientists. The old names are still in use on lunar maps.

The origin of the Moon is uncertain. The main current theory, known as the *giant impactor* or *big splash* theory, is that a

Mars-sized body collided with the newly formed Earth, and debris from the impact accreted (SEE **accretion**) to form the Moon. One piece of evidence for this is the correspondence between certain radioactive isotopes in lunar and terrestrial rocks. Subsequent large impacts produced the Moon's basins, and smaller impacts the craters. A volcanic origin for the craters has long been dismissed by most astronomers, but Clementine photographed a volcanic crater near the large farside crater Schrödinger. Mild *moonquakes* occur at depths of roughly 700 km (450 miles), and **lunar transient phenomena** appear to be associated with them; otherwise, the Moon is now geologically inactive.

Seismic measurements at Apollo landing sites yielded some information on the Moon's internal structure. The lunar crust is about 100 km (60 miles) thick in the highland regions, but only a few tens of kilometers under the mare basins. Under the largest basins the underlying mantle has bulged upward to form **mascons**. There is a small iron core at the Moon's center, accounting for about 4% of its mass. Although there is no overall magnetic field, there are distinct "magnetic areas" extending over hundreds of kilometers.

The chemical composition of material brought back from the Moon has been found to consist mainly of silica, iron oxide, aluminum oxide, calcium oxide, titanium dioxide, and magnesium oxide. Lunar rocks are igneous. The maria have a basaltic composition, while the highlands consist largely of anorthosite. As a result of the constant bombardment, crustal rocks are in the form of **breccias**, perhaps to a depth of several kilometers, and the surface is covered with a **regolith**. The Moon has only the most tenuous of atmospheres. Apollo instruments detected traces of gases such as helium, neon, and argon. The atmosphere is probably made up of **solar wind** particles retained temporarily, and atoms sputtered (knocked off) from the surface by the solar wind. Consequently the surface temperature variation is extreme: from 100 to 400 K.

Morgan–Keenan classification The categorization of a star's spectrum from its absorption features. The system was devised by William W. Morgan, Philip C. Keenan, and Edith Kellman of Yerkes Observatory for their *An Atlas of Stellar Spectra*, published in 1943; it is also known as the *MK system*, *MKK system*, or *Yerkes system*. It established formal rules for describing a star by spectral type (O, B, A, F, G, K, M, in sequence from hot to cool surface temperature) and luminosity class (I = supergiant, III = giant, V = dwarf; with interpolations Ia, Iab, Ib, II, and IV). SEE ALSO **spectral classification**.

morning star Name sometimes given to the planet Venus when it is visible in the east before sunrise.

Mount Graham International Observatory An observatory on Mount Graham, near Safford, Arizona. It has three main instruments: the Vatican Advanced Technology Telescope (also known as the Alice P. Lennon Telescope), a 1.8-m (70-inch) reflector; the 10-m (400-inch) Heinrich Hertz Submillimeter Telescope; and the **Large Binocular Telescope**.

mounting A means of supporting the weight of a telescope and aiming it at different points in the sky. Of necessity, all large professional instruments are on permanent mounts; most amateur telescopes and their mounts are portable. Permanent or portable, rigidity is a prime requirement.

In order for a telescope to be pointed at different parts of the sky, the mounting incorporates two axes, perpendicular to each other, about which the telescope can be rotated. In the **altazimuth** mount the axes are parallel and perpendicular to the horizon. The disadvantage of this mounting is that the telescope has to be moved about the two axes simultaneously in order to keep an object in view. This is overcome in **equatorial telescopes**, in which one axis (the polar axis) is parallel to the Earth's axis of rotation, and so points to the celestial pole. The telescope then only needs to be moved about the other axis, in the direction opposite to that of the Earth's rotation, to keep an object in the field of view.

Since the introduction of computer-controlled drive mechanisms, the altazimuth mount has become the choice for all large professional telescopes. A simple altazimuth platform is incorporated in the **Dobsonian telescope**, a large and portable amateur reflecting telescope.

Mount Palomar Observatory
The location of the **Palomar Observatory**.

Mount Stromlo Observatory
An observatory near Canberra, operated by the Australian National University. Founded in 1924 as a solar observatory, it changed focus to stellar astronomy after World War II. Most of its facilities, including several historic telescopes, were destroyed in a bush fire in January 2003. It also runs telescopes at **Siding Spring Observatory**.

Mount Wilson Observatory
An observatory in the San Gabriel Mountains near Pasadena, California. Its main instrument is the 2.5-m (100-inch) Hooker reflector, opened in 1917, with which Edwin **Hubble** discovered the expansion of the Universe; it was renovated and reopened in 1993. The Mount Wilson Observatory was founded in 1904 by George Ellery **Hale**, originally for solar astronomy; a 1.5-m (60-inch) reflector was opened in 1908. There are also solar telescopes and interferometers, including the CHARA optical–infrared interferometer.

moving cluster
A group of stars, often widely scattered, that share the same motion through the Galaxy. The stars in a moving cluster all follow nearly parallel paths. However, because of a perspective effect, they seem to be moving toward, or away from, a point known as the *convergent point*. This allows the distance of the cluster to be determined. First, the position of the convergent point must be established. Some clusters are very scattered, for example the Ursa Major moving cluster, which includes five stars of the Big Dipper (Beta, Gamma, Delta, Epsilon and Zeta). Sever-

al other bright stars show the same motion through the Galaxy as these five Ursa Major stars. They include Sirius, Delta Leonis, Beta Aurigae, Beta Eridani, and Alpha Coronae Borealis. In total about 100 members of the group are known, scattered widely across the sky. Their separation in space is larger than in a normal cluster. Our Sun is moving through the outskirts of the group, which is why its members appear so widely spread across the sky.

M star A star of spectral type M, characterized by absorption bands of titanium oxide molecules. M-type dwarf stars, also known as red dwarfs, define the lower end of the **main sequence**. Proxima Centauri and Barnard's Star are examples. The surface temperatures of M-type dwarfs range from 3500 to 2500 K; their masses range from 0.08 to 0.5 solar masses, and their radii from 0.1 to 0.6 times that of the Sun. The most massive M-type dwarfs are nearly one-tenth as luminous as the Sun, while those at the bottom of the main sequence have less than one-thousandth of the Sun's luminosity. Most of their radiation is emit-

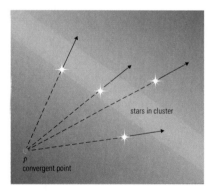

▲ **moving cluster** The velocity of stars in an open cluster can be determined if they are close enough for their parallax to be measurable over a short period. Their parallax is combined with their line-of-sight rate of motion toward or away from Earth to obtain their speed of movement at right angles to the line of sight. Combined with the measurement of the angle of movement, this allows astronomers to judge their distances.

ted in the infrared, so their visual luminosity is even fainter, from one twenty-fifth to one ten-thousandth that of the Sun. Many M-type dwarfs are **flare stars**; none are visible to the naked eye.

M-type giants and supergiants are all evolved stars with masses much greater than those of M-type dwarfs. Their surface temperatures are typically a few hundred degrees cooler than dwarfs of the same spectral type. The brightest M-type supergiants are up to a million times as luminous as the Sun. The largest of them are several thousand times the Sun's diameter – as large as the orbit of Saturn or Uranus. Even the coolest M-type giants are as large as the Earth's orbit. Such distended stars are unstable and therefore variable; examples are **Betelgeuse** and **Mira**. The high luminosity and low surface gravity of M-type giants and supergiants causes a **stellar wind** of gas in which the mass loss can be as high as one solar mass every 100,000 years.

Mullard Radio Astronomy Observatory The radio astronomy observatory of the University of Cambridge, situated at Lords Bridge near Cambridge. It was there that the technique of **aperture synthesis** was developed. The Observatory's principal instrument is the Five-Kilometre Telescope (opened in 1972), consisting of four fixed and four movable dishes 13 m (43 ft) in diameter, arranged in an east–west line. Other radio interferometers on the site are the One-Mile Telescope (opened in 1964) and the Half-Mile Telescope (opened in 1968). The Cambridge Optical Aperture Synthesis Telescope (COAST) was opened in 1995.

Müller, Johann (also known as **Regiomontanus**) (1436–76) German astronomer and mathematician. In 1463 he completed the translation into Latin of the **Almagest** begun by his tutor, Georg von Purbach. In 1471 Müller set up a printing press and became one of the first publishers of astronomical and scientific books. There is some evidence that he had the idea of a heliocentric universe, and that this reached and therefore influenced Nicholas **Copernicus**.

Multi-Element Radio-Linked Interferometer Network SEE **MERLIN**.

Multiple-Mirror Telescope SEE **MMT Observatory**.

multiple star A gravitationally connected group of more than two stars. **Castor**, for example, has six components. In multiple systems there is a recognizable hierarchy or order. In Castor, for instance, the two main pairs rotate about a common center of gravity while a pair of cool red dwarfs rotates about the same center but at a larger distance. There is no clear distinction between a multiple star with many members and a small cluster. A good example is the **Trapezium** in the Orion Nebula, which consists of four young stars enmeshed in nebulosity with a further five possible members nearby.

Musca A small but distinctive southern constellation adjoining Crux, representing a fly. Its brightest star, Alpha, is magnitude 2.7. Musca contains the globular clusters NGC 4372 and 4833.

N

nadir The point on the **celestial sphere** vertically below the observer. It is diametrically opposite the zenith.

Nagler eyepiece A telescope eyepiece with a very wide field of view, 80° or more, making it suitable for activities such as searching for comets. (After its designer, Albert Nagler, 1935– .)

Naiad One of the small inner **satellites** of Neptune discovered in 1989 during the Voyager 2 mission.

naked-eye A term applied to observations made by the eye alone, without the aid of any optical instrument, or to celestial objects thus visible.

NASA ABBREVIATION FOR **National Aeronautics and Space Administration**.

Nasmyth focus A focus of a reflecting telescope on an **altazimuth** mount, situated to one side on the altitude axis. The light beam is directed there by an extra flat mirror angled at 45°. Heavy instruments positioned at a Nasmyth focus do not have to be moved bodily as the telescope tracks an object. It is named for Scottish engineer James Nasmyth (1808–90), who incorporated it in his 6-inch (150-mm) reflector. Its use has been revived with the present generation of large, altazimuthally mounted professional instruments. SEE ALSO **coudé**.

National Aeronautics and Space Administration (NASA) The US government agency for the peaceful exploration of space, formed in 1958. Its headquarters are in Washington, DC, and it operates various field stations. These include the Goddard Space Flight Center at Greenbelt, Maryland, which handles space science research and Earth satellite tracking; the Jet Propulsion Laboratory in Pasadena, California, which is mission control for NASA space probes and manages NASA's deep-

space tracking stations around the world; the Ames Research Center, San Francisco, for aeronautics and planetary science; the John F. Kennedy Space Center at Cape Canaveral, Florida, the main US launch site; and the Lyndon B. Johnson Space Center in Houston, Texas, mission control for manned space missions.

National Radio Astronomy Observatory (NRAO) The major US radio observatory, with its headquarters at Green Bank, West Virginia. Its instruments include a 42.7-m (140-ft) radio dish and an interferometer consisting of three 26-m (85-ft) dishes plus one of 13.7 m (45 ft). The Green Bank Telescope was dedicated in 2000. With a dish measuring 100 × 110 m (330 × 360 ft), it is the world's largest fully steerable radio telescope. NRAO also operates a 12-m (39-ft) dish for millime-

▲ **Nasmyth focus** A number of large, professional telescopes have instruments at their Nasmyth foci. In the William Herschel Telescope there are two foci, one used for infrared and the other for optical observations. The third mirror (the Nasmyth flat) can be moved to direct light to either of them.

▲ **Neptune** The planet appears bluish because methane in its atmosphere absorbs red light. Ice particles in the upper cloud decks also scatter blue light.

ter-wave studies at Kitt Peak, Arizona, the **Very Large Array** in New Mexico, and the **Very Long Baseline Array**.

nautical twilight SEE **twilight**.

n-body problem SEE **many-body problem**.

neap tide SEE **tides**.

near-Earth asteroid (NEA) A small **Amor asteroid**, **Apollo asteroid** or **Aten asteroid** whose orbit brings it relatively close to the Earth. The term *near-Earth object* is also used, to include close-approaching comets. *Potentially hazardous asteroids* (PHAs) are those whose orbits bring them particularly close to the Earth. Close approaches have been made by objects as small as 9 m (30 ft) across. Objects this size striking the Earth would burn up in the atmosphere; larger objects would penetrate further, causing a powerful low-level airburst like the **Tunguska Event**, or producing a crater like **Meteor Crater**.

NEAR Shoemaker A NASA probe to the asteroid **Eros**, launched in 1996. the mission was originally the Near Earth Asteroid Rendezvous. After a fly-by of the asteroid **Mathilde**, NEAR Shoemaker became the first probe to orbit an asteroid when it entered orbit around Eros in 2000. After returning a wealth of images and geological data, the spacecraft made a successful touchdown on the surface of Eros in February 2001, at the end of the mission.

nebula (plural **nebulae**) A region of interstellar gas and dust. The word nebula is Latin for "cloud." There are three main types of nebula. **Emission nebulae** are bright diffuse nebulae which emit light and other radiation as a result of ionization (removal of electrons) and excitation of the gas atoms by ultraviolet radiation. The source of the ultraviolet is usually one or more hot stars. When gas ions recombine with free electrons, and forbidden transitions (SEE **forbidden lines**) occur in excited atoms, radiation is emitted, giving rise to emission lines in the spectrum. Examples of emission nebulae are **H II regions** such as the Orion Nebula, and **planetary nebulae** such as the Ring Nebula in Lyra. **Supernova remnants** are a form of emission nebula in which the gas is made to glow not by the ultraviolet radiation of the star within, but by frictional heating as it collides with surrounding interstellar gas; an example is the Cygnus Loop. The Crab Nebula is a type of supernova remnant that shines by **synchrotron radiation**. In contrast, the brightness of **reflection nebulae** results from the scattering by dust particles of light from nearby stars. **Dark nebulae** are not luminous: interstellar gas and dust absorb light from background stars, producing apparently dark patches in the sky. This third class includes a group known as **Bok globules**, which are dense absorption nebulae of nearly spherical shape.

Neptune The eighth major planet from the Sun, and the smallest and most distant of the four giant planets; it marks the inner edge of the **Kuiper Belt**. With a mean magnitude of 7.8, Neptune is invisible to the naked eye. Its average apparent size is 2.2 arc seconds. Through a telescope it appears as a small greenish blue disk on which very few details can be distinguished. In its size, mass, atmosphere and color, Neptune resembles Uranus. Like all the giant planets, it possesses a ring system and a retinue of satellites. The main data for Neptune are given in the table below.

Neptune was first identified by Johann **Galle**, assisted by Heinrich D'Arrest, on September 23, 1846, close to a position predicted by Urbain **Le Verrier**. Le Verrier and, independently, John Couch **Adams**, had calculated the mass, orbit, and position of an unseen planet to account for perturbations in the orbital motion of Uranus.

The fly-by of the Voyager 2 probe in 1989 provided most of our current knowledge of the planet. The upper atmosphere is about 85% molecular hydrogen and 15% helium. Its predominant blue color is due to a trace of methane, which strongly absorbs red light. Several different atmospheric features were visible at the time of the Voyager encounter. There were faint bands parallel to the equator, and spots, the most prominent of which was the oval Great Dark Spot, about 12,500 km (8000 miles) long and 7500 km (4500 miles) wide. It was about the same size relative to Neptune as the **Great Red Spot** (GRS) is to Jupiter, and like the GRS it is a giant anticyclone. White, cirrus-type clouds of methane crystals cast shadows on the main cloud deck some 50 km (30 miles) below. Neptune's atmosphere has **differential rotation**, some features taking 19 hours or more for one rotation. The atmosphere thus has retrograde rotation with respect to the interior, which rotates in just over 16 hours. There are also the highest wind speeds recorded on any planet – over 2000 km/h (1250 mile/h) in places.

Neptune's dynamic and changing atmosphere, which since the Voyager encounter has been followed by the Hubble Space Telescope, probably has to do with its internal heat source – it radiates more than twice as much heat as it receives from the Sun. The amount of white cloud also varies with the **solar cycle**, which is hard to explain for such a distant planet.

Below the visible atmosphere the composition and structure are less certain. A chemical cycle, driven by ultraviolet radiation from the Sun, might operate in which

methane is converted into other hydrocarbons and back again. At deeper levels there may be a mantle of water and ammonia about ten times the mass of the Earth, surrounding an iron–silicate core of about one Earth mass. Like Uranus, Neptune has a magnetic field in which the axis is markedly tilted (by 47°) to the axis of rotation and is displaced (by about 10,000 km/6000 miles) from the planet's center. In Neptune's case it could be that the mantle is at least partially fluid and the ammonia–water mixture is ionized, and therefore electrically conducting, and that this is the source of the magnetic field.

Of Neptune's eleven known **satellites**, **Triton** and **Nereid** were discovered before the Voyager 2 fly-by, and another (**Larissa**) was suspected from an **occultation** measurement. Triton's orbital motion is **retrograde**, and Nereid's orbit is the most eccentric of any Solar System satellite. Six inner satellites, including the second-largest, **Proteus**, lie in close, regular orbits. Three small satellites orbit at around 21 million km.

Neptune's ring system was detected from Earth during an **occultation** in 1984,

NEPTUNE DATA	
Globe	
Diameter (equatorial)	49,528 km
Diameter (polar)	48,686 km
Density	1.64 g/cm^3
Mass (Earth = 1)	17.14
Volume (Earth = 1)	57.67
Sidereal period of axial rotation	16h 07m
Escape velocity	23.5 km/s
Albedo	0.84
Inclination of equator to orbit	29° 34'
Temperature at cloud-tops	55 K
Surface gravity (Earth = 1)	0.98
Orbit	
Semimajor axis	30.06 AU = 4497 × 10^6 km
Eccentricity	0.0097
Inclination to ecliptic	1° 46'
Sidereal period of revolution	164.79y
Mean orbital velocity	5.43 km/s
Satellites	11

▲ **neutron star** A hot X-ray source within the supernova remnant IC443 is thought to be the neutron star created in the explosion.

but the measurements suggested a number of incomplete "ring arcs." Voyager revealed four continuous rings: two narrow ones, and two faint broad ones. The outermost and brightest, the Adams Ring, contains "clumps" of material that appeared as arcs in the occultation data. SEE ALSO **ring, planetary**.

Nereid Neptune's outermost **satellite**, discovered by Gerard **Kuiper** in 1949. It has the most eccentric orbit of all known satellites, taking it from 1.4 million km (850,000 miles) out to 9.7 million km (6 million miles). It is almost certainly a captured body.

neutrino astronomy The detection of neutrinos from astronomical sources. Neutrinos are atomic particles with no charge and very low (possibly zero) mass. They therefore interact hardly at all with matter. To maximize the chance of detecting one, neutrino "telescopes" consist of huge tanks filled with a suitable liquid. One type uses cleaning fluid, as an incoming neutrino has a chance of reacting with a chlorine atom to form a radioactive atom of argon, which can be detected. Another type contains water, as a neutrino may interact with an electron in a water molecule and emit a particular type of radiation. Neutrinos are emitted by nuclear reactions inside stars. Early attempts to detect neutrinos from the Sun found less than half the number that

theorists expected. This *solar neutrino problem* was resolved when it became clear that the Sum emitted different types of neutrino, and detectors were adapted to record three types. Neutrinos are also produced by the complex reactions that take place in **supernovae**, and neutrinos from **Supernova 1987A** were detected. Neutrinos are also produced in the Earth's atmosphere by cosmic rays; these can be distinguished by their different energies.

neutron star An extremely small, dense star that consists mostly of neutrons. Neutron stars are observed as **pulsars** and are also thought to be one of the components in **X-ray binaries**. They are formed when a massive star explodes as a supernova, blasting off its outer layers and compressing the core so that its component protons and electrons merge into neutrons. Neutron stars have masses comparable to that of the Sun, but diameters of only about 20 km (12 miles) and average densities of about 10^{15} g/cm^3. A neutron star's gravity is so great that objects would weigh 10^{11} times more at its surface than on the surface of the Earth, and it also produces a significant **gravitational redshift** (z about 0.2). The maximum mass for a neutron star is about 3 solar masses, beyond which it would collapse further, into a **black hole**.

Being so small and having such strong gravity, neutron stars can spin rapidly without disintegrating – up to several hundred times per second. They retain the magnetic field of the original star but compressed into a much smaller volume, so the magnetic fields of neutron stars are about a million million times stronger than that of the Earth. A neutron star's structure is complicated because of the great variation in pressure from the surface to the center. The crust is made of iron about 10,000 times denser than terrestrial iron and a million times stronger than steel. Beneath this is a liquid of neutrons, and at the very center may be a core of exotic nuclear particles.

Newcomb, Simon (1835–1909) Canadian-born American mathematical astronomer. He calculated highly accurate values for astronomical constants, such as the solar **parallax**, and prepared extremely accurate tables of the motions of the Moon and the planets. It was Newcomb who calculated the discrepancy in the advance of **Mercury**'s perihelion that was to provide a proof of **Einstein**'s Special Theory of Relativity.

new Moon The **phase** of the Moon when at conjunction. The dark side of the Moon then faces the Earth, and the illuminated side is invisible to us.

New Technology Telescope A 3.6-m (142-inch) reflector at the **European Southern Observatory** in Chile, completed in 1989. It uses a system of **active optics** to keep the main mirror in shape.

Newton The ESA X-Ray Multi-Mirror Mission satellite, launched in 1999. Newton carries X-ray telescopes with a large effective collecting area, enabling long uninterrupted exposures, and an optical monitor, the first to be flown on an X-ray observatory.

Newton, Isaac (1642–1727) English physicist and mathematician. His achievements were so important for astronomy that the phrase "Newtonian revolution" is often used. Most of his theories – on light, gravitation, and calculus, for example – he developed in basic form while in his early twenties. In 1668 he built the first reflecting telescope in order to overcome the **chromatic aberration** inherent in the lenses used in refracting telescopes. In 1684, encouraged by Edmond **Halley**, he began work on his *Philosophiae naturalis principia mathematica*. Published in 1687, and usually referred to as the *Principia*, this was one of the most epoch-making books in the history of science. In it he put forward his law of **gravitation** and laws of motion (SEE **Newton's laws of motion**), and showed that the orbit of a body moving under an **inverse-square law** of gravitation would be an ellipse. He went on to demonstrate how gravitation explained the second and third of **Kepler's laws** and also many outstanding problems in astronomy. In his *Opticks* (1704) he set out his theories of light, including how white light is made up of the colors of the spectrum. Newton's mathematical achievements were many, but his invention of calculus (independently of Gottfried Leibniz) was the greatest.

Newtonian telescope The simplest form of **reflecting telescope**. The first, built by Isaac **Newton** in 1668, had a primary concave spherical mirror of speculum metal (an alloy of copper and tin) which reflected the light up the tube of the telescope. A flat speculum mirror near the top of the tube, angled at 45°, directed the light to a focus – the *Newtonian focus* – to one side of the tube where the image of the object being observed was viewed through a small planoconvex eyepiece. This is still the basic design, although speculum has given way to glass, and for all but the smallest apertures the primary mirror is now a **paraboloid**. The Newtonian has long been a favorite with amateur astronomers.

Newton's law of gravitation SEE **gravitation**.

▲ **Newtonian telescope** In this form of reflecting telescope, light is reflected from a paraboloidal primary mirror to a flat secondary mirror placed diagonally within the tube and directed to a side-mounted eyepiece. The resulting image is inverted.

Newton's laws of motion The fundamental laws of dynamics, set out by Isaac Newton in his *Principia*:

    1. *Law of inertia*: Every body continues in its state of rest, or of uniform motion in a straight line, until acted upon by some outside force.

    2. *Law of acceleration*: The acceleration of a body is directly proportional to the force acting upon it, and is in the direction in which the force is acting.

    3. *Law of action and reaction*: To any action there is an equal and opposite reaction.

NGC Abbreviation for the *New General Catalogue of Nebulae and Clusters of Stars*, published in 1888 by J. L. E. **Dreyer**, and incorporating previous catalogs by William and John **Herschel**. The NGC contains nearly 8000 objects – including galaxies, although their real nature was not then known – and another 5000 are listed in two subsequent *Index Catalogues* (IC). Objects are still referred to by their NGC or IC numbers.

noctilucent cloud (NLC) A cloud at very high altitude (average 82 km/51 miles) in the Earth's atmosphere, visible in the summer in twilight hours (noctilucent means "night-shining") from latitudes between 50° and 70°. NLCs have a pearly or blue color and resemble delicately interwoven cirrus cloud. They are made up of water droplets that have been carried aloft and then condensed and frozen on to particles. It is not known exactly what these particles are; they may be ions produced by solar radiation, or they may be **micrometeorites**, or even products of industrial pollution. NLCs have become more common in recent years.

node The point at which one orbit cuts another. Specifically, a node is one of the two points in the orbit of a planet or comet at which it cuts the ecliptic, or at which the orbit of a satellite cuts that of its primary. The *ascending node* is the node at which a celestial body passes from south to north; the *descending node* is the node at which a celestial body passes from north to south.

The *line of nodes* joins the ascending and descending nodes. The *regression of the nodes* is the backward motion of the Moon's nodes around its orbit. They move slowly westward so that the nodal line makes one complete revolution of the orbit in 18.6 years. The regression is caused by the Sun's gravitational attraction.

non-gravitational force Any force acting on a celestial body which is not gravitational, in particular one which produces similar effects to gravity. The term is most commonly used in the context of comets which, on their approach to perihelion, lose volatile material through one or several vents in their crust, producing a jet or jets in a particular direction. This can alter their course in much the same way as the gravitational **perturbations** to which they are prone. The **Poynting–Robertson effect** results from another type of non-gravitational force. SEE ALSO **magnetohydrodynamics**.

Norma A small southern constellation introduced by **Lacaille** between Ara and Lupus, representing a surveyor's level. It contains no star above magnitude 4.0.

North America Nebula A large emission nebula about 3500 light years distant in the constellation Cygnus. It is also known as NGC 7000. An associated region of nebulosity, IC 5070, is the *Pelican Nebula*.

Northern Cross A popular name sometimes given to the cross-shaped arrangement of the brightest stars in the constellation Cygnus.

North Polar Sequence A list of stars with accurately measured magnitudes within 2° of the north celestial pole. It includes stars from magnitude 2 down to magnitude 20. Comparison with the North Polar Sequence enables the magnitudes of other stars to be determined.

nova (plural **novae**) An existing star, usually quite faint, whose brightness suddenly increases by 10 magnitudes or more before slowly returning to its prenova state. A typical nova rises to maximum in a few days, and declines thereafter in brightness by a factor of about ten in 40 days, although cases of both slower and faster declines are known. The diagram shows the light curve of the brightest nova (as opposed to **supernova**) of modern times, Nova Aquilae in 1918. About 40 novae are estimated to explode in the Galaxy each year, although we see only a small percentage of them.

Novae occur in close binary systems where one component is a white dwarf and the other is a giant. Matter drains from the surface of the giant, where gravity is weak, to the surface of the white dwarf, where gravity is strong. Nuclear reactions involving the infalling material cause a violent explosion on the surface of the white dwarf. The energy of a typical explosion is about the same as the Sun emits in 10,000 years. Clouds of gas are ejected at speeds around 1500 km/s (900 mile/s) during the outburst. The total amount of material lost is estimated to be about one ten-thousandth the mass of the white dwarf. Some stars, known as **recurrent novae**, undergo explosions repeatedly. Given time, all novae probably recur. SEE ALSO **dwarf nova**.

Nozomi A Japanese space probe to Mars, launched in 1998 with an international payload of instruments. Its main purpose is to study the structure of the Martian atmosphere, and the effect upon it of the solar wind. Problems with its main engine delayed its arrival at Mars, now due in 2004.

Nubecula Major, Nubecula Minor
Obsolete names for the Large and Small **Magellanic Clouds**.

nucleosynthesis The production of chemical elements from other chemical elements via naturally occurring nuclear reactions. These reactions are of two types: *fusion*, in which heavier elements are built up from lighter ones, and *fission*, in which heavier elements are broken down into lighter ones. Fusion created most of the Universe's helium from hydrogen in the first few minutes following the **Big Bang**, and fusion processes such as the **carbon–nitrogen cycle** operate inside stars to create helium and heavier elements which are carried into interstellar space by stellar winds. Both fusion and fission occur in the extreme conditions of novae and supernovae to produce many different elements which are then ejected into the surrounding space.

nucleus (1) The solid part of a **comet**.

nucleus (2) The central region of a **galaxy**.

nutation The periodic oscillation in the precessional motion of the Earth's axis of rotation (SEE **precession**). It was discovered by James **Bradley** in 1747. It is in effect a slight "nodding" of the Earth's axis, caused by the combined effect of the gravitational attractions of the Sun and Moon upon the Earth. The effect is constantly varying as the relative positions of the three bodies are continuously changing. *Lunar nutation*, which causes the Earth's axis to describe an ellipse, has a period of 18.6 years. *Solar nutation* has a period of 0.5 of a tropical year. The *fortnightly nutation* has a period of 15 days.

▶ **nova** The light curve of a nova typically shows a steep rise in brightness by a factor of many thousands followed by a slow, irregular, decline. The rapid rise in brightness is caused by explosive hydrogen burning on the surface of the white dwarf.

O

OB association A region of space in which recent star formation has produced high-mass O and B stars. Such associations are less tightly bound than a normal star cluster and are slowly dispersing. Since O and B stars are very luminous, OB associations are recognizable at great distances, even through intervening dust. Tens or hundreds of these massive stars can occur in a single association. Fainter, less massive stars may also exist, but are harder to see at great distances. The Orion Nebula is part of an OB association that is relatively close to us, so young, low-mass stars are also detectable within it. OB associations are a feature of the gas-rich spiral and irregular galaxies, but not of elliptical galaxies, in which no young stars are found. COMPARE **T association**.

Oberon One of the five main **satellites** of Uranus, discovered in 1787 by William Herschel. It is 1520 km (945 miles) in diameter and consists largely of a mixture of rock and water ice, with a density of 1.58 g/cm^3. Oberon has many craters, some with **rays** and some with dark floors, and mountains, one of which appears to be 20 km (12 miles) high. The high density of craters is evidence of a world unaltered by geological activity.

objective The main light-gathering element of an optical instrument. In a refracting telescope it is the lens or lens system (called the *object glass* in older texts) that collects and focuses light from a celestial body to form an image of it.

objective prism A large, narrow-angle prism mounted across the aperture of a telescope so that the image of each star in the field of view is transformed into a low-resolution spectrum. From photographs taken by this means it is possible to rapidly assign spectral classifications to the stars in a given area.

oblateness (ellipticity) A measure of how far a **spheroid** deviates from a true sphere, given by the ratio of the difference between the equatorial and polar radii to the equatorial radius. The oblateness of a celestial body is an indication of the speed at which it is rotating.

oblate spheroid SEE **spheroid**.

obliquity of the ecliptic The angle between the plane of the ecliptic and the celestial equator. Alternatively, it may be defined as the angle between the axis of rotation of a planet and the pole of its orbit. The Earth's obliquity is currently about 23° 26′, and is decreasing by 0.47′ per year. The angle will start to increase again in about 1500 years; the maximum range is 21° 55′ to 28° 18′, in a period of 40,000 years. The obliquity is responsible for the seasons. It represents the greatest angular distance that the Sun can lie north and south of the equator.

observatory A structure that is built and equipped for making celestial observations and measurements, acting principally as a housing for a telescope. The best location for astronomical purposes is on a mountain or high plateau where there are many cloudless or near-cloudless nights. The most favorable locations include Hawaii, Chile, western parts of the USA, and the Canary Islands.

Professional optical and infrared telescopes are installed in buildings with a rotating roof, usually dome-shaped, although square buildings are now becoming fashionable, and in some cases the entire building may rotate. The dome has a slit, closed by sliding shutters when not in use, through which observations are made. In addition there are control rooms, laboratories, photographic darkrooms, workshops, living accommodation for personnel, and other facilities. There are over 200 professional optical observatories throughout the world.

Sites for radio telescopes are less critical than for optical observatories because radio waves are not affected so much by the atmosphere. Radio telescopes consisting of an array of antennas do, however, require a considerable area of land.

The latest observatories are those being launched into Earth orbit, entirely above the absorbing and distorting effects of the Earth's atmosphere. In addition to optical and radio studies these craft carry out measurements in the infrared, ultraviolet, X-ray and gamma-ray regions of the spectrum which do not penetrate the Earth's atmosphere and so cannot be studied from ground-based observatories. SEE ALSO the names of individual observatories.

occultation The temporary cutting off of the light from one celestial body as another, nearer one passes in front of it. For example, the Moon may pass in front of a star; a planet may do the same and may also occult its own satellites, as in the case of Jupiter. The rings of **Uranus** were first detected when the planet occulted a star in 1977. Strictly speaking, a solar eclipse is an occultation of the Sun by the Moon. Accurate timings of lunar occultations are valuable for accurately determining the Moon's position and can also provide information about the occulted body. *Grazing occultations*, in which the star appears to skim the Moon's limb, are of particular interest.

Octans A faint constellation, notable only because it contains the south celestial pole. The nearest naked-eye star to the pole is Sigma Octantis, magnitude 5.4. Octans was introduced by **Lacaille** and represents an octant, the forerunner of the sextant. The brightest star in the constellation is Nu, magnitude 3.7.

ocular Another word for **eyepiece**.

Olbers, Heinrich Wilhelm Matthäus (1758–1840) German doctor and amateur astronomer. He is best known for having posed the question now known as **Olbers' paradox**. He made the second and fourth asteroid discoveries, of Pallas (1802) and Vesta (1807). Olbers invented a method for calculating cometary orbits, and discovered several comets. He also proposed the theory that the tails of comets are highly rarified matter expelled from the head of the comet by pressure of some kind from the Sun, thus anticipating the concept of **radiation pressure**.

Olbers' paradox The paradox that arises in attempting to answer the question, "Why is the sky dark at night?" If space is infinite and uniformly filled with galaxies, then an observer would end up looking at the surface of a star or galaxy at every point in the sky, so the entire night sky should be bright. In fact, space is not infinite. The Universe is believed to have been born in the **Big Bang** about 15,000 million years ago, and so light from the most distant objects has not yet reached us. In addition, the light from distant galaxies is weakened by the **redshift** caused by the expansion of the Universe. The paradox is named for Heinrich **Olbers**, who drew attention to it in 1823.

Olympus Mons An extinct volcano on Mars, and the largest known volcano in the Solar System, rising to 24 km (15 miles) above the planet's mean surface level. It is about 700 km (450 miles) across, including the surrounding area of eroded volcanic rock. The caldera (summit crater) is over 80 km (50 miles) across.

▲ **Oberon** This is the closest image obtained of Uranus's satellite Oberon during the Voyager 2 fly-by of January 1986. Oberon's surface shows several impact craters surrounded by bright rays.

Omega Centauri The brightest globular cluster in the sky, visible to the naked eye as a fuzzy star of magnitude 3.7 in the constellation Centaurus. Designated NGC 5139, it is 17,000 light years away, one of the nearer globular clusters. It contains several hundred thousand stars in a volume 200 light years across.

Omega Nebula An emission nebula in the constellation Sagittarius, designated M17 or NGC 6618. Other names for it are the Horseshoe Nebula and the Swan Nebula. It lies 6800 light years away.

Oort, Jan Hendrik (1900–1992) Dutch astronomer. He carried out research on the structure and dynamics of stellar systems, especially the Galaxy, whose rotation he discovered in 1927. He was also a pioneer of radio astronomy, in particular with respect to the 21 cm radiation (SEE **twenty-one centimetre line**) of interstellar hydrogen, collaborating with Hendrik van de **Hulst** and Bart **Bok** to map the Galaxy at this wavelength. In 1950 he proposed the existence of what has come to be called the **Oort Cloud** of comets.

Oort Cloud (Oort–Öpik Cloud) The region of space surrounding the Solar System in which **comets** are thought to reside. It is spherical, extending from about

▲ **Omega Nebula** Also known as the Swan Nebula, M17 in Sagittarius is part of a huge complex of emission nebulosity. It contains about 800 solar masses of material.

6000 AU from the Sun to halfway to the nearest star, with the greatest concentration of comets in a torus-shaped region about 40,000 AU from the Sun. The size and structure of the Oort Cloud have been worked out from statistical studies of comets' orbits, although there is no direct evidence for its existence. The idea was put forward by Ernst Öpik (1893–1985) in 1932, and first worked out in detail by Jan **Oort** in 1950. SEE ALSO **Kuiper Belt**.

open cluster A group of young stars in the spiral arms of our Galaxy, containing from a few tens of stars to a few thousand. Open clusters were formerly also known as *galactic clusters*. They are usually several light years across. Examples are the **Hyades** and **Pleiades** in Taurus, and **Praesepe** in Cancer. Related to open clusters are *associations*, loose groupings of stars of common origin. Often one of more open clusters are found in the central parts of associations. SEE ALSO **star cluster**, **OB association**, **T association**.

open universe In cosmology, a theoretical "model" universe in which there is not enough matter to halt the expansion triggered in the **Big Bang**. It would thus continue to expand for ever. In an open universe the **curvature of space** is negative, the **deceleration parameter** is less than 0.5, and the **density parameter** is less than 1. COMPARE **closed universe**.

Ophelia One of the small inner **satellites** of Uranus discovered in 1986 during the Voyager 2 mission. It and Cordelia act as **shepherd moons** to the planet's Epsilon Ring.

Ophiuchus A large constellation spanning the celestial equator, commemorating Aesculapius, a great healer. Its brightest star, Alpha (Rasalhague), is magnitude 2.1 and lies near Alpha Herculis. RS Ophiuchi, a **recurrent nova**, flared up to naked-eye brightness in 1898, 1933, 1958, 1967, and 1985. Rho Ophiuchi is a celebrated multiple star, and 70 Ophiuchi is a well-known binary, of magnitudes 4.2 and 6.0, period

88 years. The constellation contains several fairly bright globular clusters: M9, M10, M12, M14, M19, M62, and M107. It also contains **Barnard's Star**.

opposition The alignment of the Earth and Sun with another body in the Solar System so that the body lies exactly opposite the Sun in the sky, or the time when this occurs. The term is used mostly for alignments with the major planets, although it is also applied to asteroids. Only a **superior planet** can come to opposition (SEE the diagram at **conjunction**). Opposition is the most favorable time to observe a body, for it is then closer to the Earth than at any other point in its orbit.

optical double Two stars that appear to be close together, but are not components of a binary system. They appear close together because they happen to be in almost the same line of sight, but are in fact separated by great distances.

optical interferometer An instrument for studying a celestial object by **interferometry** at optical wavelengths. The first was developed by Albert Michelson in the 1920s. It consisted of mirrors at either end of a steel beam placed across the aperture of the 2.5-m (100-inch) telescope at Mount Wilson Observatory, and with it he measured the diameters of a few large stars by examining the interference pattern formed in the eyepiece. Present-day systems working on the same principle use separate telescopes linked either electronically or by laser beams, and **photometers** in place of the human eye. Another type is the *speckle interferometer*, which uses an image enhancer or a CCD to capture a succession of very-short-exposure, high-magnification pictures. These individual pictures, which "freeze" the effects of atmospheric turbulence, are made up of a large number of "speckles." Computer image-processing combines these pictures and yields an image from which the effects of the atmosphere are electronically subtracted. Other types, such as the **Fabry–Pérot interferometer**, exploit

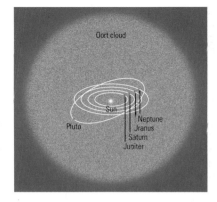

▲ **Oort cloud** The spherical Oort cloud of cometary nuclei surrounds the Solar System. Nuclei from this vast reservoir may be perturbed into orbits that carry them close to the Sun.

interference to capture high-resolution spectra. SEE ALSO **radio interferometer**.

orbit The path of a celestial body in a gravitational field. The path is usually a closed one about the focus of the system to which it belongs, as with those of the planets about the Sun, or the components of a binary system about their common center of mass. Most celestial orbits are elliptical, although the eccentricity can vary greatly. It is rare for an orbit to be parabolic or hyperbolic. To define the size, shape, and orientation of the orbit, seven quantities must be determined by observation. These are known as *orbital elements* (SEE the diagram on page 164). In the case of a planetary orbit, only six are needed. They are as follows:

a	the *semimajor axis*, in AU,
e	the *eccentricity*,
i	the *inclination* of the orbital plane to the plane of the ecliptic,
Ω	the *longitude of the ascending node*,
ω	the *longitude of perihelion*,
T	the *epoch* (the time of perihelion passage).

For the orbit of a binary star system a seventh element is needed, if the mass is not known; this is the *period* (P), or alternatively the *mean motion* (n).

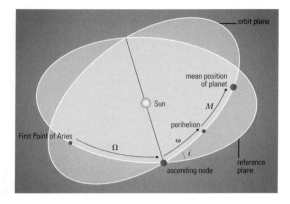

◀ **orbital elements** The orbit of a planet around the Sun can be defined by a number of characteristics, including the inclination (*i*) relative to the reference plane of the ecliptic, the longitude (Ω) of the ascending node measured in degrees from the First Point of Aries, and the argument of perihelion (ω) measured in degrees from the ascending node.

In general, the same considerations apply in determining the orbit of a satellite as in the case of determining the orbit of a planet, except the inclination is usually referred to the equatorial plane of the primary planet, instead of to the plane of the ecliptic. SEE ALSO **ellipse**, **parabola**, **hyperbola**.

orbital element One of the quantities needed to define an **orbit**, or the position of a body in its orbit at a given time.

Orion Perhaps the most splendid of all the constellations, representing a hunter. Its brightest stars are **Betelgeuse** and **Rigel**. Gamma (Bellatrix) is a giant of type B2, magnitude 1.6, 243 light years away. Orion is so distinctive as to be unmistakable. A line of three stars – Zeta (Alnitak), magnitude 1.8; Epsilon (Alnilam), magnitude 1.7; and Delta (Mintaka), magnitude 2.2 – make up Orion's Belt. The celestial equator passes closely north of Mintaka. From the belt hangs Orion's Sword, marked by the **Orion Nebula**. Within the nebula lies Theta, a multiple star known as the **Trapezium**. The **Horsehead Nebula** is a dark nebula near Zeta Orionis.

Orion Arm A spiral arm of the **Galaxy**, also known as the *local arm*. The Sun is located on the inner side of a part of this arm known as the *Orion Spur*.

Orionids A **meteor shower** that occurs in October. Its radiant is near the border of the constellations Orion and Gemini, close to the star Gamma Geminorum. Like the Eta Aquarids, the Orionids are associated with Halley's Comet.

Orion Nebula A gaseous emission nebula, visible to the naked eye as a diffuse glow marking Orion's Sword. It is known as M42, and is separated from a smaller nebula, M43, by a dark nebula called the Fish's Mouth. It lies about 1500 light years away and is 25–30 light years across. It is an **H II region** lit up by a group of young stars called the Trapezium, probably less than one million years old. The Trapezium stars have destroyed most of the volatile dust in the nebula, etching out a roughly spherical hollow that is still growing. The visible Orion Nebula is therefore a hole in a much larger dark nebula. Within the densest portions of the dark cloud, star formation is still under way. The most recent stars to be formed lie concealed within it.

orrery A mechanical model of the Solar System. Orreries varied from simple ones with just the Earth, Moon, and Sun to highly complicated representations of the whole Solar System, where the planets with their satellites not only revolve in their orbits but also rotate on their axes. The name is from Charles Boyle, fourth Earl of Orrery (1676–1731), who commissioned one from John Rowley in 1712, although this was not the first to be made.

orthoscopic eyepiece A standard telescope lens with good definition and **eye relief**. It is a two-element lens, one of which is a cemented triple achromatic (SEE **achromatic lens**).

oscillating universe A variant of the **Big Bang** theory in which it is suggested that the Universe passes through successive cycles of expansion and contraction (or collapse). At the end of the collapse phase, with the Universe packed into a small volume of great density, it is possible that a "bounce" would occur. The Universe would thus oscillate between Big Bang and "Big Crunch" episodes and so be infinite in age. However, for this to happen the density of the Universe would have to be above a certain value (the **critical density**), which is not thought to be the case.

osculating orbit The path that an orbiting body, such as a planet or comet, would follow if it were not subject to any perturbations, but were governed solely by the **inverse-square law** of attraction by the Sun or other central body. The orbital elements, in this case, are called *osculating elements*, and are used in calculating perturbations.

O star A star of spectral type O, characterized by absorption lines of ionized helium in its spectrum. O stars are the hottest and brightest stars on the main sequence, with surface temperatures in the range 30,000 to 50,000 K and luminosities from 30,000 up to 2 million times that of the Sun. They are also the heaviest stars, ranging from about 15 solar masses upward (the upper limit is unknown, and may be more than 120 solar masses). O stars are rare – there are only about 300,000 in the Galaxy at present – because they remain on the main sequence for less than a million years. However, there has been time for 10,000 generations of O stars during the estimated life of the Galaxy. As they return most of their mass to space via stellar winds and supernova explosions, they have been an important source of heavy

elements (SEE **metals**). Together with the hotter B stars, they form **OB associations**.

Most O and B stars form in binary systems, many with components of similar mass, an example being **Plaskett's Star**. When one star explodes as a supernova, it may release its companion from orbit at high velocity. As many as 20% of O stars may be such runaways. If the binary survives the supernova explosion intact, the result will be a neutron star or black hole orbiting through the stellar wind from its companion, producing an X-ray binary such as **Cygnus X-1**. Examples of O stars include Zeta Puppis, Delta Orionis, and Zeta Orionis.

outer planets The planets Jupiter, Saturn, Uranus, Neptune and Pluto, which orbit beyond the main asteroid belt.

Owl Nebula A planetary nebula, M97, in Ursa Major, of 11th magnitude and apparent diameter 3 arc minutes. It takes its name from two dark patches, like eyes, that give it the appearance of an owl's face. It lies 2600 light years away.

▲ **Orion Nebula** A color composite of infrared images taken with the Very Large Telescope (VLT) at Cerro Paranal shows the central region of the Orion Nebula. The four bright, young Trapezium stars are prominent in the middle of this view.

P

PA ABBREVIATION FOR **position angle**.

Pallas Asteroid no. 2, the second to be discovered, by Heinrich **Olbers** in 1802. It is the second-largest, with a diameter of 580 × 470 km (360 × 290 miles).

Palomar Observatory An observatory located on Palomar Mountain, near San Diego, California, site of the 5-m (200-inch) Hale Telescope, a reflector opened in 1948, and the 1.2-m (48-inch) Oschin Telescope, a Schmidt camera. The Schmidt was used for the *Palomar Observatory Sky Survey*, an all-sky photographic survey. Palomar Observatory was founded by George Ellery **Hale** and is now run by the California Institute of Technology (Caltech).

Pan A small **satellite** of Saturn, identified in 1991 by Mark Showalter from images obtained during the Voyager missions. It orbits within the **Encke Division** in Saturn's rings.

Pandora A small inner **satellite** of Saturn discovered in 1980 during the Voyager missions. Pandora and Prometheus are the two **shepherd satellites** of Saturn's F Ring.

parabola An open curve. In mathematics the parabola is a type of conic section, so called because it is one of the intersections of a plane with a cone, and is formed when a plane cuts a cone parallel to its slope. The **eccentricity** of a parabola is 1. In astronomy, it is a type of open **orbit**, perhaps most familiar in the context of cometary orbits.

Half of all comets observed have an orbital eccentricity of approximately 1. Since these comets are observable over only a short arc of their orbits near perihelion, it is not possible to determine their orbital elements sufficiently accurately to distinguish between exactly parabolic orbits and extremely elongated elliptical orbits.

paraboloid The geometrical figure obtained by rotating a **parabola** about its axis. It is the shape most often used for the primary mirror of reflecting telescopes, because a paraboloidal reflector, unlike a spherical one, brings an incoming beam of light to a single focus and is thus free from spherical **aberration** (2).

parallax (symbol π) The angular distance by which a celestial object appears to be displaced with respect to more distant objects, when viewed from opposite ends of a baseline, used as a measure of the object's distance. If the parallax is determined by direct means, using the principle of triangulation and a known baseline, it is known as **trigonometric parallax**. If the parallax is deduced by examining a star's spectrum, it is known as **spectroscopic parallax**.

In trigonometric parallax, the choice of baseline is made according to the remoteness of the object. For fairly close objects, such as members of the Solar System, the Earth's equatorial radius is used; this gives a *geocentric* (or *diurnal*) *parallax*. For more remote objects the radius of the Earth's orbit – the **astronomical unit** (AU) – is used; this gives the *annual* (or *heliocentric*) *parallax*. The parallax of an object is therefore the angular size of the radius of the Earth or of its orbit as it would appear from that object.

The *solar parallax* is the Sun's geocentric parallax, defined as the angular size of the Earth's equatorial radius from a distance of 1 AU. It has been historically very important in establishing the scale of the Solar System.

parhelion (mock Sun, sundog) A round patch of light at the same altitude as the Sun. Parhelia are images of the Sun produced when sunlight is refracted by ice crystals in cirrus cloud in the upper atmosphere. They often occur in pairs, one either side of the Sun.

Parkes Observatory The National Radio Astronomy Observatory of Australia, in New South Wales. Its main instrument is a 64-m (210-ft) dish aerial, completed in 1961, which is now part of the **Australia Telescope**.

parsec (symbol pc) The distance at which a star would have a **parallax** of one second of arc; equivalent to 3.2616 light years, 206,265 astronomical units, or 30.857×10^{12} km.

Parsons, William SEE **Rosse**, Third Earl of.

partial eclipse SEE **lunar eclipse, solar eclipse**.

Pasiphae One of Jupiter's four small outermost **satellites**, discovered in 1908 by P. J. Melotte. It is in a retrograde orbit – SEE **retrograde motion** (1) – and may well be a captured asteroid.

patrol A systematic survey aimed at recording certain transient astronomical phenomena. In a meteor patrol, for example, networks of cameras are used to photograph the sky to maximize the chance of recording meteors. In a supernova patrol, images of selected galaxies are obtained on as many nights as possible in the hope of discovering a supernova.

Pavo A southern constellation, representing a peacock; its brightest star, Alpha, magnitude 1.9, is called Peacock. Kappa is a **Cepheid variable** with a magnitude range from 3.9 to 4.8 in a period of 9.1 days.

Pegasus A conspicuous constellation of the northern sky, representing the flying horse of Greek mythology. Its most famous feature is the *Square of Pegasus*, made up of four stars, although one of these stars is actually part of Andromeda. Beta (Scheat) is a red giant irregular variable with a small magnitude range (2.3 to 2.7). Pegasus is not a rich constellation but it does contain a major globular cluster, M15, of magnitude 6.

Pelican Nebula SEE **North America Nebula**.

penumbra (1) The outer part of the shadow cast by a celestial body illuminated by an extended source such as the Sun. An observer in the penumbral region of a shadow sees only part of the source obscured, as, for example, in a partial **solar eclipse**. SEE ALSO **lunar eclipse**. COMPARE **umbra** (1).

penumbra (2) The lighter zone surrounding the dark central part of a sunspot. COMPARE **umbra** (2).

penumbral eclipse SEE **lunar eclipse**.

Penzias, Arno Allan (1933–) German–American physicist. In 1964 he was working with Robert **Wilson**, using a small radio telescope to measure noise that might interfere with satellite communications. They detected a strange signal that, after consultation with Robert Dicke and James Peebles, they realized was the **cosmic microwave background** – radiation left over from the Big Bang, as had been predicted by George **Gamow** and others in the 1940s. Penzias and Wilson shared the 1978 Nobel Prize for Physics for what was the most important discovery in modern cosmology.

▲ **penumbra** (1) Mir cosmonauts took this photograph of the shadow cast by the Moon on to the Earth during the solar eclipse of August 11, 1999. The dark central umbra, under which the eclipse was total, is surrounded by a lighter outer penumbra.

peri- A prefix referring to the point in an object's orbit at which it comes closest to its primary, as in *perigee* and *perihelion*.

periastron The point of nearest approach of the components of a binary star in their orbit.

perigee The point of nearest approach to Earth by the Moon or an artificial satellite.

perihelion The point in the orbit of a planet or comet at which it is nearest to the Sun.

periodic comet A common term for a short-period **comet** – one whose period is less than 200 years.

period–luminosity law A relationship between the period of light variation and the mean absolute magnitude (luminosity) of a **Cepheid variable** star: the absolute magnitude increases as the period increases. Once the period of a Cepheid is known, its absolute magnitude can be deduced from the period–luminosity law. Furthermore, the absolute magnitude taken in conjunction with the apparent magnitude then enables the distance of the Cepheid to be calculated. It was discovered by Henrietta **Leavitt** in 1912.

Perseids A **meteor shower** that occurs at the end of July and the beginning of August. Its radiant starts in the constellation Cassiopeia, and moves through the north of the constellation Perseus. The shower is associated with Comet **Swift–Tuttle**. The Perseids are one of the strongest, best known, and most long lived of all the meteor showers. In the late Middle Ages they were known as the *tears of St Lawrence*, who was martyred in AD 258, on August 10, close to the shower's maximum.

Perseus A prominent northern constellation, representing the Greek hero who rescued Andromeda. Alpha Persei (Algenib or Mirfak) is magnitude 1.8. Beta (**Algol**) is the prototype eclipsing binary and the brightest in the sky; the second-brightest is Gamma, whose first eclipse was observed only in 1990 (components are magnitude 3.2 and 4.4, period 15 years). Rho is a red giant semiregular variable with a magnitude range from 3.3 to 4.0 and a very rough period of 50 days. Perseus is a rich constellation, crossed by the Milky Way; there are several open clusters, including the famous **Double Cluster** and M34.

personal equation A correction factor applied to visual observations made by a particular person. An individual observer may record timings of, for example, transits that are consistently early or late. Such systematic errors cannot be measured

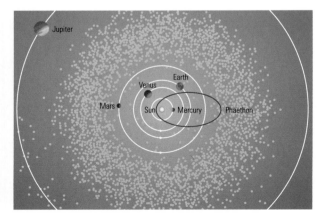

◀ **Phaethon** Phaethon's orbit (red) carries it from the inner part of the main asteroid belt at aphelion, to a perihelion closer to the Sun than Mercury. As an Apollo asteroid, Phaethon has an orbit that cuts across that of Earth.

▶ **phase** The changing appearance of the Moon's illuminated disk results from its orbital motion around the Earth. At new (right), the Moon lies in line with the Sun, and the hemisphere presented to Earth is completely dark. At full, the Moon lies opposite the Sun in Earth's sky and is completely illuminated. Between these two extremes, varying degrees of illumination are seen as shown.

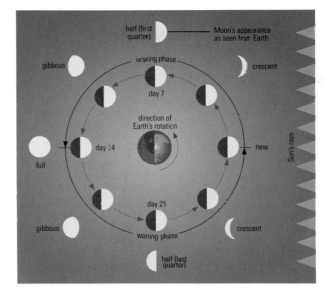

directly, but in any program of work carried out by several observers a personal equation can be found by carrying out a statistical analysis of all the results. It can then be applied to improve the accuracy of future observations.

perturbation An irregularity in the orbital motion of a celestial body, brought about by the gravitational attraction of other bodies. Attractions by planets and satellites cause significant displacements of bodies from where they would be in a purely elliptical orbit.

Periodic perturbations are small oscillations with periods similar to the orbital period of the perturbed body. *Secular perturbations* are slow continuous changes in orbital motion, for example the slow rotation of the orbit of a planet (SEE **perihelion**), and in the orbit's eccentricity and inclination, that arise from commensurabilities (SEE **commensurable**). It is such perturbations that produce **precession**. Perturbations are responsible for dramatic changes in the orbits of **comets**, converting long-period orbits into short-period ones. SEE ALSO **non-gravitational force**.

Phaethon Asteroid no. 3200, discovered in 1983 by Simon Green and John Davies, using the **Infrared Astronomical Satellite**. Phaethon is a 5-km (2.5-mile) diameter **Apollo asteroid** with a perihelion within the orbit of Mercury. Its orbit coincides with the **Geminid** meteor shower, so it seems likely that it is an extinct comet.

phase The proportion of the illuminated hemisphere of a body in the Solar System (in particular the Moon or an inferior planet) as seen from Earth. The phase of a body changes as it, the Sun, and the Earth change their relative positions. All the phases of the Moon (SEE the diagram) – new, crescent, half, gibbous, and full – are observable with the naked eye. The inferior planets, Mercury and Venus, show phases from a slender crescent to a fully illuminated disk when observed through a telescope. Of the superior planets, Mars can show quite a marked gibbous phase, and Jupiter a very slight gibbous phase, but with Saturn and the planets beyond no phase is ever discernible. Phase is sometimes expressed as the percentage of the visible disk that is illuminated.

▲ **Phobos** Mars's larger moon, Phobos, is seen in this image obtained by the Mars Global Surveyor spacecraft in 1998. Believed to be a captured asteroid, Phobos is heavily cratered and shows linear structures resulting from impact shockwaves.

Phobos The larger of the two **satellites** of Mars, discovered in 1877 by Asaph **Hall**. It is a dark, irregular body, measuring 26 × 21 × 18 km (16 × 13 × 11 miles), and may well be a captured asteroid. It has two large craters, named Stickney and Hall.

Phoebe The outermost **satellite** of Saturn, discovered in 1898 by William Pickering. Phoebe is quite small (diameter 230 × 210 km/145 × 130 miles), moves in a distant, inclined and retrograde orbit, and has a non-**synchronous rotation** period of 9.4 hours, all of which suggests that it is a captured body. Its similarities with **Chiron** indicate a possible common origin.

Phoenix A southern constellation representing a phoenix. Alpha (Ankaa), magnitude 2.4, is the brightest star. Zeta is an eclipsing binary with a range from magnitude 3.9 to 4.4 in a period of 1.67 days.

photoelectric magnitude The **apparent magnitude** of an object measured by a **photoelectric photometer** attached to the telescope. The measurement of brightness is made through a number of colored filters in turn. SEE ALSO **UBV system**.

photoelectric photometer An instrument that measures the brightness of a celestial object by generating an electrical signal that is directly proportional to the intensity of the light from that object falling on it. There are various types of detector. A *photomultiplier tube* is a device in which electrons that are given off when light falls upon a special screen liberate more electrons, so producing a measurable electric current. A **CCD** can be used instead (strictly speaking, it is then not a photoelectric device, but it performs the same function). The output from the detector is fed into a computer. The brightness thus measured is an object's **photoelectric magnitude**. An accuracy of 0.01 magnitude is possible, even with fairly basic equipment, which is far better than can be achieved by photographic methods. The photoelectric photometer is therefore used for accurate tracking of the variations of variable stars, and also for measuring the brightness variations of asteroids, and thus their rotation periods. The light falling on the detector can be confined to a particular part of the visible spectrum, or infrared or ultraviolet wavelengths can be selected.

photographic magnitude (symbol m_{pg}) The **apparent magnitude** of an object as measured on a traditional photographic emulsion which is more sensitive to blue light than the human eye is. Hence the photographic magnitude differs significantly from the visual magnitude.

photographic zenith tube (PZT) A fixed telescope directed at the local zenith and used to photograph stars as they pass across its field of view. The photographs are used to determine the zenith distance and time of transit of **fundamental stars**. These measurements yield highly accurate values of the stars' positions, from which can be calculated the *clock error* – a very small nonuniformity in the Earth's rotation. The photographic zenith tube has a much greater degree of accuracy than was possible with traditional **transit instruments**.

photometer SEE **photoelectric photometer**.

photosphere ("sphere of light") The visible surface of the Sun. It is a layer of highly luminous gas about 500 km (300 miles) thick and with a temperature of about 6000 K, falling to 4000 K at its upper level, where it meets the **chromosphere**. As this is the region of the Sun from which all its output of visible light is emitted, the photosphere is the source of the Sun's visible spectrum. The lower, hotter gases produce the continuous emission spectrum, while the higher, cooler gases absorb certain wavelengths and so give rise to the dark **Fraunhofer lines**. This absorption contributes to **limb darkening**. Sunspots and other visible features of the Sun are situated in the photosphere.

photovisual magnitude (symbol m_{pv}) The **apparent magnitude** of an object measured photographically using emulsions and filters that mimic the spectral response of the human eye.

Piazzi, Giuseppe (1746–1826) Italian astronomer and monk. He discovered the first asteroid, Ceres, on January 1, 1801, during a lengthy program of taking accurate measurements of star positions (published as a catalog in two parts, in 1803 and 1814). Piazzi proposed the name "planetoid," but "asteroid" has prevailed.

Pic du Midi Observatory An observatory in the French Pyrenees, at an altitude of nearly 2900 m (9400 ft), noted for its solar and planetary observations. Its main instrument is a 2-m (78-inch) reflector.

Pickering, Edward Charles (1846–1919) American astrophysicist, who was director of the Harvard College Observatory. Most of his work was on stellar photometry and the spectral classification of stars. He built a meridian photometer and began a program of accurate stellar magnitude determinations, published in the *Revised Harvard Photometry*. Pickering discovered spectroscopic binary systems, and invented a method of obtaining the spectra of several stars on one photographic plate. The spectroscopic work he instituted culminated in the *Harvard system* of **spectral classification** and the *Henry Draper Catalogue* of stellar spectra. This work was largely carried out by a team of women, originally hired as assistants but who grew to be able astronomers in their own right. They included Annie Jump **Cannon**, Williamina Fleming, and Antonia C. Maury. Edward's younger brother, **William Henry Pickering** (1858–1938), was a lunar and planetary astronomer who instituted the search for **Pluto**. He also discovered Saturn's satellite Phoebe, in 1898.

Pictor An unremarkable constellation in the southern sky introduced by **Lacaille**, representing a painter's easel. Its leading star is Alpha, magnitude 3.3, and the only other star above magnitude 4 is *Beta Pictoris*, magnitude 3.8, which in 1984 was discovered to have a disk of gas and dust around it, believed to be a planetary system in the process of formation.

▲ **Pictor** This Very Large Telescope (VLT) false-color image of the protoplanetary disk around Beta Pictoris, in Pictor, shows that it is not even in composition. The visible kink in the disk is about 20 AU from the star, which corresponds roughly to the distance of Uranus from the Sun.

Pinwheel Galaxy A large, face-on spiral galaxy in Ursa Major, also known as M101 and NGC 5457. It lies about 25 million light years away. The name is also sometimes used for the galaxies M33 in **Triangulum**, and M99 in Coma Berenices, both also face-on spirals.

Pioneer A series of US space probes launched between 1958 and 1978. The first four were failed Moon probes. Pioneers 5 to 9 were put into orbit around the Sun to study the solar wind and solar flare activity. Pioneer 10, launched in March 1972, was the first probe to travel through the asteroid belt; it passed Jupiter in December 1973, sending back photographs. Pioneer 11 followed it, flying past Jupiter in December 1974 and moving on to Saturn, which it reached in September 1979. Both probes are now on their way out of the Solar System; contact was maintained until 1997. The last two in the series were the **Pioneer Venus** probes.

Pioneer Venus Two NASA space probes to the planet Venus, launched in May and August 1978. Pioneer Venus 1 went into orbit around the planet in December 1978, making a radar map of

▲ **Pinwheel Galaxy** This face-on spiral galaxy, M33 in Triangulum, has loosely wound spiral arms surrounding a relatively small nucleus. M33 belongs to the Local Group, which also includes the Milky Way and the Andromeda Galaxy.

the surface and photographing the planet's clouds; it finally burned up in the planet's atmosphere in October 1992. Pioneer Venus 2 dropped five small probes into the atmosphere of Venus in December 1978, which sent back data on the planet's clouds and atmospheric conditions.

Pisces A constellation of the zodiac, representing a pair of fishes. It is very obscure, consisting of a chain of rather faint stars south of Pegasus. Its brightest star, Eta, is magnitude 3.6. Alpha (Alrescha) is a close binary, magnitudes 4.2 and 5.2, with a period of over 900 years.

Piscis Austrinus The constellation of the southern fish. First-magnitude **Fomalhaut** is its only star brighter than magnitude 4.

plage A bright cloud of hot gas in the Sun's **chromosphere**. Plages are intimately connected with **faculae**, which occur below them in the photosphere, and are found in active regions, where the local magnetic field is enhanced.

planet A large non-stellar body in orbit around a star, shining only by reflecting the star's light. In our **Solar System** there are nine **major planets**, as opposed to the thousands of small bodies known as **asteroids** or minor planets. SEE ALSO **extrasolar planet, Kuiper Belt, Planet X**.

planetarium (plural **planetaria**) A structure in which a representation of the stars and planets as visible in the night sky is projected on to the inside of a dome for an audience seated below. They vary in size and sophistication from small portable projectors used in inflatable 3-m (10-ft) domes to highly complex installations in permanent theatres seating hundreds of people. In all planetaria the positions and motions of celestial bodies can be simulated. The first planetarium, built by the Carl Zeiss company, began operation in Munich in 1923. From then until the early 1980s, planetarium projectors consisted of many individual optical projectors mount-

▲ **Pioneer** Pioneer 10, shown here in an artist's impression, was launched on March 2, 1972. It passed Jupiter on December 3, 1973, at a distance of 132,000 km (82,500 miles).

ed on a complicated rotating frame. In the new generation of projectors, a computer-generated image is projected through a single, stationary fish-eye lens.

planetary nebula An emission nebula formed when a red giant or supergiant sheds its outer layers, leaving a hot core. The core's stellar wind compresses the gas, and its ultraviolet radiation ionizes it and makes it shine. Their sizes range from about that of our Solar System to a light year or so across. They were so named because the first ones to be discovered gave

the impression of planetary disks when viewed in small telescopes; the Ring Nebula (M57) in Lyra is a well-known example. However, only 10% of the 1000-plus now known have a circular shape, and about 70% have two lobes.

The central stars of planetary nebulae are hot and blue, with temperatures of 30,000 to over 100,000 K. They form a transitional stage at the end of a star's evolution between the giant stars and the white dwarfs. Ultraviolet light from the central stars make planetary nebulae fluoresce, as in other emission nebulae. In addition to hydrogen and helium, planetary nebulae contain small proportions of oxygen, nitrogen, and neon, which accounts for their various colors as seen on photographs.

Planetary nebulae are thought to result from stellar winds blown out from red giants and supergiants, leaving the red giant's core as the central star (SEE **RV Tauri star**). Over time, this star cools to become a white dwarf. Because red giants are so big the gravity at their surfaces is very low, and storms on the surface can easily throw off material. A star up to about 8 solar masses can lose three-quarters of its mass into space before it turns into a white dwarf of about 1.4 solar masses. The visible nebu-

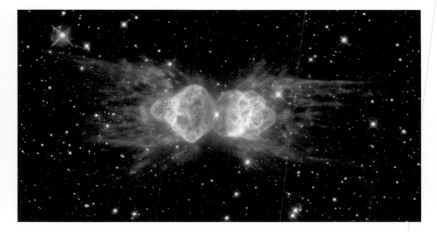

▲ **planetary nebula** This Hubble Space Telescope image shows the Ant Nebula (Mz 3), a planetary nebula in the southern constellation Norma. The twin-lobed structure has been produced by ejection of gas from the aged star at the nebula's center. The Ant Nebula lies about 3000 light years away.

la is simply the central region of a much larger object, the outer parts of which are not lit up by the central star. All planetary nebulae are expanding, at speeds of around 20 km/s (12 mile/s).

The curious shapes of planetary nebulae are probably linked to the magnetic field and rotation of the stars. The magnetic field lines might control the outflow, channeling it in particular directions, and the rotation of the star could add spiral twists. The double-lobed structure of most planetary nebulae must be connected with the star's direction of spin or magnetic axis. Some planetary nebulae, such as the Eskimo and Saturn nebulae, have a bright central nebula surrounded by fainter extensions, as though the central star has ejected successive shells. Since they are expanding as the central stars fade, planetary nebulae are relatively transitory objects with lifetimes of tens of thousands of years. Several new planetary nebulae form each year in our Galaxy.

planetary precession The smaller of the two components of **precession**. It results from the gravitational effects of other planets upon the Earth. It moves the equinoxes eastward by about 0.12 arc seconds per year.

planetesimal In theories of the origin of the Solar System (and other planetary systems), a body, between a few millimeters and a few kilometers in size, that condensed out of a cloud of gas and dust (the solar nebula), away from the center where the Sun was forming. Once they had reached a few kilometers in size, the gravitational attraction of planetesimals would allow them to combine with one another, by a process of **accretion**, to form **protoplanets**. The term was first used for the very small planetary bodies supposed to have condensed from matter torn out of the Sun by a passing star in the "planetesimal theory" proposed by Forest Moulton and Thomas Chamberlin in the early 1900s.

Planet X The name given to a hypothetical tenth major planet orbiting the Sun beyond **Pluto**, the X representing both "ten" and "unknown." It was originally the name given by Percival **Lowell** to the planet whose gravitational **perturbations** he supposed were causing irregularities in the orbits of Uranus and Neptune. Pluto's mass was calculated accurately in 1978 and found to be far too small to account for these perturbations, reviving interest in Planet X. A trans-Plutonian planet was also invoked to explain how comets and objects like **Chiron** become diverted into the region of the major planets. But the perturbations of Uranus and Neptune are now almost certainly explained away by errors in observing their positions, and the inward diversion of cometary bodies is now believed to be caused by *chaos* in the outer Solar System – in other words, orbits way beyond the major planets lack long-term stability. Several long-term searches, including one by Clyde **Tombaugh** that continued for 13 years after he discovered Pluto, have turned up no sizable body. Some of the **Kuiper Belt** objects found from 1992 onward are quite large but nowhere near the size of a major planet. There is no longer any reason to invoke a Planet X, and it is highly unlikely that one exists.

planisphere A simple device for showing the stars on view from a certain latitude at any time and on any date. It consists of two disks held together by a pivot at their centers. On the lower disk is printed a map of the night sky, with the celestial pole at the pivot, and into the upper disk is cut an oval window. The planisphere is set by rotating one disk with respect to the other so that the correct date, marked on the rim of one, is aligned with the correct time, marked on the rim of the other. There is some distortion of constellation shapes, particularly near the rim, but the planisphere is nevertheless a useful and easily portable guide.

Plaskett's Star An exceptionally massive binary star in the constellation Monoceros. In 1922 the Canadian astronomer John Plaskett (1865–1941) showed it to be a spectroscopic binary with a period of

14 days. Each component is a blue super-giant of about 50 solar masses. It is the most massive binary known.

Platonic year The time taken for the celestial pole to describe a circle around the pole of the ecliptic, as a result of **precession**. Its value is about 25,800 years. (After the Greek philosopher Plato, *c.* 427 – *c.* 347 BC.)

Pleiades A young open cluster in the constellation Taurus, popularly called the *Seven Sisters*. Although only six or seven stars are visible to the naked eye, there are in fact over a thousand embedded in a reflection nebula. The brightest member is Alcyone, a B-type giant over 300 times as luminous as the Sun. Another member of the cluster, Pleione, is a slightly variable **shell star** that throws off shells of gas as a result of its fast rotation. The other named Pleiades are Maia, Atlas, Electra, Merope, and Taygeta. The cluster lies 410 light years away.

plerion SEE **supernova remnant**.

Plössl eyepiece A telescope eyepiece with good **eye relief** and a wider field of view (about 40°) than an orthoscopic lens. The most common form has two identical achromatic doublets (SEE **achromatic lens**), but some quite different designs have been called "Plössls." (After Austrian optician Simon Plössl, 1794–1868.)

Plough Popular British name for the group of stars in Ursa Major known as the *Big Dipper*.

Pluto The smallest and outermost planet of the Solar System. It is in a much more inclined and eccentric orbit than are the other major planets, and for a 20-year period around perihelion (as, for example, from 1979 to 1999) it comes within the orbit of Neptune. However, a **resonance** between the two planets' orbits (their periods are in the ratio 3 : 2) means they never get close. Pluto's small size and great distance give it a maximum apparent diameter of 0.1 arc seconds, and it is never brighter than magnitude 14.0. The main data for Pluto are given in the table.

Following the discovery of Neptune (SEE **Adams**, John Couch), it seemed that there were still unexplained irregularities in the motion of Uranus. Independently, William H. Pickering (SEE **Pickering**, Edward) and Percival **Lowell** concluded that an undiscovered planet beyond Neptune was responsible, and made calculations of its size, orbit, and position in the sky. The planet was eventually located by Clyde **Tombaugh**, within 5° of Lowell's predicted position, on photographic plates taken in January and February 1930. The discovery was publicly announced on March 13, 1930.

Pluto has yet to be visited by a space probe. Most of our knowledge of it dates from the discovery in 1978 of its satellite, **Charon**, which allowed, among other things, Pluto's mass and unusual axial tilt to be determined. In 1988 Charon's orbit around Pluto was edge-on to the line of sight from Earth. A series of mutual **occultations** of planet and moon from 1985 until 1990 gave information on their rela-

▶ **Pluto** Pluto (left) and its satellite Charon (right) are sufficiently similar in size to be regarded as a "double planet." They are locked in synchronous rotation about the barycenter between them.

tive brightnesses and sizes, and a very rough brightness map of Pluto's surface. The Hubble Space Telescope has enabled these results to be refined further.

Although there is still some uncertainty in their measurements, the Pluto–Charon pair is known to be unique in many ways. Charon is by far the largest satellite with respect to its primary, with between 8 and 16% of Pluto's mass, and from Pluto's surface its apparent size is 4°. Charon has both a **synchronous orbit** and **synchronous rotation**, so it keeps the same face turned toward Pluto, and hangs motionless in Pluto's sky.

The coupling of Pluto and Charon could well have brought about sufficient tidal heating (SEE **tides**) to cause **differentiation** in both bodies, giving them rocky cores and icy mantles. Pluto seems to have a mottled surface with light and dark regions, and signs of polar caps. The surface is covered with icy deposits consisting of 98% nitrogen, with traces of methane, and also probably water, carbon dioxide, and carbon monoxide. For a 60-year period in each orbit, when Pluto's high orbital eccentricity brings it closer to the Sun, the surface warms to above 50 K, sufficient to release a thin, temporary atmosphere of nitrogen. Interaction with the **solar wind** probably draws some of this atmosphere into a comet-like tail. Pluto therefore loses material at each perihelion passage; some of it may be swept up by Charon.

Pluto is certainly not the massive planet whose supposed perturbations of Uranus and Neptune led to Pluto's discovery. The realization of just how small Pluto is has led some astronomers to question its status as a major planet. Pluto and Charon may be alternatively regarded as the largest members of the **Kuiper Belt** so far detected.

Poincaré, (Jules) Henri (1854–1912) French mathematician. In astronomy he studied celestial mechanics in general and the **three-body problem** in particular. He showed that the **many-body problem** can never be solved exactly, and in the process discovered chaotic orbits – ones that lack any long-term stability. Poincaré established many new directions in celestial mechanics, with many innovative mathematical treatments. He also made important early studies of the Special Theory of Relativity.

Pointers The stars Alpha and Beta Ursae Majoris (Dubhe and Merak), which point toward Polaris – the Pole Star and the brightest star in Ursa Minor.

polar caps Deposits of ice in the polar regions of a planet or satellite. The Earth's permanent polar caps are of water ice, and show only a little seasonal variation. Mars has caps of water ice and carbon dioxide that show pronounced seasonal variation because of the planet's eccentric orbit. The axial inclination of Neptune's moon **Triton** with respect to the Sun produces a migration of its nitrogen-ice cap from one pole to the other over the course of the planet's 165-year orbital period.

polar distance The angular distance of a celestial body from the celestial pole. It is the complement of **declination** (i.e. polar distance plus declination equals 90°).

PLUTO: DATA	
Globe	
Diameter	2320 km
Density	2.0 g/cm^3
Mass (Earth = 1)	0.0021
Volume (Earth = 1)	0.0058
Sidereal period of axial rotation (retrograde)	6.387d
Escape velocity	1.1 km/s
Surface gravity (Earth = 1)	0.03
Albedo	0.9
Inclination of equator to orbit	122° 27′
Surface temperature (average)	45 K
Orbit	
Semimajor axis	39.54 AU = 5914 × 10^6 km
Eccentricity	0.249
Inclination to ecliptic	17° 09′
Sidereal period of revolution	248.54y
Mean orbital velocity	4.74 km/s
Satellites	1

Polaris The star Alpha Ursae Minoris, lying within 1° of the north celestial pole. **Precession** is bringing Polaris closer to the pole, and it will be at its closest around AD 2100. Polaris is a supergiant of spectral type F, magnitude 2.0, distance 431 light years, luminosity nearly 3700 times that of the Sun. It is a **Cepheid variable** whose amplitude steadily decreased during the 20th century, stabilizing at 0.03 magnitude in the mid-1990s. Its period is 3.97 days. It is an optical double, with an 8th-magnitude companion 18 arc seconds away.

pole The points on a sphere that are 90° from the equatorial plane. The *celestial poles*, the points where the Earth's axis of rotation intersects the celestial sphere, are 90° north and south of the celestial equator. Similarly, the *ecliptic poles* are 90° from the plane of the ecliptic, and the *galactic poles* are 90° from the galactic plane.

pole star The nearest naked-eye star to either celestial pole. In the northern hemisphere it is **Polaris**, magnitude 2.0; in the south it is 5th-magnitude Sigma Octantis. Because of **precession** the positions of the celestial poles, and hence of the pole stars, are continually changing. In 3000 BC the north pole star was Thuban, in Draco; by AD 10,000 it will be Deneb, and by AD 14,000 it will be Vega.

Pollux The star Beta Geminorum, magnitude 1.16, distance 34 light years, luminosity 40 times that of the Sun. It is a giant of type K0.

Population I star A relatively young star containing a high abundance of metals, found especially in the spiral arms of galaxies. The Sun is a Population I star. SEE ALSO **stellar populations**.

Population II star An old star containing a lower abundance of metals than a Population I star. Population II stars, which include RR Lyrae stars and subdwarfs, are found in the halo of our Galaxy, especially in globular clusters, and in the nuclei of spiral galaxies. The oldest of them

are sometimes called *halo population stars*. SEE ALSO **stellar populations**.

pore A very small and short-lived **sunspot**.

Portia One of the small inner **satellites** of Uranus discovered in 1986 during the Voyager 2 mission.

position angle (PA) The orientation in the sky of one celestial body with respect to another, measured from 0 to 360° from north via east. For a binary system it is the angle between the direction of the celestial north pole and the direction of the line joining the stars.

post-nova A **nova** after its outburst and return to its original magnitude.

Poynting–Robertson effect A **non-gravitational force** produced by the action of solar radiation on small particles in the Solar System which causes them to spiral inward toward the Sun. When particles in orbit around the Sun absorb energy from the Sun and then re-radiate it, they lose kinetic energy and their orbital radius shrinks slightly. The effect is most marked for particles between approximately a micrometer and a centimeter in size. A similar effect makes particles in planetary ring systems slowly spiral inward. The effect is named for John Poynting (1852–1914), who first described it in 1903, and Howard Robertson (1903–61), who proved its reality in 1937 by derivation from the theory of relativity.

Praesepe A large open cluster in the constellation Cancer, known also as the *Beehive* or *Manger*; its number on Messier's list is M44. It was known to the Greeks as a hazy naked-eye patch, but binoculars and telescopes show about 60 stars. The cluster is believed to be about 700 million years old, and lies 525 light years away.

preceding (abbreviation *p*) Describing the leading edge, feature, or member of an astronomical object or group of objects.

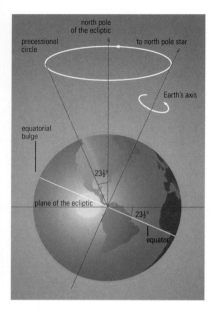

▲ **precession** Over the course of a 25,800-year cycle, the direction in which Earth's axis of rotation points describes a 47° diameter circle relative to the star background. One result of this is that different stars will, over time, become Pole Star.

describes a circle on the sky of radius $23\frac{1}{2}°$ (the inclination of the Earth's axis), centered on the pole of the ecliptic, in about 25,800 years. Because of precession, the positions of all stars (including, of course, the pole star) are continuously changing. Precession also has the effect of moving the **equinoxes** around the ecliptic, and is therefore sometimes called the *precession of the equinoxes*. The First Point of Aries, for example, is now in Pisces. Star catalogs give positions for a given **epoch**, e.g. 2000, so these positions must be corrected for precession to the date of observation.

For example, the *preceding limb* of the Moon is the edge facing toward its direction of motion; and the *preceding spot* (or *p-spot*) of a group of sunspots is the first of the group to be brought into view by the Sun's rotation.

precession The circular motion of the celestial poles around the poles of the ecliptic, and the associated westward movement of the equinoxes with respect to the background stars. The main component of precession is caused by the gravitational pull of the Sun and Moon on the Earth's equatorial bulge. This is called the *lunisolar precession*. A smaller, similar effect caused by the gravitational influence of the other planets is called the *planetary precession*, but it acts in the opposite direction. The total effect, called the *general precession*, amounts to about 50.3 arc seconds per year. The celestial pole therefore

primary (1) The largest of a system of celestial bodies, around which the others orbit. The term is used for the more massive component of a binary or multiple star, and for a planet with respect to its moons.

primary (2) **(primary mirror)** The largest mirror in a **reflecting telescope**, which collects light from the object being observed.

prime focus The point at which the objective lens or primary mirror of a telescope brings light to a focus in the absence of any other optical component, such as a secondary mirror. In large telescopes, special instruments are sometimes mounted at the prime focus. In the **Schmidt camera** the photographic plate is placed at the prime focus.

In *prime focus photography*, a technique which can be used with small telescopes, the lens of a camera is removed and the camera back attached to the telescope via an adapter, in place of the eyepiece; photography does not, despite the name, actually take place at the prime focus.

primeval fireball The explosion which, according to the **Big Bang** theory, gave the present expanding Universe.

primordial black hole SEE **black hole**.

Procyon The star Alpha Canis Minoris, the eighth-brightest star in the sky, magnitude 0.40, distance 11.4 light years, luminosity seven times that of the Sun. It is a main-sequence star of type F5 and has a white dwarf companion of 11th magnitude that orbits it every 41 years.

prograde SEE **direct motion** (1).

prolate spheroid SEE **spheroid**.

Prometheus One of the small inner **satellites** of Saturn discovered in 1980 during the Voyager missions. Prometheus and Pandora are the two **shepherd moons** of Saturn's F Ring.

prominence A cloud of matter extending outward from the Sun's chromosphere, into the corona. Prominences are regions of higher density and lower temperature than the surrounding corona. Ones near the Sun's limb are visible as bright protrusions into the corona during total eclipses or with the aid of an instrument such as a **coronagraph** or a special filter. Away from the limb, they are silhouetted against the bright photosphere as dark **filaments**. *Quiescent prominences*, which extend for tens of thousands of kilometers, can last for weeks or months. *Active prominences* are short-lived, high-speed, flamelike eruptions that

can reach heights of up to 700,000 km (450,000 miles) in just 1 hour. All prominences have spectra showing lines of neutral hydrogen, helium, and ionized calcium. Their forms and behavior are very varied, and their direction of motion is controlled by the Sun's magnetic field.

proper motion (symbol μ) The apparent motion of a star on the celestial sphere, as a result of its movement relative to the Sun. About 300 stars are known to have proper motions greater than 1 arc second per year, but most annual proper motions are smaller than 0.1 arc seconds. **Barnard's Star** has the largest known proper motion, 10.3 arc seconds per year. Proper motion is determined by comparing the star's position on images obtained on widely separated occasions, usually many years or decades apart, perhaps with the aid of a **comparator**. Considerably improved proper motions for over 100,000 stars have been obtained by the position-measuring satellite **Hipparcos**.

proplyd A contraction of the term "protoplanetary disk," a disk of dust and gas around a young star which is thought to represent the initial phase of planet for-

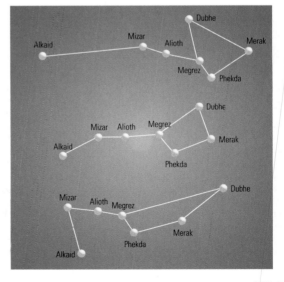

▶ **proper motion** Five of the stars making up the familiar pattern of the Big Dipper are members of a cluster showing a common proper motion relative to the more distant stellar background. The two non-members, Alkaid and Dubhe, have different proper motions, and over the course of hundreds of thousands of years, the pattern will become distorted. The view in the center shows the stars' relative positions today, with an earlier epoch represented at top, and the configuration in the distant future at bottom.

mation. The disks are thought to form around **protostars** as they grow by the gravitational collapse of interstellar matter in molecular clouds. As the disks spread out and become gravitationally unstable, clumps of matter within them form the progenitors of planets. More than a hundred proplyds have been observed in the **Orion Nebula**.

Proteus Neptune's second-largest **satellite**, discovered in 1989 during the Voyager 2 mission. It is a squarish, dark body, heavily cratered, and its most prominent feature is the large crater Pharos, 250 km (150 miles) across.

proton–proton reaction The nuclear reaction thought to be the main source of energy for main-sequence stars with masses equal to or less than the Sun's. In more massive stars the main energy source is the **carbon–nitrogen cycle**. The proton–proton reaction is a three-stage process in which hydrogen is converted into helium with the liberation of an enormous amount of energy. In the first stage two protons (hydrogen nuclei) combine to form a deuterium nucleus, releasing of a positron, a neutrino, and radiation. In the second stage the deuterium nucleus combines with a proton to form an isotope of

▲ **Proteus** Discovered by Voyager 2 during its 1989 fly-by, Proteus is the largest of the inner satellites of Neptune. Proteus is notably dark, with an albedo of only 0.07.

helium (^{3}He), again releasing radiation. In the third stage two ^{3}He nuclei combine to form the normal helium nucleus (^{4}He), releasing two protons and radiation. Overall, four hydrogen nuclei have combined to form one helium nucleus and give out radiation. The reaction requires a temperature of about 10 million K.

protoplanet In theories of the origin of the Solar System (and other planetary systems), a body formed by the accretion of **planetesimals**. Once the protoplanets attained the size of, say, the Moon, they could grow no further by accretion. Instead, collisions between them caused some to grow by accumulating fragments. The major planets of our Solar System are the result of this second stage of growth, while the collisional fragments that were not captured survive as bodies ranging in size from asteroids down to **micrometeoroids**.

protoplanetary disk SEE **proplyd**.

protostar An embryonic star, formed as dense clumps of gas and dust collapse within molecular clouds. Protostars are the earliest, optically invisible stage of star formation. They are very red and can be observed at infrared wavelengths. The Sun's protostellar stage lasted about 0.1 to 1 million years. SEE ALSO **stellar evolution**.

Proxima Centauri The nearest star to the Sun, slightly closer than the nearby star **Alpha Centauri**. It is magnitude 11.0, and its distance is 4.22 light years. Proxima Centauri is an M-type dwarf 20,000 times fainter than the Sun. It is a **flare star**. It is thought to be part of the Alpha Centauri system, though it may be an unrelated star making a close approach.

Ptolemy (Claudius Ptolemaeus) (2nd century AD) Egyptian astronomer and geographer. His chief astronomical work, known by its title in Arabic translation, the *Almagest*, was largely a compendium of contemporary knowledge, including a star catalog, drawing heavily on the work of

◀ **Puck** This distant view of Uranus's inner satellite Puck was taken from Voyager 2 during its fly-by of January 1986. Puck has a diameter of about 154 km (96 miles).

Hipparchus of Nicaea. The *Ptolemaic system* is based on the geocentric world system of the ancient Greeks, with the Earth fixed at the center, and the Moon, Mercury, Venus, the Sun, Mars, Jupiter, and Saturn revolving about it. Beyond these planets lies the sphere of fixed stars. Each of the bodies moves around a small circle called the *epicycle*, the center of which in turn revolves around the Earth on a larger circle called the *deferent*. Ptolemy added to each orbit two more points, the *eccentric* and the *equant*, equally spaced either side of the Earth. He made the epicycle revolve around the eccentric rather than the Earth, and the planet have uniform motion with respect to the equant. The resulting model reproduced the apparent motions of the planets, including retrograde loops, so well that it remained unchallenged until the revival of the heliocentric theory by Nicholas **Copernicus** in the 16th century. Ptolemy's *Geography* remained a definitive text for a similar length of time.

Puck The largest of the small inner **satellites** of Uranus discovered during the Voyager 2 mission. It was first detected at the end of 1985, shortly before the encounter itself in January 1986. It is cratered and approximately spherical.

pulsar An object emitting radio waves in pulses of great regularity. They were first noticed in 1967 by Jocelyn **Bell**, working at the Mullard Radio Astronomy Observatory, Cambridge. The first pulsar to be discovered pulsed every 1.337 seconds.

Pulsars are believed to be rapidly rotating **neutron stars**, the aftermaths of **supernova** explosions. A beam of radio waves emitted by the rotating pulsar sweeps past the Earth and is received in the form of pulses, in the same way that a lighthouse is seen to flash. One theory supposes that the beam is emitted along the magnetic axis of the star. With the Crab Pulsar, in the **Crab Nebula**, two pulses are observed for each rotation, which suggests that the magnetic axis is pointed almost directly at us so that we receive a beam from each pole as the star spins. The Crab Pulsar was the first pulsar to be seen flashing optically, in 1969. In some pulsars, the pulse amplitude and shape can change with time, and the pulsation can fade out temporarily.

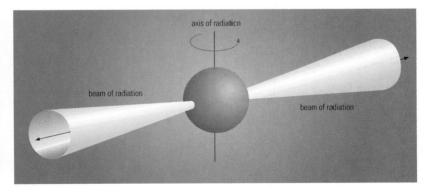

▲ **pulsar** This schematic illustration of a pulsar shows the two beams of radiation directed from the collapsed star along its magnetic poles, which need not coincide with the axis of rotation. If the rotational axis is so aligned that the radiation beams sweep across our line of sight, the rapidly pulsing radio signal characteristic of a pulsar will be detected.

▲ **pulsar** This Chandra X-ray image shows the heart of the Crab Nebula, the site of a pulsar remnant from the supernova explosion of 1054. A thin jet almost 20 light years long emerges from the pulsar's south pole toward lower left.

Pulsars are gradually slowing down, but some, such as the Vela Pulsar, occasionally increase their spin rate abruptly; such an event is called a **glitch**. Over a thousand pulsars have been found, flashing at rates from 1.5 milliseconds (SEE **millisecond pulsar**) to nearly 5 seconds. The most common period is just under 1 second. Pulsars are concentrated toward the disk of our Galaxy, which is where supernovae most commonly occur. Most pulsars are thought to be over a million years old, so that the supernova remnants that once surrounded them would have long since dispersed and faded from view. SEE ALSO **binary pulsar**.

pulsating star A type of variable star whose fluctuating brightness arises from a regular pulsation of the star's size. **Cepheid variables** and **RR Lyrae stars** are examples of pulsating stars.

Puppis A southern constellation representing the stern of the ship Argo Navis. Its brightest star is Zeta (Naos), magnitude 2.2. L2 is a red giant semiregular variable with a range from 2.6 and 6.2, and a period of 140 days or so. Puppis is a rich constellation, with several bright open clusters, including M46, M47, M93, NGC 2451, and NGC 2477.

Purkinje effect The shift in the maximum color sensitivity of the human eye, from yellow toward green, that takes place as the level of illumination falls, during **dark adaptation**. It can affect the observer's perception of star colors. (After Czech physiologist Jan Purkinje, 1787–1869.)

Pyxis A southern constellation adjoining Vela and Puppis, representing a ship's compass, introduced by **Lacaille**. There is no star above magnitude 3.7, but there is one object of particular interest, the **recurrent nova** T Pyxidis.

PZT ABBREVIATION FOR **photographic zenith tube**.

Q

Quadrantids A **meteor shower**, one of the most prolific, which occurs in January. Its radiant lies in the constellation Boötes, near the borders with Hercules and Draco. The shower gets its name from a constellation called Quadrans Muralis which once occupied this region, but is no longer officially recognized.

quadrature The position of the Moon or a planet when it makes an angle of 90° with the Sun (SEE the diagram at **conjunction**). The Moon is at quadrature at first and last quarter.

quasar A compact object which is an extremely luminous source of energy, radiated over a wide range of wavelengths from X-rays through optical to radio wavelengths. The name is a shortened form of *quasi-stellar object* (QSO), which is what they were first called because of their starlike appearance on photographs. Quasars were discovered in 1963 when astronomers identified optical counterparts of certain strong radio sources. However, only a small proportion of quasars are actually radio sources.

Every quasar shows a very large **redshift** which, if it results from the expansion of the Universe, indicates that quasars are the oldest and most distant objects in the Universe. In some cases, galaxies that lie between us and a quasar cause **gravitational lens** effects. Quasars seem to inhabit the centers of very remote galaxies. Since they vary in brightness with periods of a few days, they can be no larger than a few hundred astronomical units wide. A surrounding galaxy has been seen around the nearest quasars, almost lost in the glare from the brilliant center. Quasars are thus extreme examples of **active galactic nuclei**.

Quasar activity is thought to be caused by a **black hole** accreting material at the center of the host galaxy. The most luminous quasars would need to be powered by black holes of 100 million solar masses,

swallowing stars at a rate of about one every year. Quasars are bright ultraviolet sources, and this hot radiation seems to come from a swirling **accretion disk**. Also associated with the inner regions of quasars are emission lines of hydrogen, helium, and iron that indicate motions as rapid as 5000 km/s, as would be expected near a massive black hole.

Quasar-like activity may be possible within every galaxy, but at any given time the black hole is accreting matter in only a small proportion of galaxies. Activity in the gas-free elliptical galaxies may produce a **radio galaxy** or **BL Lacertae object**. In galaxies such as our own, where the black hole is small, even at its most active it is swamped by the light of surrounding stars.

Because quasars are the most luminous objects in the Universe, they can be seen to greater distances, and hence further back in time, than anything else. At redshifts near $z = 4$ we see the Universe as it was about one-tenth of its present age. Therefore quasars give us a way of examining the Universe in its youth.

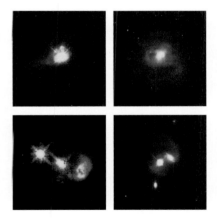

▲ **quasar** This set of quasar images shows the bright nuclear activity that lies at the heart of these galaxies. Each of these quasars is probably fueled by material falling into a massive central black hole.

R

RA ABBREVIATION FOR **right ascension**.

radar astronomy The use of radar techniques to study bodies in space, by transmitting continuous waves or pulses and observing the reflections, as opposed to radio astronomy, which involves only the reception of radio waves emitted naturally by objects in the Universe. Because of the restrictions of signal strength, radar astronomy is limited to objects within the Solar System.

Radar astronomy began with studies of ionized **meteor** trails in the Earth's atmosphere after World War II. These studies revealed the existence of daytime meteor showers, and demonstrated that sporadic meteors are moving in orbit around the Sun and do not originate from interstellar space, as many astronomers had maintained.

Radar echoes from the Moon were first obtained by John H. De Witt of the US Army Signals Corps on January 10, 1946, followed the next month by Zoltan Bay in Hungary. Radar reflections from Venus were received in 1961 at NASA's Goldstone tracking station and Jodrell Bank. The resulting measurements provided an accurate new value of the solar **parallax** and hence of the scale of the Solar System. In 1965 the giant dish at **Arecibo Observatory** was used to establish the true rotation rates of Venus and Mercury.

As the power of radars grew, techniques were developed for radar mapping of planetary surfaces, which proved particularly important for cloud-covered **Venus**. Radar altimeters have subsequently been carried aboard space probes to map Venus in even greater detail. In 1991, radar mapping of Mercury from the Earth revealed what could be metallic deposits or perhaps water ice in craters near the poles. Other objects in the Solar System studied by radar include asteroids, Saturn's rings, planetary satellites and the nuclei of comets.

radial velocity The velocity of a body in the line of sight, either toward (positive) or away from (negative) an observer. The radial velocity is determined from the **Doppler effect** on lines in the object's spectrum. SEE ALSO **redshift**.

radiant The point on the celestial sphere from which a **meteor shower** seems to radiate. Most meteor showers are identified by the constellation in which their radiant is situated.

radiation belts (radiation zones) Doughnut-shaped regions of a planetary **magnetosphere** in which charged parti-

◀ **radar astronomy** This Magellan orbiter radar image shows Maat Mons on Venus. Radar observations have allowed planetary astronomers to penetrate Venus's dense atmosphere and obtain an idea of the planet's surface topography.

▶ **radiant** Meteors from a common source, occurring during a shower such as the Perseids, enter the atmosphere along parallel trajectories, becoming luminous at altitudes of around 80–100 km (50–60 miles). If an observer on the ground plots the apparent positions of meteors on a chart of the background stars, they will appear to diverge from a single area of sky, the radiant. The radiant effect is a result of perspective.

meteor layer

observer

observer's field of view

cles become trapped by the planet's magnetic field. The Earth's radiation belts are the **Van Allen Belts**. All the giant planets have radiation belts; Jupiter's are huge in extent and 10,000 times as intense as those of the Earth.

radiation pressure The pressure exerted by **electromagnetic radiation**. A beam of electromagnetic radiation can also be considered as a beam of particles called photons. When a beam of photons strikes a surface, the photons transfer their momentum to the surface and so exert a pressure. It is radiation pressure that keeps a star's outer layers from collapsing inward under its own gravity. In interplanetary space, small particles are subjected to considerable radiation pressure from the **solar wind** (which to some extent counteracts the **Poynting–Robertson effect**). For example, the tail of a comet always points away from the Sun.

radio astronomy The study of radio emissions, **electromagnetic radiation** with wavelengths from about 1 mm to many meters, reaching the Earth from space. Observations down to about 2 cm can be made at sea level without serious interference from the atmosphere. At shorter wavelengths it is necessary to observe from high mountains to avoid atmospheric absorption, mainly by water vapor. At the long-wavelength end, from 10 to 20 m, radio waves are absorbed and scattered in the ionosphere. Between these extremes, interfer-

ence from terrestrial communications is the main difficulty. International agreements set aside certain wave bands for the sole use of radio astronomy.

Radio noise from the Milky Way was discovered in 1931, by Karl **Jansky**. The first radio telescope purposely built to study such emission was made in the US by Grote Reber, who made a radio map of the sky. In 1942, James Stanley Hey in England found that the Sun was an emitter of radio waves. The subject grew rapidly after World War II as it was realized that significant information about the Universe could be obtained in this region of the spectrum.

The radio emissions first detected from the Milky Way were produced by the process of **synchrotron radiation** – the motion of high-speed electrons in the Galaxy's magnetic field. A number of localized radio sources were also found. Some of these turned out to be **supernova remnants** within our Galaxy, but optical identifications of sources led to the realization that most were far off in the Universe. These sources are now known as **radio galaxies** and **quasars**. The number of radio sources increases with distance, demonstrating that the Universe has been evolving with time. This, combined with the discovery at radio wavelengths of the **cosmic microwave background**, is strong evidence in favor of the **Big Bang** theory of the origin of the Universe. In 1967 a new class of individual radio sources was discovered in our Galaxy: the fast-flashing

pulsars, which are ultra-dense neutron stars formed when massive stars die.

Individual spectral lines can also be detected at radio wavelengths, similar to optical spectral lines. In 1951 the emission from hydrogen at 21 cm (SEE **twenty-one centimeter line**) was detected, and it has proved to be an especially useful means of studying of the structure of the Galaxy. In 1965 the first **interstellar molecule** was discovered at radio wavelengths: hydroxyl (OH). Spectral lines at radio wavelengths have now been observed from over a hundred interstellar molecules. The study of these molecular lines has stimulated the extension of radio observations into the centimeter and millimeter range.

radio galaxy A galaxy that is an abnormally high emitter of radio waves, around a million times stronger than the weak emissions from galaxies such as our own. Radio galaxies are usually giant ellipticals, such as M87 in the constellation Virgo. The emission from a typical radio galaxy is concentrated in two *lobes* well outside the galaxy, often located symmetrically either side, which appear to have been ejected from the galaxy in explosions. The biggest radio galaxies have lobes around 15 million light years across, comparable in size to a typical cluster of galaxies. The extended radio lobes are almost invariably connected to the main galaxy by jets which seem to originate from the galactic nucleus. Strong, point-like radio sources exist in the nuclei of many of these objects.

It is not understood why some elliptical galaxies emit such enormous amounts of radio energy. However, it does seem clear that the source of the activity lies in the galaxy's nucleus, and in this respect radio galaxies are similar to radio **quasars** and **BL Lacertae objects**. A radio galaxy's enormous energy output comes from **synchrotron radiation**. The central "engine" produces vast quantities of energetic electrons traveling near the speed of light. When these electrons encounter a magnetic field, they spiral around and, in so doing, lose energy. It is this energy that we detect as radio waves.

radio interferometer A type of radio telescope that operates on the same principle as an **optical interferometer**, to achieve far higher resolution than a single dish. Single-dish telescopes typically have beamwidths of around one arc minute, which means in effect that they can only resolve angular sizes larger than this and measure positions to a few tens of arc seconds. A radio interferometer consists of two or more antennas which are spaced along a baseline and receive radio waves from the same source. The output signals from pairs of antennae are fed to a common radio receiver. The length of the baseline can be changed by varying the distance between the pairs of antennas. A small dish can be very effective as an interferometer in conjunction with a second dish of large collecting area. In **very long baseline interferometry** the radio telescopes can be on different continents, thereby providing a baseline the size of the Earth and achieving resolutions of a thousandth of an arc second or so.

radio telescope An instrument used to collect and record radio waves from space. The basic design is the large single dish or parabolic reflector, similar to an optical reflector, but much larger – 100 m (330 ft) or more in diameter. The smoothness and accuracy of the reflecting surface determine the shortest wavelength that can be detected, since surface irregularities must be no more than a small fraction of this wavelength. However, for longer wavelengths, such as around 21 cm, it is perfectly adequate to use open wire mesh for the reflector. Radio waves are reflected by the dish via a secondary reflector to a focus, where they are converted into electrical signals. The signals are amplified and sent to the main control room, where there is further amplification before analysis and recording. A single-dish telescope does not produce a radio "photograph" of a source: it only collects radio energy in a narrow beam along its axis. To build up a picture of a radio source, the dish must be scanned to and fro, and the signals processed by computer.

A radio telescope needs to have high *sensitivity* (the ability to detect faint signals), and high *resolution* (the ability to see fine detail). The largest individual dishes, such as the **Effelsberg Radio Telescope**, and those at **Jodrell Bank** and **Parkes Observatory**, are very sensitive. They collect radio energy at a rate of only one hundred-million-millionth of a watt. Resolution depends on wavelength and aperture. Meter wavelengths are a million or so times longer than the wavelengths of light, and resolution is correspondingly poorer. At 1 m wavelength, large radio telescopes have a resolving power of around 1°, about 20 times inferior to that of the naked eye. At a few centimeters their resolution equals that of the eye, which is still much poorer than optical telescopes.

Better resolution can be obtained by linking two or more dishes to form a **radio interferometer**, or by combining many of them to form an **aperture synthesis** array, which gives resolutions of a few hundredths of an arc second, 10 to 100 times better than that of the largest optical tele-scopes on Earth. The sensitivity remains that of the individual dishes added together. An example of such a large array is the **Very Large Array**.

radius vector The imaginary straight line between an orbiting celestial body and its primary, such as the line connecting a planet and the Sun. SEE ALSO **Kepler's laws**.

Ramsden eyepiece A basic telescope eyepiece consisting of two simple elements. It suffers from **chromatic aberration**, so the **Kellner eyepiece** is usually preferred. (After English optician Jesse Ramsden, 1735–1800.)

Ranger The first series of US Moon probes. After several failures, Rangers 7, 8 and 9, in 1964 and 1965, transmitted thousands of photographs of the lunar surface as they approached predetermined impact points. The photographs showed details of boulders and craters as small as 1 m (3 ft), a much higher resolution than with Earth-based telescopes.

▲ **radio telescope** Just like an optical reflector, a radio telescope depends on collection of electromagnetic radiation by a large aperture, parabolic reflector, often in the form of a dish. Radio waves are brought to a focus on the detector. Most radio telescopes can be steered in altitude and azimuth.

rays Bright streaks radiating from a crater. They consist of *ejecta* – material ejected by the impact that produced the crater. They are bright because the shock wave accompanying the impact is sufficient to vaporize rock, which resolidifies in a glassy form. Ray craters are numerous on the Moon. The most prominent is Tycho, whose rays are seen at full Moon to extend across much of the Moon's disk. Ray craters are also found on other bodies in the Solar System, such as Jupiter's moon **Ganymede**.

R Coronae Borealis star

A member of a class of variable stars that are subject to sudden, unpredictable fadings. They normally spend most of their time at maximum, sometimes pulsating slowly by about half a magnitude in periods of five to seven weeks. The fadings can be anywhere between one and nine magnitudes and occur completely at random; the resultant deep minima can last for a few weeks or months, or for more than a year. There is no way of determining in advance how far the star will fade or how long it will remain at minimum. Decline is usually rapid, and the subsequent rise back to maximum is usually very much slower, with many fluctuations.

▲ **rays** This Galileo image of the Moon's northeast quadrant, was taken as the spacecraft sped past Earth en route to Jupiter. A bright ray is seen crossing the oval Mare Serenitatis at left.

The type star, R Coronae Borealis, was first found to be variable in 1795. As shown in the diagram, it is normally at 6th magnitude, where it may remain for as long as ten years before suddenly fading. The minimum may be anywhere between 7th and 15th magnitude and can last for a few weeks or several years.

Only about three dozen stars of this sort are known, although none of the others is as bright at maximum as R Coronae Borealis itself. They are thought to be stars of 1 solar mass which have evolved past the red giant stage and have ejected their hydrogen-rich outer layers to expose a core which is rich in carbon but depleted in hydrogen. Their declines are due to obscuration by clouds of carbon particles they puff off. Each cloud covers only a few per cent of the star's surface, and we see a fading when a cloud is ejected in our line of sight.

recurrent nova A **nova** that is seen to erupt more than once. Recurrent novae are binary stars in which the primary is a late-type (G, K, or M) giant filling its **Roche lobe** and transferring material to a white dwarf secondary. At intervals of years to tens of years this material, which has accumulated on the surface of the white dwarf, explodes in a thermonuclear reaction, causing the brightness of the star to increase by 7 to 10 magnitudes for about 100 days. The longer the time between outbursts, the brighter the outbursts are. Examples are T Coronae Borealis, which underwent major outbursts in 1866 and 1946, and T Pyxidis (1890, 1902, 1922, 1944, and 1966). All novae probably recur, given time.

reddening A phenomenon exhibited by starlight as it is selectively absorbed and scattered on passing through clouds of interstellar dust and the Earth's atmosphere. Blue light is scattered more than red light, so an unscattered beam of starlight that continues on its way is effectively reddened. The longer the path and the greater the density of the clouds of dust, the greater the amount of reddening. A correction factor for interstellar reddening must be intro-

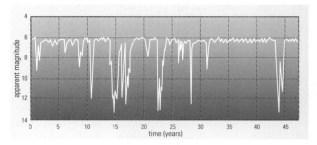

► **R Coronae Borealis star** The light curve of a typical R Coronae Borealis star shows long intervals spent at maximum brightness, punctuated by irregular episodes of abrupt dimming caused by condensation of carbon in the stellar atmosphere.

duced when spectra are used for estimating the distances of remote stars, otherwise they will seem to be more distant than they are. This reddening is not to be confused with **redshift**.

red dwarf A star at the lower end of the main sequence, of spectral type K or M. Red dwarfs have masses of between 0.8 and 0.08 of a solar mass. They are of small diameter, relatively low surface temperature (between 2500 and 5000 K), and low absolute magnitude. Barnard's Star and Proxima Centauri are examples.

red giant A giant star of spectral type K or M, having a surface temperature of less than 4700 K, a diameter 10 to 100 times that of the Sun, and a luminosity 100 to 10,000 times that of the Sun. A red giant is a star in a late stage of evolution, having exhausted the hydrogen fuel in its core. Red giants lie in the upper right hand part of the **Hertzsprung–Russell** diagram. Examples are Arcturus and Aldebaran.

Red Planet Popular name for the planet Mars, which has a distinctly red color.

redshift (symbol z) A lengthening of the wavelength of light or other electromagnetic radiation from a source, caused either by the source moving away (the **Doppler effect**) or by the expansion of the Universe. It is defined as the change in the wavelength of a particular spectral line, divided by the unshifted, or rest, wavelength of that line. For example, a redshift of $z = 0.1$ means that all the wavelengths of a source are lengthened by 10%, for $z = 0.2$ they

are lengthened by 20%, and so on. The speed (v) at which a source is receding can be obtained from its redshift by the formula $v = z \times c$, where c is the speed of light. For speeds greater than about one-third that of light, **Einstein**'s Special Theory of Relativity gives a less simple relation between z and v, since it is not possible to travel at the speed of light. An object with a redshift of $z = 2$, for example, is not receding from us at twice the speed of light. Instead, a redshift of 2 corresponds to a speed of 80% of the velocity of light, and a redshift of 4 corresponds to 92% of the velocity of light. By early 2003, quasars had been detected out to a redshift of $z = 6.4$. The present experimental (and possibly theoretical) limit is around 7.0.

Redshifts caused by the expansion of the Universe, called *cosmological redshifts*, have nothing to do with the Doppler effect. The Doppler effect results from motion through space, whereas cosmological redshifts are caused by the expansion of space itself, which literally stretches the wavelengths of light that is traveling toward us. The longer the travel time of light, the more its wavelength is stretched, as embodied in the **Hubble law**. The most highly redshifted photons are those in the **cosmic microwave background** radiation. These photons have been traveling toward us since matter and radiation decoupled from each other some 300,000 years after the **Big Bang**. They exhibit a redshift of $z = 1000$, meaning that their wavelengths have been stretched a thousandfold. Since the most remote quasars have redshifts of over 5, there is a vast region of space in which no objects have yet been seen.

▲ **reflection nebula** The nebulosity surrounding the Pleiades star cluster is a well-known example of a reflection nebula. A small part of the nebula can be seen in this Hubble Space Telescope image. The star Merope is just out of frame at top right.

Gravitational redshift is a phenomenon predicted by Einstein's General Theory of Relativity. Light emitted by a star has to do work to overcome the star's gravitational field. As a result there is a slight loss of energy and a consequent increase in wavelength, so that all the spectral lines are shifted toward the red. The effect was first observed in 1925 in the spectrum of the white dwarf companion of Sirius.

redshift survey A mapping of the three-dimensional distribution of galaxies in space, with the aim of determining the large-scale structure of the Universe. Current redshift surveys, such as the **Sloan Digital Sky Survey**, employ bundles of optical fibers, each precisely aimed at a different galaxy, to enable the spectra – and hence the redshifts – of several hundred galaxies to be obtained at one pointing of the telescope.

reflecting telescope (reflector) A telescope that forms an image by the reflection of light from a primary concave mirror (usually parabolic) at the back of the tube, via a secondary mirror, to a focus whose position depends on the design of the instrument. The chief optical designs for smaller, amateur reflectors are the **Newtonian** and **Schmidt–Cassegrain**. Larger, professional instruments include the **Cassegrain**, **coudé** and **Schmidt camera** configurations. Many large professional reflectors incorporate two or more different optical configurations, and with some it is possible to site instruments at the **prime focus**. Reflectors have the advantage of being free from **chromatic aberration**.

A reflector's mirrors are usually made of low-expansion glass or a ceramic material and coated with a thin layer of aluminum, which is highly reflective. Reflectors can be made larger than refractors because the primary mirror can be supported at the back, whereas a lens can be supported only around its rim, and is consequently liable to distort under its own weight. Also, new ways of building and controlling mirrors, as in the segmented mirrors of the **Keck Telescopes**, and the technique of **active optics**, have made possible further increases in the size and performance of big reflectors.

The Scottish mathematician James Gregory was the first to describe a reflecting telescope, in 1663; the first to build one was Isaac **Newton**, in 1668. In the late 18th century William **Herschel** demonstrated the instrument's potential by making large, accurately figured mirrors. SEE ALSO **Dobsonian telescope**, **Maksutov telescope**, **Ritchey–Chrétien telescope**.

reflection nebula A nebula in which starlight is reflected off the dust particles mixed in with the gas. Reflection nebulae generally appear blue in photographs; they are always bluer than the star that illuminates them because blue light is scattered more efficiently than red light. Some reflection nebulae, notably the infrared cirrus discovered by the Infrared Astronomical Satellite (SEE **infrared astronomy**), are lit not by an individual star but by the combined light of the whole Galaxy.

refracting telescope (refractor) A telescope that forms an image by the refraction of light by a **lens** at the front of the tube. This lens, called the *objective*, is a compound lens, of perhaps more than one type of high-quality glass in order to minimize **chromatic aberration**, and it may be coated to reduce reflection and so increase the amount of light transmitted to the eyepiece. The tube is long so as to reduce spherical aberration. Refractors are still the first choice of some amateur observers for subjects such as the planets, but the difficulties of making a large, optically perfect lens and supporting its weight mean that all large professional instruments are now **reflecting telescopes**.

The origins of the refracting telescope are uncertain. The Dutch spectacle-maker Hans Lippershey applied for a patent on one in 1608, although others were making similar instruments around the same time, and telescopes of some sort may have been in use at the end of the 15th century. **Galileo** was among the first to use the refractor in astronomy. Improvements were made by a succession of opticians, including John **Dollond** and Joseph von **Fraunhofer**, who devised means of reducing aberration, and Christiaan **Huygens** and Jesse Ramsden (SEE **Ramsden eyepiece**), who devised better eyepieces. The largest refractor, at the Yerkes Observatory, has an aperture of 1 m (40 inches) and was built in 1897.

refraction SEE **atmospheric refraction, refracting telescope**.

Regiomontanus SEE **Müller**, Johann.

regolith The surface dust and debris covering the surface of the Moon and other bodies in the Solar System. It is produced by the disintegration of crustal rocks caused by the impact of meteorites.

regression of the nodes The slow westward motion of the **nodes** of the Moon's orbit, due to the gravitational pull of the Sun. A full circuit takes 18.6 years.

Regulus The star Alpha Leonis, magnitude 1.36, distance 77 light years, luminosity 135 times that of the Sun. It is a main-sequence star of type B7.

relative sunspot number (symbol R) An index of sunspot activity which reduces the sunspot counts of different observers to a common, statistical basis. For a particular day, it is given by

$$R = k \, (f + 10g)$$

where g is the number of groups of sunspots, irrespective of the number of spots each contains, and f is the total number of spots in all the groups; k is a factor based on the estimated efficiency of observer and telescope. The work of recording sunspots in this way was begun by Swiss astronomer Rudolf Wolf at the Zurich Federal Observatory, and the terms *Wolf number* and *Zurich relative sunspot number* were both formerly used.

relativity SEE **Einstein**, Albert.

▲ **refracting telescope** The optical layout of a refractor is shown in this schematic cutaway. Light is collected by the main, objective lens at the front end of the telescope and brought to a focus. The observer views a magnified image of the focused light through a smaller lens, or set of lenses, making up the eyepiece. The telescope is shown on a German equatorial mount.

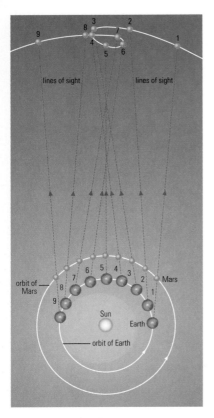

▲ **retrograde motion** (2) The cause of the apparent reversal of superior planets in their motion along the ecliptic close to the time of opposition is explained in the diagram. Here, Earth (inner orbit) is seen catching up Mars (outer orbit). Before opposition, Mars appears to move eastward (top). As Earth, moving more rapidly on its smaller orbit, catches up, Mars's eastward motion apparently slows, then reverses relative to the background stars as Earth overtakes. After opposition, the apparent eastward motion of Mars resumes.

residuals The differences between predicted and observed values, frequently found on analysis of astronomical observations. Residuals in, say, observations of the position of a planet, if collected over a sufficiently long period of time and then analyzed statistically, may reveal some long-term trend. They may also result from inaccurate observations.

resolution (1) SEE **resolving power**.

resolution (2) The size of the smallest detail distinguishable by an imaging system at a particular distance, or of the smallest detail visible in an image (of, for example, a photograph taken by a space probe).

resolving power (resolution) The ability of an optical system, such as a telescope or the human eye, to distinguish close but separate objects such as the components of a close binary star, or a single small object. It is measured in angular units. The resolving power of a telescope in seconds of arc is found by dividing 110 by the aperture in millimeters (or 4.56 by the aperture in inches). This measure is called *Dawes' limit*, after the English amateur astronomer William Rutter Dawes (1799–1868). The resolving power of the human eye is about one minute of arc. SEE ALSO **radio telescope**.

resonance An effect produced by the gravitational interaction between two bodies orbiting the same primary in orbits whose periods are **commensurable**. Resonance between bodies of similar size can lock bodies into commensurable orbits, as has happened with the **Galilean satellites** of Jupiter. Between bodies of highly dissimilar size, however, it can prevent small bodies from occupying commensurable orbits. This is what keeps clear the **Kirkwood gaps** in the asteroid belt and the various gaps in the rings of **Saturn**.

Reticulum A small but distinctive southern constellation, introduced by **Lacaille**, representing a reticle used for measuring star positions. Its leading star is Alpha, magnitude 3.3, which forms a compact pattern with Beta, Gamma, Delta, and Epsilon.

retrograde motion (1) Orbital or rotational motion in the opposite sense to the motion of the Earth: clockwise, as opposed to counterclockwise, as viewed from above the Sun's north pole. Some satellites, such as the outermost ones of Jupiter, have a retrograde orbital motion, and so do certain comets, notably Halley's Comet. Venus, Uranus, and Pluto all have retrograde rotation. Orbital or rotational motion is retrograde if the orbital or axial inclination is more than 90°. COMPARE **direct motion** (1).

retrograde motion (2) The temporary east-to-west motion traced out by a superior planet or asteroid on the celestial sphere as the Earth catches up and overtakes it. This reversal is called a *retrograde loop*. The places at which the planet reverses direction are termed *stationary points* (points 4 and 6 in the diagram). COMPARE **direct motion** (2).

revolution The movement of a planet or other celestial object around its orbit, as distinct from the rotation of the object on its axis.

Rhea Saturn's second-largest **satellite**, diameter 1528 km (950 miles), discovered by Giovanni Cassini in 1672. It is a typical heavily cratered icy world, but with few large craters, the largest being the 250-km (150-mile) diameter Izanagi. Some resurfacing seems to have taken place in the past, but there is little sign of any more recent geological activity, although wispy streaks poorly imaged by Voyager 1 may be signs of tectonic activity.

rich-field telescope (richest-field telescope)

(RFT) A low-power telescope equipped with a wide-angle eyepiece (such as a **Nagler eyepiece**) designed to show a wide field of view at a relatively low magnification. Such an instrument is ideal for studying starfields and hunting for novae.

Rigel The star Beta Orionis, the brightest star in the constellation Orion, and marking Orion's left leg. It is of magnitude 0.18 (the seventh-brightest star in the sky), and a supergiant of spectral type B8. Its distance is 773 light years, and its luminosity 40,000 times that of the Sun.

right ascension (RA, symbol α) A coordinate used to define the position of a celestial object. It is the angular distance measured eastward from the vernal equinox (also known as the First Point of Aries) to the point where the hour circle of an object meets the celestial equator. RA is usually expressed in hours, minutes, and seconds, although it may be expressed as an angle, 1 hour of RA being equivalent to 15°. SEE ALSO **celestial sphere**.

Rigil Kentaurus Another name for **Alpha Centauri**.

rille A cleft on the surface of the Moon. Rilles that are curved or straight are geological faults. The winding *sinuous rilles* are most likely collapsed lava tubes. Lava was able to flow on the Moon for hundreds of kilometers. The top and sides of a flow can solidify, insulating the molten lava inside which continues to flow. When such a tube has been drained, the roof may collapse, leaving what we see today as a sinuous rille. There are similar features on Uranus's satellite **Ariel**.

ring, planetary A narrow ring or thin disk of large or small particles, or both, orbiting in the equatorial plane of a planet. **Jupiter**, **Saturn**, **Uranus**, and **Neptune** all have systems of rings. Saturn's, by far the most prominent and extensive, were

▲ **Rhea** This Voyager 1 image of Saturn's satellite Rhea was obtained from a distance of 128,000 km (79,500 miles) in November 1980. Features as small as 2.5 km (1.5 miles) in diameter are visible on Rhea's heavily cratered surface.

discovered telescopically in the 17th century. Jupiter's were found in 1979 by Voyager 1; those of Uranus and Neptune were first observed during **occultations**, and were confirmed by Voyager 2 in the 1980s.

Several important factors governing ring systems were established in the 19th century. Saturn's rings were found in 1849 to lie inside the **Roche limit**; any sizable satellite inside this limit would be broken up by gravitational forces, and any smaller bodies would be prevented from accreting into a larger one. In 1857 the Scottish physicist

PLANETARY RINGS		
Ring or gap	Distance from centre (thousand km)	Width (km)
Jupiter		
Halo	100–122.8	22800
Main	122.8–129.2	6400
Gossamer	129.2–214.2	85000
Saturn		
D	67–74.5	7500
C (Crepe)	74.5–92	17500
Maxwell Gap	88	270
B	92–117.5	25500
Cassini Division	117.5–122.2	4700
A	122.2–136.8	14600
Encke Gap	133.6	325
Keeler Gap	136.5	35
F	140.2	30–500
G	165.8–173.8	8000
E	180–480	300000
Uranus		
1986U2R	37–39.5	2500
6	41.8	1–3
5	42.2	2–3
4	42.6	2–3
Alpha	44.7	7–12
Beta	45.7	7–12
Eta	47.2	0–2
Gamma	47.6	1–4
Delta	48.3	3–9
1986U1R	50	1–2
Epsilon	51.1	20–100
Neptune		
Galle	41.9–43.6	1700
Le Verrier	53.2	15
Lassell/Arago	53.2–59.1	5900
Adams	62.9	50

James Clerk Maxwell showed that on theoretical grounds the rings had to be made up of many separate particles, and this was proved to be so in 1895 by the American James Keeler, who found that Doppler shifts in the spectra of sunlight reflected from the inner and outer parts of the ring system were different.

The ring systems around the four giant planets show similarities and dissimilarities. For example, Saturn's main rings are made up of bright, icy particles, whereas the other three planets have dark rings. Some rings are very narrow and distinct, others are continuous sheets, but still a kilometer or less thick, and in the equatorial plane, while others are large, tenuous, doughnut-shapes haloes. Small **shepherd moons** keep some rings in check (in Neptune's case, creating clumps of material – ring arcs), while larger satellites create gaps. Particles range from micrometer-sized in Jupiter's main ring to as much as 10 m (30 ft) in Saturn's B Ring. Continual collisions between ring particles make rings spread inward and outward; particles can be moved outward by a magnetic field (as in Saturn's spokes), or inward by the **Poynting–Robertson effect**.

ring galaxy A galaxy having the form of an elliptical ring around the nucleus, like the rim of a wheel. Ring galaxies are thought to form as the result of a collision when a small galaxy passes through a larger one.

Ring Nebula The planetary nebula M57, 4100 light years away in the constellation Lyra, also known as NGC 6720. It appears as a hollow ellipse; in three dimensions it is shaped not like a spherical shell but like a toroid, or ring doughnut. The nebula, centered on a blue star of 0.2 of a solar mass, is expanding at 19 km/s (12 mile/s) and is estimated to be about 5500 years old.

Ritchey–Chrétien telescope A modified form of **Cassegrain telescope**, designed originally by George Ritchey (1864–1945) and Henri Chrétien (1879–1956). The primary and secondary mirrors are hyperboloidal, although there are variants with near-hyperboloidal, paraboloidal, or ellipsoidal components. The system is free from **coma** (1) over a wide field, though there is some astigmatism and the field is curved.

Roche limit (symbol R) The minimum distance at which a planetary satellite can remain in orbit without being torn apart by gravitational forces; alternatively, the minimum distance at which smaller bodies in orbit round a planet can combine by **accretion** to form a larger one. A planet's Roche limit is about 2.5 times its radius (r). It depends, though not strongly, on the densities of planet (d_p) and satellite (d_s):

$$R = 2.44 \ r^3 \ \sqrt{(d_p/d_s)}$$

Actually the Roche limit applies only to "fluid" satellites. Rocky bodies with a high structural integrity, such as Neptune's innermost satellites, can exist inside the Roche limit. All planetary ring systems lie inside their planet's Roche limit. The concept was originated by the French astronomer and mathematician Edouard Roche (1820–83).

Roche lobe A surface defining the maximum size of a star in a binary system. If one star's surface overflows this size as it evolves, the companion's gravity will pull matter off it.

ROE ABBREVIATION FOR **Royal Observatory, Edinburgh**.

Römer, Ole Christensen (1644–1710) Danish astronomer. He was the first to measure the speed of light. Working at Paris in 1675, he found that the predictions of the times of eclipses of the **Galilean satellites** made by Giovanni Domenico **Cassini** did not match the observed times. Intervals between successive eclipses decreased as the orbital motions of the Earth and Jupiter brought them closer together, and increased as they moved further apart. He deduced that the differences were caused by light taking time

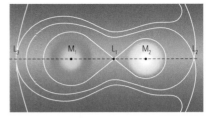

▲ **Roche lobe In** a binary star system, the stars (here marked M_1 and M_2) have their own gravitational sphere of influence, defined by the Roche lobe. If one component fills its Roche lobe, material can escape through the inner Lagrangian point, L_1, on to its companion. If both stars fill their Roche lobes, material can escape entirely from the system through the outer Lagrangian points, L_2.

to travel the intervening distance. His timings gave the speed of light as 200,000 km (120,000 miles) per second, two-thirds of the true value. Römer also invented the first **transit instrument**.

Roque de los Muchachos Observatory A Spanish observatory on La Palma in the Canary Islands, the home of telescopes owned by several European nations, foremost among them the UK's **William Herschel Telescope**. Other major instruments include the 2.5-m (100-inch) Isaac Newton Telescope, moved from England in 1984, and the 2.5-m Nordic Optical Telescope, opened in 1989. The 3.5-m (138-inch) Italian **Galileo National Telescope** opened in 1997, and the 10-m (394-inch) Spanish Gran Telescope Canarias is expected to open in 2004.

Rosalind One of the small inner **satellites** of Uranus discovered in 1986 during the Voyager 2 mission.

Rosat A German satellite containing an X-ray telescope and a British wide-field camera that made the first survey of the sky at extreme ultraviolet wavelengths. It was launched in 1990, and operated until 1998.

Rosetta A European Space Agency probe planned for launch in 2003 but delayed until 2004/5. The original aim was to place Rosetta in orbit around Comet Wirtanen in 2012, but a different objective may be chosen.

Rosette Nebula A diffuse nebula 4900 light years away in the constellation Mono-ceros, surrounding the open cluster NGC 2244. The stars of the cluster were born within the past half million years from the Rosette Nebula and now illuminate it. Radiation pressure from the stars has created a cavity about 12 light years across in the central part of the nebula. This cavity will expand through the nebula, which will thus disperse in a few million years.

Rosse, Third Earl of (William Parsons) (1800–1867) Irish amateur astronomer. In the grounds of Birr Castle, near Parsonstown in Ireland, and aided only by estate laborers whom he trained, he built a 72-inch (1.83-m) reflector. Completed in 1845, it was the largest telescope in the world for the rest of its lifetime. Lord Rosse studied in particular the hazy cloud-like objects then known collectively as nebulae. In 1845 he discovered the spiral structure of what is now called the **Whirlpool Galaxy** and other "spiral nebulae" (recognized as other galaxies in the 1920s). He was able to resolve certain nebulae into groups of stars. He also discovered the ring structure of planetary nebulae. His son **Laurence Parsons** (1840–1908), the fourth earl, continued the work at Birr Castle; another son, **Charles Algernon Parsons** (1854–1931), was an outstanding engineer who continued the telescope-making business of Howard **Grubb**.

rotation The turning of a celestial body on its axis, as distinct from its orbital revolution.

Royal Greenwich Observatory (RGO) The former UK national astronomical observatory, founded at Greenwich, London, in 1675. After World War II the observatory moved to Herstmonceux in Sussex, and in 1990 it moved again, to Cambridge. It was closed in 1998, and its functions were transferred to the Royal Observatory, Edinburgh.

Royal Observatory, Edinburgh (ROE) An observatory in Edinburgh which acts as the UK's national astronomy administrative center, a role it inherited in 1998 following the closure of the **Royal Greenwich Observatory**. The ROE operates British research telescopes at overseas sites, chiefly at the **Roque de los Muchachos Observatory**. Its two main departments are the Institute for Astronomy, run with the University of Edinburgh, and the UK Astronomy Technology Centre, which plans, designs, and manufactures telescopes and instrumentation.

RR Lyrae star A pulsating variable star of a class also known as *short-period Cepheids* or *cluster variables* since the first were found in globular clusters. However, RR Lyrae, the type star, was the first to be found outside a globular cluster. RR Lyrae stars are Population II giants of spectral type between A and F. Their periods range from 0.2 to 1.2 days and their amplitudes from 0.2 to 2.0 magnitudes, although most have periods of between 9 and 17 hours.

◄ RR Lyrae variable
This is the light curve of a typical RR Lyrae star. These stars show a rapid rise to peak light followed by a slower decline. They are commonly found in globular clusters.

► **RV Tauri star** The light curve of a typical RV Tauri star shows alternating deep and shallow minima. The deep minima result from different modes of pulsation in the star's outer layers coming into phase with each other.

Their light curves differ from those of the so-called classical Cepheids; that of RR Lyrae itself is shown in the diagram. Most rise quickly to maximum in no more than a tenth of their total period. Their minima are comparatively long, so that for a few hours their light remains roughly constant. The absolute magnitude of all RR Lyrae stars is about +0.5, so they can be used as "standard candles" for distance finding. However, they are too faint to be seen beyond the nearest galaxies. SEE ALSO **Cepheid variable**.

runaway star A star of spectral type O or B which has an unusually high velocity in space. Such stars are former members of massive binaries, one member of which has exploded as a supernova, flinging off the other like a slingshot. Three runaway stars with a common origin in Orion – 53 Arietis, AE Aurigae, and Mu Columbae – may result from a supernova in what was a quadruple star system three million years ago. It has been estimated that as many as 20% of O stars are runaways.

Russell, Henry Norris (1877–1957) American astronomer. In 1910 he produced a diagram plotting the absolute magnitude of stars against their spectral type, which showed the division between main-sequence stars and giant stars. Unknown to him, Ejnar **Hertzsprung** had done the same some years previously. The

diagram, now called the **Hertzsprung–Russell diagram**, is the starting-point for all modern theories of stellar evolution, although Russell first thought, erroneously, that stars evolved along the main sequence. In 1928 he determined the composition of the Sun's atmosphere from a study of its spectrum.

RV Tauri star A member of a class of pulsating variable stars with alternating primary and secondary minima. RV Tauri stars are luminous supergiants of spectral type between F and M, with periods ranging from 30 to 150 days and amplitudes of up to 4 magnitudes. The diagram shows the light curve of AC Herculis, which has a constant mean magnitude, but with some other stars of this class, such as DF Cygni, the mean varies by up to 2 magnitudes over a period of 600 to 1500 days. RV Tauri stars are undergoing pulsations of more than one period, and appear to be giving off material in a strong **stellar wind** in what may be the early stages of formation of a **planetary nebula**.

Ryle, Martin (1918–84) English radio astronomer. In 1946 he discovered **Cygnus A**, the first radio galaxy to be found. In 1955 he built the first **radio interferometer**, and in 1960 developed the technique of **aperture synthesis**. For his pioneering work in radio astronomy he received a share of the 1974 Nobel Prize for Physics.

S

Sagan, Carl Edward (1934–96) American astronomer. He studied many aspects of the Solar System, including the physics and chemistry of planetary atmospheres and surfaces, particularly of Venus and Mars. Sagan investigated the origins of terrestrial life (discovering in 1963 that ATP, an important biochemical, is formed when chemicals thought to be present on the early Earth are mixed together) and the possibility of life existing elsewhere. He did much to popularize astronomy and science in general.

Sagitta A small constellation representing an arrow, adjoining Aquila. Its brightest star is Gamma, magnitude 3.5. The arrow shape is completed by the 4th-magnitude stars Alpha, Beta, and Delta.

Sagittarius The southernmost of the zodiacal constellations, representing a centaur with a bow, and notable for containing the dense star clouds that mark the direction of the center of our Galaxy. The outline of its main stars is sometimes likened to a teapot. Sagittarius's brightest star is Epsilon (Kaus Australis), magnitude 1.8, followed by Sigma (Nunki), 2.0. There are plenty of variable stars, including the Cepheids X (range 4.3–5.1) and W Sagittarii (range 4.2–4.9). The Milky Way is at its richest in Sagittarius, and there are no fewer than 15 Messier objects – globular clusters, open clusters, and gaseous nebulae, including the **Omega Nebula**, the **Lagoon Nebula**, and the **Trifid Nebula**.

Sagittarius A A strong radio source at the center of our Galaxy. The exact center of the Galaxy is thought to be marked by a small part of this source, referred to as *Sagittarius A**, less than 4 AU across. It is most likely a black hole surrounded by an **accretion disk**. Estimates of the hole's mass range from as high as 5 million solar masses to as low as 100 solar masses.

Sagittarius Dwarf Galaxy The closest companion galaxy of the Milky Way, 50,000 light years beyond our Galaxy's nucleus, first detected in 1994. In 100 million years it could be absorbed into the Galaxy by **cannibalism**.

Saha, Meghnad (1894–1956) Indian astrophysicist and nuclear physicist. His theoretical work on spectra, the solar corona, radiation, and ionization resulted in the 1920s in an equation now known as the *Saha equation*. This relates the degree of ionization of atoms (i.e. how many electrons they have lost) in a star's atmosphere to the temperature. It provides an important key to the differences in the spectra of stars of different spectral types, and allows their temperatures to be deduced from the lines present in their spectra. Saha's later work in nuclear physics led to the founding of an Indian institute for the study of the subject.

Salpeter process SEE **triple-alpha process**.

Sandage, Allan Rex (1926–) American astronomer. Initially an assistant to Edwin **Hubble**, he has continued to work toward establishing the value of the **Hubble constant** and thus fixing the age of the Universe. In 1960 Sandage made the first identification at optical wavelengths of what would prove to be a **quasar**, and five years later he found the first radio-quiet quasar.

saros A cycle of lunar and solar eclipses lasting just over 18 years. A saros consists of the same sequence of eclipses repeated at nearly the same time intervals. This is because 223 **lunations** (6585.32 days) are almost equal to 19 **eclipse years** (6585.78 days), after which time the Sun, the Moon, and the Moon's **nodes** return to very nearly the same positions relative to one another. This period was known to many ancient peoples, including the Babylonians and the

Mayans. Knowledge of it enabled astronomers to predict eclipses, since, if an eclipse of the Sun or the Moon is observed, then after a complete saros there will be a similar eclipse. The difference of 0.46 of a day between the number of lunations and eclipse years in one saros means that each eclipse in the sequence will recur about 165° further west. In predicting a date, allowance must be made for leap days. SEE ALSO **Metonic cycle**.

satellite (moon) A celestial body in orbit around a planet. Of the major planets in the Solar System, Mercury and Venus have no known satellite. The Earth has one satellite, Mars two, Jupiter sixty, Saturn thirty-two, Uranus twenty-one, Neptune eleven, and Pluto one (as of early 2003). There are almost certainly further small satellites of the giant planets awaiting discovery. Some asteroids, such as **Ida**, are known to have a satellite.

The satellites of the Solar System vary enormously in their size, orbit, surface features, and supposed origin. One, Jupiter's Ganymede, is bigger than Mercury, and seven are bigger than Pluto, while the smallest are tiny bodies a few kilometers across. Their orbits vary just as much, from Phobos, which circles Mars's equator more than three times a day, to the outer satellites of Jupiter, which take more than two Earth years to complete their inclined, eccentric orbits. In their surface features they are remarkably diverse, ranging from dead, densely cratered worlds to ones which show a fascinating variety of geological processes. Indeed, Jupiter's Io is the most geologically active body known. As to origin, some, such as Jupiter's **Galilean satellites**, may have condensed out of the same part of the solar nebula as their parent planet, while several small ones are believed to be captured asteroids. The **Moon** is currently thought to have formed out of debris flung out when a Mars-sized body collided with the Earth.

Data for the principal planetary satellites are given in the table on pages 200–201, and each has an entry of its own. SEE ALSO **artificial satellite**.

Saturn The sixth major planet from the Sun, and the second-largest of the four giant planets. Saturn was the remotest planet known in the pretelescopic era. Viewed through a telescope it appears as a flattened golden yellow disk encircled by white rings. The disk has surface markings and belts similar to, but fainter than, those on Jupiter. Because of Saturn's axial inclination, the angle at which the ring system is presented to us depends on the planet's position in its orbit, and this has a considerable effect on its brightness. At perihelic opposition with the rings fully open, Saturn is magnitude −0.3; at aphelic opposition with the rings edge-on, it is magnitude 0.8. The planet's apparent size varies from 15 to 21 arc seconds (the rings vary from 35 to 48 arc seconds). The main data for Saturn are given in the table.

Saturn's magnificent system of rings is its dominant feature. They were first observed by **Galileo** in 1610, but Christiaan Huygens was the first to identify them as rings, in 1656. (For further historical details and for the dimensions of the rings, SEE **ring, planetary**.) As seen from the

SATURN: DATA	
Globe	
Diameter (equatorial)	120,000 km
Diameter (polar)	107,100 km
Density	0.69 g/cm³
Mass (Earth = 1)	95.2
Volume (Earth = 1)	764
Sidereal period of axial rotation (equatorial)	10h 14m
Escape velocity	35.5 km/s
Albedo	0.70
Inclination of equator to orbit	26° 43′
Temperature at cloud-tops	95 K
Surface gravity (Earth = 1)	1.19
Orbit	
Semimajor axis	9.539 AU = 1427 × 10⁶ km
Eccentricity	0.056
Inclination to ecliptic	2° 29′
Sidereal period of revolution	29.46y
Mean orbital velocity	9.65 km/s
Satellites	32

PRINCIPAL SATELLITES OF THE MAJOR PLANETS				
	Diameter (km)	Distance from center of planet (thousand km)	Orbital period (days)	Mean opposition magnitude
Earth				
Moon	3,476	384	27.32	−12.7
Mars				
Phobos	26 × 21 × 18	9.4	0.32	11.3
Deimos	15 × 12 × 10	23.5	1.26	12.4
Jupiter				
Metis	40	128	0.29	17.5
Adrastea	25 × 20 × 15	129	0.30	19.1
Amalthea	270 × 165 × 150	181	0.50	14.1
Thebe	110 × 90	222	0.67	15.7
Io	3,630	422	1.77	5.0
Europa	3,138	671	3.55	5.3
Ganymede	5,262	1,070	7.15	4.6
Callisto	4,800	1,883	16.69	5.6
Leda	16	11,094	238.7	20.2
Himalia	186	11,480	250.6	14.8
Lysithea	36	11,720	259.2	18.4
Elara	76	11,737	259.6	16.8
Ananke	30	21,200	631 R	18.9
Carme	40	22,600	692 R	18.0
Pasiphae	50	23,500	735 R	17.0
Sinope	36	23,700	758 R	18.3
Saturn				
Pan	20	134	0.58	19
Atlas	36 × 28	138	0.60	18.0
Prometheus	150 × 90 × 70	139	0.61	15.8
Pandora	110 × 85 × 60	142	0.63	16.5
Epimetheus	140 × 120 × 105	151	0.69	15.7
Janus	200 × 180 × 150	151	0.69	14.5
Mimas	398	186	0.94	12.9
Enceladus	498	238	1.37	11.7
Tethys	1,058	295	1.89	10.2
Telesto	30 × 16	295	1.89	18.5
Calypso	30 × 16	295	1.89	18.7
Dione	1,120	377	2.74	10.4
Helene	32	377	2.74	18.4

R indicates retrograde orbit *continued*

Earth, the ring system consists of a few principal components, separated by a number of gaps. These were discovered gradually, as telescopes improved, over a 300-year period from Huygens' time until just before the first fly-by, by Pioneer 11 in 1979. The three main rings, starting from the outside, are Ring A, which is grayish-white, Ring B, which is bright white, and Ring C, which is fainter and blue-gray. Ring D, the innermost, was discovered in 1969; Rings E, F and G, which lie outside Ring A, were discovered by space probes. The most prominent gap is the Cassini Division, between Rings A and B; the Encke Gap in Ring A is also quite prominent. These gaps may be produced by the perturbing effect of Saturn's inner satel-

PRINCIPAL SATELLITES OF THE MAJOR PLANETS (continued)

	Diameter (km)	Distance from center of planet (thousand km)	Orbital period (days)	Mean opposition magnitude
Saturn (cont.)				
Rhea	1,528	527	4.52	9.7
Titan	5,150	1,221	15.95	8.3
Hyperion	370 × 225	1,481	21.28	14.2
Iapetus	1,440	3,561	79.33	10.2–11.9
Phoebe	230 × 210	12,952	550.5 R	16.4
Uranus				
Cordelia	26	50	0.34	24.1
Ophelia	32	54	0.38	23.8
Bianca	44	59	0.43	23.0
Cressida	66	62	0.46	22.2
Desdemona	58	63	0.47	22.5
Juliet	84	64	0.49	21.5
Portia	108	66	0.51	21.0
Rosalind	58	70	0.56	22.5
Belinda	68	75	0.62	22.1
1986 U10	40	77	0.64	23.3
Puck	154	86	0.76	20.2
Miranda	480 × 466	129	1.41	16.3
Ariel	1,158	191	2.52	14.2
Umbriel	1,170	266	4.14	14.8
Titania	1,580	436	8.71	13.7
Oberon	1,520	583	13.46	13.9
Caliban	60	7,169	579 R	22
Sycorax	120	12,214	1,289 R	20
Neptune				
Naiad	58	48	0.29	24.7
Thalassa	80	50	0.31	23.8
Despina	150	53	0.33	22.6
Galatea	160	62	0.43	22.3
Larissa	208 × 178	74	0.55	22.0
Proteus	436 × 416 × 402	118	1.12	20.3
Triton	2,706	355	5.88 R	13.5
Nereid	340	5,513	360.1	18.7
Pluto				
Charon	1,186	20	6.39	16.8
R indicates retrograde orbit				

lites, in the same way that Jupiter's gravitational influence maintains the **Kirkwood gaps** in the asteroid belt. The rings are made up of particles ranging from dust to objects a few meters in size, all in individual orbits around Saturn. The main rings are only a kilometer or so thick, although the very tenuous Rings G and E may be up to 1000 km (600 miles) thick. Ring A's diameter is about 275,000 km (170,000 miles); this is the extent of the ring system as seen from Earth. Following Pioneer 11, the two Voyager probes flew past Saturn, Voyager 1 in 1980 and Voyager 2 in 1981. They revealed the ring system to be made up of thousands of separate ringlets. The gaps were not empty, but simply regions with a lower ring density. Some rings were

◄ **Saturn** In this high-resolution image of Saturn from the Hubble Space Telescope, the belts in the atmosphere are clearly visible. The open angle of the rings means that some of the finer details can be seen.

eccentric, not circular. Ring F has a braided structure, being made up of intertwined ringlets and controlled by **shepherd moons**. *Spokes*, dark radial features in Ring B, apparently mark the trajectories of ring particles being moved outward along magnetic field lines.

Saturn's disk shows features similar to Jupiter's, but rather less dynamic and more muted in color because of a high-altitude haze layer. There are dark and light yellow bands parallel to the equator, termed *belts* and *zones*, as on Jupiter (and with a similar nomenclature – SEE the table at **Jupiter** – except that there are no "north north" or "south south" components). Generally the bands are more straight-edged and less turbulent than Jupiter's; there is no equivalent of the Great Red Spot, although short-lived white spots occasionally appear. The most prominent erupt near the equator, and with a 57-year period: there were major eruptions in 1876, 1933, and 1990. The last of these grew until it was spread around much of the equator, and was referred to as the Great White Spot. There is **differential rotation** on Saturn: features at the equator rotate in about $10\frac{1}{4}$ hours, those near the poles in nearly $10\frac{3}{4}$ hours (the rotation period of the interior). Winds at the equator can reach speeds of 1800 km/h (1100 mile/h).

Saturn has an internal heat source, which probably drives its weather systems (as on Jupiter, aurorae have been observed). Its density, 0.7 g/cm^3, is much lower than for any other planet (and its oblateness is higher than for any other), and it also has much more of its mass concentrated at the center. It is therefore assumed to be composed predominantly of hydrogen, and to have an iron–silicate core about five times the Earth's mass, surrounded by an ice mantle of maybe 25 Earth masses. Near the mantle the hydrogen would be in metallic form (so compressed it behaves as a conducting metal); above this it would exist as liquid molecular hydrogen. The upper atmosphere contains 97% hydrogen and 3% helium, with traces of other gases including methane, ethane, ammonia, and phosphine.

The magnetic field of Saturn is 500 to 1000 times the strength of the Earth's, and unlike the other giant planets' fields the magnetic axis is nearly coincident with the axis of rotation. The magnetic field generates a **magnetosphere** intermediate between those of the Earth and Jupiter in its shape, extent, and the intensity of its radiation belts of trapped particles.

As of early 2003 Saturn has 32 known **satellites**. They include **Titan**, the only satellite in the Solar System to have a significant atmosphere.

Saturn Nebula The planetary nebula NGC 7009 in Aquarius, just 160 light years away. Handle-like protuberances give it the appearance of Saturn. The nebula is small, of 8th magnitude, and its appearance probably indicates an internal structure consisting of a series of shells ejected by the central star.

scattering The deflection of light or other types of electromagnetic wave by particles. Where the particles are very much larger than the wavelength, the scattering consists of a mixture of reflection and **diffraction**, and the amount of scattering depends very little on wavelength. Where the particles are very much smaller than the wavelength, the amount of scattering (d) is inversely proportional to the fourth power of the wavelength (λ): $d \propto 1 / \lambda^4$. Thus blue light is scattered by small particles ten times as much as red light. Scattering of sunlight by atoms and molecules in the atmosphere is what makes the sky blue. It is the main cause of **atmospheric extinction**, making the Sun appear red at sunrise and sunset by preferentially scattering blue light out of the line of sight.

Schiaparelli, Giovanni Virginio

(1835–1910) Italian astronomer. He was an assiduous observer of the inner planets, Mars in particular. He prepared a map of the surface of Mars, introducing a nomenclature for the various features, that remained standard until the planet was mapped by space probes. His use of the term *canali* (SEE **canals**) was the cause of much subsequent controversy. He also prepared maps of Venus and Mercury, believing – mistakenly – that they had **synchronous rotation**. Schiaparelli discovered the connection between **meteor streams** and comets, and also studied double stars.

Schmidt, Maarten (1929–) Dutch–

American astronomer. He is known for his work on **quasars**, in particular for his discovery in 1963 of the immense **redshift** of the lines in the spectrum of the quasar designated 3C 273. This discovery, and his subsequent finding that the number of quasars increases with distance, provided important support for the Big Bang theory. Schmidt also ascertained the spiral structure of the Galaxy and the distribution of mass within it.

Schmidt camera (Schmidt telescope) A wide-field **catadioptric** tele-

scope used mainly for photography, developed in 1930 by the Estonian instrument-maker and astronomer Bernhard Schmidt (1879–1935).

The large telescopes of the time were reflectors whose primary mirrors were paraboloidal in order to overcome spherical **aberration** (2). This meant, however, that their effective field of view was limited, sometimes to as little as a few minutes of arc, because away from the center of the field the image became rapidly affected by **coma** (1), especially in instruments having the small **focal ratio** necessary for good images of extended objects. Schmidt's solution was to use a spherical primary mirror but with a correcting plate, placed at the mirror's center of curvature, figured so as to minimize optical aberrations in the image. The image is formed on a curved focal surface in front of the mirror, necessitating the use of a special plate-holder to curve the photographic plate or film, and is sharp over a diameter of 10° or more.

The Schmidt camera and its numerous variations enabled whole-sky photographic surveys to be carried out, and has enabled significant advances to be made in photographic astronomy. A typical instrument is the UK Schmidt Telescope at the Anglo-Australian Observatory. Its aperture of 1.2 m (48 inches) is the diameter of the correcting plate; the primary mirror measures 1.8 m (72 inches) across. Modifications to the Schmidt camera design

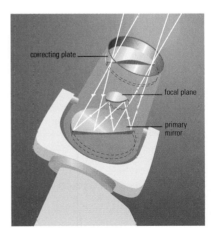

correcting plate

focal plane

primary mirror

▲ **Schmidt camera** This type of reflecting telescope gives a wide field of view. They are used only for astrophotography and are useful in survey work.

that have made popular amateur telescopes have been the **Maksutov telescope** and, more recently, the **Schmidt–Cassegrain telescope**.

Schmidt–Cassegrain telescope

(SCT) A short-focus telescope combining features of the **Schmidt camera** and the **Cassegrain telescope**. The SCT has the Schmidt's spherical primary mirror and specially figured correcting plate, but light is reflected back down the tube by a convex secondary mirror mounted behind the plate and then through a hole in the primary to a Cassegrain focus. This makes for a highly compact and portable telescope that since the 1970s has become very popular with amateur astronomers.

Schmidt telescope SEE **Schmidt camera**.

Schröter effect A difference between the observed phase of Venus near dichotomy and the phase it should theoretically show. At western elongation dichotomy comes a day or two late, while at eastern elongation it is a day or two early. The effect, the cause of which is unknown, was first observed in the 1790s by the German astronomer Johann Schröter (1745–1816).

Schwarzschild, Karl (1873–1916)

German astronomer. He worked on methods of measuring stellar magnitudes from photographs, and studied stellar atmospheres and motions. In the last year of his life he solved equations in **Einstein**'s General Theory of Relativity that dealt with the gravitational field of a point mass, from which he developed the concepts of what are now called the **Schwarzschild radius** and the **black hole**. His son, **Martin Schwarzschild** (1912–), became a naturalized American and an astrophysicist, working on stellar evolution.

Schwarzschild radius The critical radius at which a very massive body under the influence of its own gravitation becomes a **black hole**. It is the radius of

the event horizon of a black hole, from which nothing can escape, not even light. The radius is given by the expression

$$R = 2GM/c^2$$

where G is the gravitational constant, M is the mass of the body, and c is the speed of light. The Schwarzschild radius for the Sun is about 3 km, and for the Earth it is about 1 cm. (After Karl Schwarzschild.)

scintillation The twinkling of the stars. When observed with the naked eye, stars appear to change in brightness and color, particularly if they are low down in the sky; through the telescope, the positions of stars are seen to undergo short rapid changes. The effect is caused by the non-uniform density of the Earth's atmosphere, which produces uneven refraction of starlight. Planets do not exhibit this phenomenon (except that Mercury and Mars when in certain phases may do so slightly when close to the horizon) because the scintillations from different points on the surface are not in phase, and the fluctuations are lost in the general illumination. For astronomical purposes the inconvenience of the effects of scintillation is reduced by siting observatories at high altitudes, and it is almost entirely avoided by using telescopes carried by balloons, rockets or artificial satellites.

It has also been found that radio sources scintillate. In this case it is the non-uniformity of the refractive indices of the interstellar medium, the interplanetary medium, and the ionosphere that causes the strength of radio waves to fluctuate. SEE ALSO **seeing**.

Scorpius A constellation of the zodiac, in mythology representing the scorpion that killed Orion. Its long line of bright stars has the appearance of a scorpion and its sting, with its heart marked by first-magnitude **Antares**, the constellation's brightest star. The "sting" is Lambda (Shaula, magnitude 1.6). Both Mu and Zeta are optical doubles. In addition to being immersed in a very rich part of the Milky

Way, Scorpius contains some magnificent clusters, notably the globular M4 near Antares, and M6 and M7, open clusters both visible with the naked eye.

Scorpius X-1
The first cosmic X-ray source, discovered in 1962, and the brightest known apart from occasional transients (SEE **X-ray burster**). It is thought to be a low-mass X-ray binary with an orbital period of 19.2 hours, one component being a neutron star; the nature of the other is unknown. X-rays are thought to originate from the neutron star and from a thin **accretion disk** around it.

SCT
ABBREVIATION FOR **Schmidt–Cassegrain telescope**.

Sculptor
A barren southern constellation representing a sculptor's studio, introduced by **Lacaille**. It contains no star above magnitude 4.3, but is rich in faint galaxies, including a dwarf member of our Local Group.

Scutum
A small constellation adjoining Aquila, representing a shield. Alpha, magnitude 3.9, is the only star above magnitude 4, but it contains the splendid open cluster M11, nicknamed the **Wild Duck Cluster**, and is crossed by the Milky Way. Delta Scuti is the prototype of a class of variable stars that pulsate in periods of 0.01 to 0.2 days with very small amplitudes, usually only several hundredths of a magnitude.

season
One of a number of periods making up a cycle of changes in the surface conditions of a planet or satellite (average temperature and "weather") as its axial tilt gives different parts of its surface longer or shorter periods of sunlight over the course of one revolution around the Sun. The Earth's seasons are defined by the different positions of the Sun with respect to the ecliptic. In the northern hemisphere, spring is reckoned from the vernal equinox (March 21) to the summer solstice (June 21); summer from the summer solstice to the autumnal equinox (September 23); fall from the autumnal equinox to the winter solstice (December 22); and winter from the winter solstice to the vernal equinox. Spring in the northern hemisphere corresponds to fall in the southern hemisphere, summer to winter, fall to spring, and winter to summer.

Of the other planets, Mars has a similar axial inclination to that of the Earth (25° 11′ compared to 23° 27′), and shows seasonal variations such as changes in its **polar caps**. The polar cap on Neptune's moon Triton migrates from pole to pole over the course of Neptune's 165-year orbital period.

Secchi, (Pietro) Angelo (1818–78)
Italian astronomer and priest. He was a pioneer of spectroscopy, which he used to study the Sun and the stars. In the 1860s he carried out the first spectroscopic survey of stars, cataloguing the spectra of over 4000 and dividing them into four classes

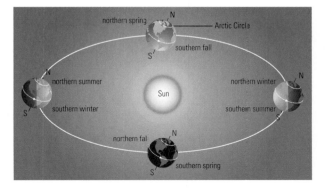

▶ **season** Because the Earth's axis is inclined relative to its plane of orbit, the Earth experiences seasons. The side that is tilted toward the Sun gets more light and has longer days and so experiences summer.

according to their color and spectral characteristics (SEE **spectral classification**).

secondary (1) Any of the smaller members of a system of celestial bodies that orbit around the largest, the **primary** (1).

secondary (2) **(secondary mirror)** The small mirror in a **reflecting telescope** that diverts the converging beam of light from the primary mirror toward the eyepiece.

second contact In a total **solar eclipse**, the moment when the leading edge of the Moon's disk covers the last visible part of the Sun's disk, and totality begins. In a **lunar eclipse**, it is the moment when the trailing edge of the Moon enters the umbra of the Earth's shadow. The term is also used for the corresponding stage in eclipses involving other bodies.

second of arc SEE **arc minute, second**.

secular acceleration The gradual long-term acceleration of the Moon's apparent motion around the Earth. Examination of ancient lunar eclipse records has revealed that the Moon's period is less now than in former times, and is decreasing by about 10 arc seconds per century. The acceleration is caused by **perturbations** by other planets, and tidal friction (SEE **tides**) between the Earth and the Moon which is slowing the Earth's rotation and transferring angular momentum from the Earth to the Moon.

secular parallax The apparent displacement of a celestial body in the sky over time as a result of the Sun's motion in space.

seeing The quality of the image produced by a telescope, as affected by the steadiness of the air. Seeing depends on the calmness of the atmosphere immediately overhead, and also on air movements at ground level, and even in the telescope tube itself. The evaluation of seeing is largely subjective, though values assigned according to the

Antoniadi scale can help achieve a degree of uniformity in the reporting of conditions under which observations are made.

seleno- Prefix referring to the Moon, as in selenography, the description and mapping of the Moon's surface features. (From *Selēnē*, Greek for Moon.)

semimajor axis (symbol *a*) Half the longest diameter of an **ellipse**. The semimajor axis of an elliptical orbit is the average distance of an object from its primary, and is one of the elements used when defining an **orbit**.

semiregular variable A pulsating giant or supergiant star of late spectral type with a period from 20 to 2000 days or more and an amplitude from a few hundredths of a magnitude to several magnitudes. Some semiregular variables show a definite periodicity, interrupted at times by irregularities. Others differ little from long-period variables except in having smaller amplitudes.

separation The angular distance between two members of a visual binary or multiple star system. It is measured in arc seconds.

Serpens A constellation in two parts, Caput (the head) and Cauda (the body), separated by the serpent-bearer Ophiuchus; Serpens represents a snake coiled around Ophiuchus. Alpha (Unukalhai), its brightest star, is magnitude 2.6. Theta (Alya) is a wide, easy pair with components of magnitudes 4.6 and 5.0. Delta is also an easy telescopic double, magnitudes 4.2 and 5.2. The Mira-type variable R Serpentis can reach magnitude 5.2 (minimum 14.4, period 356 days). There are two bright Messier objects: M5, a globular cluster, and M16, an open cluster embedded in the **Eagle Nebula**.

SETI Abbreviation for "search for extraterrestrial intelligence," a term used for a variety of projects seeking signs of intelligent life elsewhere in the Universe. The first was Project Ozma, which in 1960 began to search for radio signals. Current SETI projects scan the microwave region. None has met with success.

Seven Sisters Popular name for the **Pleiades**, an open star cluster in Taurus.

Sextans An obscure constellation between Leo and Hydra representing a sextant. Its brightest star is only magnitude 4.5.

Seyfert galaxy A class of galaxies which have extremely bright, compact **active galactic nuclei** and whose spectra show strong emission lines. About 1% of all galaxies are Seyferts; M77 (NGC 1068) is one of the brightest. Seyferts emit strongly at ultraviolet and infrared wavelengths, and exhibit a degree of short-term variability. Some are strong X-ray sources but few are particularly strong radio emitters. Most are spirals. The gas clouds in Seyfert galaxies move at several thousand kilometers per second, probably due to the gravitational influence of a massive black hole at the nucleus. They are named for American astrophysicist Carl Seyfert (1911–60), who first studied them.

shadow bands Gray ripples seen moving across light-colored surfaces in the minutes either side of a total eclipse of the Sun. Close to totality the bands are more ordered and show higher contrast. They are caused by turbulence in the atmosphere affecting light from the very thin crescent of the Sun, just as it affects starlight to produce **scintillation**.

Shapley, Harlow (1885–1972) American astronomer. In 1914 he explained how the variability of **Cepheid variables** is caused by their pulsations. In the years that followed he calibrated the period–luminosity law discovered by Henrietta **Leavitt**, using Cepheid variables in globular clusters, which enabled him to make the first accurate estimate of the size of the Galaxy and the position in it of the Solar System. He also made estimates of the sizes

▲ **Seyfert galaxy** M77 (NGC 1068) is the prototype type 2 Seyfert galaxy. Its spectrum indicates that there is high-velocity gas in its inner regions. Visually, Seyfert galaxies are characterized by their bright, small nuclei.

of eclipsing binaries, and worked on photometry and spectroscopy.

shell star A hot B-type star with an equatorial disk thrown out by rapid rotation. Strong shell spectra, which include emission lines from the surrounding disk, are seen in B0e to B3e giants and dwarfs. Examples are Gamma Cassiopeiae, BU Tauri (Pleione), and 48 Librae (FX Librae).

shepherd moon A minor moon that through its gravitational influence keeps in check the particles in a planetary ring. Shepherd moons often act in pairs. Examples are Prometheus and Pandora, orbiting either side of Saturn's F Ring, and Cordelia and Ophelia, orbiting either side of Uranus's Epsilon Ring.

Shoemaker–Levy 9, Comet A **comet** discovered in 1993 jointly by Eugene and Carolyn Shoemaker, David Levy, and Philippe Bendjoya. It was in a temporary orbit around Jupiter, and a close approach to the planet in 1992 had disrupted its nucleus into over 20 separate fragments. Over a six-day period in July 1994 these fragments collided with Jupiter, creating fireballs and leaving temporary dark scars at the impact sites. This was a unique event in the history of observational astronomy, and was observed worldwide by amateur and professional astronomers, and by the **Galileo** space probe.

shooting star A popular (chiefly British) name for a **meteor**.

short-period comet A **comet** whose apparitions are sufficiently frequent to permit correlation of orbital data; formally defined as one whose period is less than 200 years.

sidereal Of or pertaining to the stars; measured or determined with reference to the stars.

sidereal day The interval between two successive passages of a given star, or the vernal equinox, across the observer's meridian. It is equal to 23h 56m 4.091s of mean solar time, and divided into 24 *sidereal hours*.

sidereal month One revolution of the Moon around the Earth relative to a fixed star. The sidereal month is equal to 27.32166 mean solar days.

sidereal period The orbital period of a planet or other celestial body with respect

to a background star. It is the true orbital period. COMPARE **synodic period**.

sidereal time Local time reckoned according to the rotation of the Earth with respect to the stars. The time is 0h when the vernal equinox crosses the observer's meridian. The sidereal day is 23h 56m 4s of mean solar time – nearly 4 minutes shorter than the mean solar day. Sidereal time is equal to the **right ascension** of an object on the observer's meridian.

sidereal year The time required for the Earth to complete one revolution around the Sun relative to the fixed stars. It is equal to 365.25636 mean solar days.

siderostat A mirror mounted equatorially and driven so as to counteract the apparent movement of a star or other celestial object and direct its light into a fixed telescope. A more sophisticated version is the **coelostat**.

Siding Spring Observatory An astronomical observatory in the Warrumbungle mountains of New South Wales, founded in 1962 by the Australian National University, which operates several telescopes there including the 2.3-m (90-inch) Advanced Technology Telescope opened

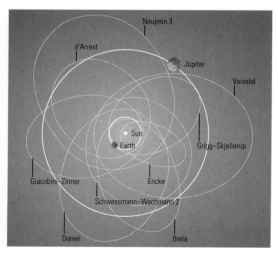

◄ **short-period comet** Most short-period comets are thought to have originated in the Kuiper Belt and to have been perturbed into the inner Solar System by the gravitational influence of the gas giants. Encke has an orbital period of only 3.3 years, Grigg–Skjellerup 5.1 years and Giacobini–Zinner 6.5 years. Biela's Comet suffered the probable fate of all short-period comets: on one orbit, its nucleus was seen to have broken into at least two pieces and it never reappeared.

in 1984. The site is shared by the **Anglo-Australian Observatory**.

singularity A point at which the known laws of physics break down. For example, the singularity at the center of a **black hole** is a point at which matter is compressed into an infinitely small volume and so has an infinitely large density. In the theory of the **Big Bang**, the Universe was born from a singularity of infinite density and temperature.

Sinope One of Jupiter's four small outermost **satellites**, discovered in 1914 by Seth Nicholson. It is in a retrograde orbit and may well be a captured asteroid.

sinuous rille SEE **rille**.

Sirius The star Alpha Canis Majoris. It is the brightest star in the sky, magnitude −1.44, distance 8.6 light years, luminosity 22 times that of the Sun. Sirius is a main-sequence star of spectral type A0, and is the sixth-closest star system to us. It has a binary companion, Sirius B, whose existence was deduced by Friedrich **Bessel** in 1844 from its gravitational perturbations on Sirius, which affected the proper motion of Sirius. The companion, magnitude 8.4, was discovered optically by Alvan G. Clark in 1862. It is a **white dwarf** of 1.03 solar masses, but with a diameter only 0.92 that of the Earth. Its orbital period is 50 years. Sirius B was the first star to be recognized as a white dwarf.

SIRTF ABBREVIATION FOR **Space Infrared Telescope Facility**.

Sixty-one Cygni (61 Cygni) The first star to have its parallax measured (by Friedrich **Bessel** in 1838). It lies 11.4 light years away and is a binary consisting of two K-type main-sequence stars, magnitudes 5.2 and 6.0, with a period of 660 years.

Slipher, Vesto Melvin (1875–1969) American astronomer. On the death of Percival **Lowell** he became director of the Lowell Observatory. He was a specialist in

astronomical spectroscopy, especially of galaxies. In 1912 he became the first to obtain the spectrum of another galaxy, the Andromeda Galaxy, and in 1920 he discovered the **redshifts** in galactic spectra. By 1925, following Edwin **Hubble**'s finding that the redshifts of galaxies increase with distance, he had measured the radial velocities of over 40 galaxies. Slipher also showed that light from certain nebulae was reflected from stars embedded in them, thus discovering reflection nebulae.

Sloan Digital Sky Survey (SDSS) A survey, including a **redshift survey**, begun in 1998, aimed at covering a quarter of the sky. In five years it is hoped to plot the distribution of galaxies in a volume of space over 30 times as large as in any previous survey. The SDSS has a dedicated 2.5-m (100-inch) reflector at **Apache Point Observatory**. It is named for Alfred Pritchard Sloan, Jr. (1875–1966), whose foundation financed the project.

Small Magellanic Cloud SEE **Magellanic Clouds**.

SNC meteorites A group of achondritic meteorites (SEE **achondrites**) believed to have originated on Mars. Their ages (mostly 1.3 billion years or less) indicate that they originated on a sizable planet that has been geologically active, and gases trapped in them are consistent with analyses of the Martian atmosphere made by spacecraft. It is unclear how they could have emerged as intact fragments from an impact of sufficient force to accelerate them to Mars's escape velocity. The name (pronounced "snick") comes from the three subtypes: *shergottites*, *nakhlites*, and *chassignites*.

SNR ABBREVIATION FOR **supernova remnant**.

SOFIA ABBREVIATION FOR **Stratospheric Observatory for Infrared Astronomy**.

SOHO ABBREVIATION FOR **Solar and Heliospheric Observatory**.

Solar and Heliospheric Observatory (SOHO) A European Space Agency probe, operated jointly with NASA, to study the corona of the Sun, oscillations of the Sun's surface, and the solar wind. It was launched in December 1995, and placed at the L_1 **Lagrangian point** 1.5 million km (900,000 miles) from Earth in the direction of the Sun in March 1996. A main aim of the mission is to probe the Sun's interior structure. SOHO discovered many **sungrazer** comets.

solar apex, antapex SEE **apex**.

solar constant A measure of the amount of solar energy received by a body a certain distance from the Sun. For the Earth, the solar constant is defined as the solar power received per unit area, at the top of the atmosphere, at the average Earth–Sun distance of 1 AU; its value is about 1.35 kW/m^2. The "constant" varies from day to day with sunspot activity, and also in the longer term with the **solar cycle**.

solar cycle The periodic fluctuation in the number of **sunspots** and the level of other kinds of solar activity; the cycle lasts about 11 years. Over the course of a cycle, sunspots vary both in number and in latitude in a way which is neatly illustrated in the **butterfly diagram**. The variation in

latitude is described by **Spörer's law**. The diagram here shows the cycle of sunspot activity going back to the 17th century, as reckoned by the **relative sunspot number**. At *solar maximum*, when sunspot numbers are greatest, astronomers refer to the *active Sun*; *solar minimum* is also called the *quiet Sun*. There is some variation in the level of activity at solar maximum, and in the period up to 1715 – during the **Maunder minimum** – activity was low for about 70 years.

solar day SEE **day**.

solar eclipse An **eclipse** of the Sun by the Moon. Since the Moon's orbital plane is inclined to the plane of the ecliptic, a solar eclipse can occur only when the Moon is at conjunction (i.e. at new Moon) and at the same time is at or near one of its **nodes**; the Sun, Moon, and Earth are then very nearly in a straight line. In addition, the Sun's angular distance from one of the Moon's nodes at conjunction will determine whether an eclipse can or cannot occur (SEE **ecliptic limits**).

The apparent diameters of the Sun and the Moon as seen in the sky are almost the same, but they do vary slightly, particularly the Moon's, whose maximum and minimum distances from the Earth differ by about 10%. A *total solar eclipse* is seen at

▲ **solar cycle** The number of sunspots observed peaks roughly every 11 years, although the peak intensity can vary widely, as can be seen in this diagram. During the second half of the 17th century there were very few spots; this period was known as the Maunder minimum.

places where the umbra of the Moon's shadow-cone falls on and moves over the Earth's surface; at the same time the eclipse will appear *partial* to observers on either side of the central track of totality. Shortly before and after totality the phenomena of **shadow bands** and **Baily's beads** are observable. During the brief period of totality, the **corona** and any prominent **prominences** become visible. When the Moon is near apogee, the tip of its shadow-cone does not reach to the Earth and there is an *annular eclipse*, in which a rim of light is seen around the darkened disk of the Moon.

The overall duration of a solar eclipse from **first contact** to **fourth contact** can be as much as 4 hours; totality, from **second contact** to **third contact**, lasts at most $7\frac{1}{2}$ minutes. There are from two to five solar eclipses each year; if there are five, they will all be partial. The total number of solar and lunar eclipses in a year varies from two to seven; if there are only two, they will both be solar.

solar mass The mass of the Sun, used as a unit of mass for other stars and celestial objects.

solar parallax SEE **parallax**.

Solar System The Sun and all the celestial bodies which revolve around it: the nine **planets**, together with their **satellites** and ring systems, the thousands of **asteroids** and **comets**, meteoroids, and other interplanetary material. The boundaries of the Solar System extend beyond the orbit of Pluto, which orbits the Sun at an average distance of 30 AU, to include the **Kuiper Belt**, which may extend to 100 AU, and the inferred **Oort Cloud** of comets, some of which may have aphelion distances of as much as 100,000 AU ($1\frac{1}{2}$ light years).

All the planets have direct orbits, i.e. they revolve around the Sun in the same direction as the Sun itself rotates; the orbital motion of most of the satellites is direct too. The Sun is by far the most massive component of the Solar System, its

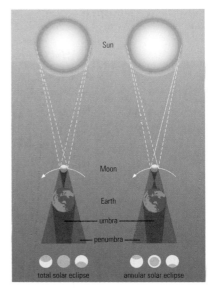

▲ **solar eclipse** In a total solar eclipse (left) the Sun's photosphere is completely obscured by the Moon's disk in the area covered by the cone of shadow. In an annular solar eclipse (right) the Moon is slightly farther away and so the cone of shadow does not reach the Earth.

mass being 330,000 times that of the Earth. The nine planets have a total mass equal to 448 times that of the Earth, of which Jupiter accounts for 70%. Interplanetary space contains cosmic dust and extremely tenuous ionized gas (particles of the **solar wind**).

The Solar System came into being nearly 5000 million years ago, as the end-product of a contracting cloud of interstellar gas and dust, the *solar nebula* (SEE **cosmogony**). Similar systems centered on other stars, referred to as "solar systems," without the capital S's (SEE **extrasolar planet**), and the existence of disks of matter surrounding stars (SEE **proplyd**), suggest that planets are by-products of star formation.

solar wind The steady flow of charged particles (mainly protons and electrons) from the solar corona into interplanetary space, controlled by the Sun's magnetic field. Solar wind particles are too energetic to be held back by the Sun's gravity. Some of them get trapped in planetary magnetic fields. At the Earth, some are trapped in the outer **Van Allen Belt**; others reach the Earth's upper atmosphere in the region of the magnetic poles and cause aurorae. The solar wind carries away about 10^{-13} of the Sun's mass per year. Its intensity increases during periods of solar activity. SEE ALSO **stellar wind**.

solstice The extreme northern or southern position of the Sun in its yearly path among the stars. At the solstices the Sun reaches its greatest declination, $23\frac{1}{2}°$N or $23\frac{1}{2}°$S. The *summer solstice* occurs around June 21 and the *winter solstice* is on or about December 22.

solstitial colure The **great circle** that passes through the north and south celestial poles and the winter and summer **solstices**.

Sombrero Galaxy A spiral galaxy in the constellation Virgo, also known as M104 or NGC 4594. The galaxy is seen nearly edge-on, and has the appearance of a sombrero hat. It is crossed by a thick dark lane of interstellar dust, and there is an unusually large central bulge containing numerous globular clusters. It lies about 30 million light years away.

South African Astronomical Observatory An observatory at Sutherland in the Karoo semidesert northeast of Cape Town, founded in 1972. Its main instrument is a 1.88-m (74-inch) reflector which was originally at the Radcliffe Observatory, Pretoria; there is also a 1-m (40-inch) reflector. When completed in 2004, the 11-m (36-ft) Southern African Large Telescope will be the largest single-mirror instrument in the southern hemisphere.

Southern Cross The popular name for the southern-hemisphere constellation **Crux**.

Space Infrared Telescope Facility (SIRTF) A spaceborne telescope developed by NASA, planned for launch in 2003, to study the sky at infrared wavelengths between 3 and 180 μm. With its 0.85-m (33-inch) telescope SIRTF will take images and spectra. It will be placed into a solar orbit trailing behind the Earth.

space probe An unmanned space vehicle sent to investigate the Moon, planets, or interplanetary space.

spacetime A unified dimensional framework in which the three dimensions of space (length, breadth, and height) and the dimension of time are linked together. An event in spacetime is specified by the three space coordinates and the time coordinate. The concept arose from **Einstein**'s Special and General Theories of Relativity. In the General Theory, gravity is a distortion of spacetime by matter.

space velocity The true velocity of a star in space with respect to the Sun. It is the hypotenuse of the right-angled triangle formed by its **tangential velocity** (obtained from observation of its proper motion) and its **radial velocity** (obtained by measurement of the Doppler shift in its spectrum).

Spacewatch A project to monitor the sky for **near-Earth asteroids** (NEAs), using the 0.9-m (36-inch) and 1.8-m (72-inch) telescopes at **Steward Observatory**. With most NEAs there is no risk of collision, but around a hundred come close enough to be classified as *potentially hazardous asteroids* (PHAs). The Spacewatch project has discovered several of these objects, as well as many NEAs and main-belt asteroids.

speckle interferometer SEE **optical interferometer**.

spectral classification The categorization of stars based on the characteristics of their spectra. In the 1860s Angelo **Secchi** made the first spectral classifica-

SPECTRAL CLASSIFICATION: MAIN SPECTRAL TYPES		
Type	Surface	Main features temperature (K)
O	40,000–25,000	Blue-white stars. Lines of ionized helium. Weak hydrogen lines.
B	25,000–11,000	Blue-white stars. Neutral helium lines, with hydrogen lines strengthening from B6 to B9.
A	11,000–7500	White stars. Very strong hydrogen lines at A0, decreasing toward A9. Lines of ionized calcium increase in strength from A0 to A9.
F	7500–6000	Yellow-white stars. Ionized calcium continuing to increase in strength, and hydrogen weakening. Lines of other elements begin to strengthen.
G	6000–5000	Yellow stars. Strong lines of calcium, hydrogen weaker. Lines of iron become prominent.
K	5000–3500	Orange stars. Strong metallic lines. Molecular bands of CH and CN become prominent.
M	3500–3000	Orange-red stars. Strong absorption bands of titanium oxide and large numbers of metallic lines.

SPECTRAL CLASSIFICATION: LUMINOSITY CLASSES	
Ia	Supergiants of high luminosity
Ib	Supergiants of lower luminosity
II	Bright giants
III	Normal giants
IV	Subgiants
V	Main-sequence stars (dwarfs)
VI	Subdwarfs

tion of stars by dividing them into four groups according to color and spectral lines. As spectroscopy improved, a more comprehensive system was developed at Harvard College Observatory. This was embodied in the **Henry Draper Cata-**

logue of stellar spectra published in 1918–24. The sequence of spectral types was arranged according to the prominence or absence of certain lines in the spectra. There were seven spectral types, designated O, B, A, F, G, K and M (an ordering which results from revising an earlier alphabetical sequence). This sequence is a temperature sequence, from the hottest stars of types O and B, which appear blue-white, to type M, the coolest stars, which appear orange-red. As a result of this ordering, O, B and A stars are called **early-type stars**; K and M stars are called **late-type stars**. Three new types were added when it was found that some cool stars had strong absorption bands not usually seen in other stars of the same color. These were classes R and N with strong bands of molecular carbon, and class S with bands of zirconium oxide.

Stellar spectra can be classified into even finer divisions within these seven types, so decimal subdivisions were introduced. G5, for example, indicates a star midway in type between G0 and K0. Further refinements include the use of additional letters as suffixes to the spectral type, giving more information about the star, for example the existence of emission lines (e), metallic lines (m), broad lines due to rotation (n and nn), or a peculiar spectrum (p).

However, the Harvard system could not deal with stars of different luminosities at a given temperature (that is, dwarfs, giants, and supergiants). In 1943 William Morgan, Philip Keenan and Edith Kellman of Yerkes Observatory redefined the spectral types and added a classification scheme for the luminosity (absolute magnitude) of stars. This is now known as the **Mor-**

gan–Keenan system, shortened to *MK system*, and is used universally. For example, a star classified in the MK system as O9.5 IV–V has a spectral type (and therefore a temperature) midway between that of an O9 and a B0 star, and a luminosity between that of a dwarf and a subgiant. The spectral types and luminosity classes of the MK system are listed in the tables on page 213.

The MK system is applicable to stars of normal chemical composition, which is to say about 95% of all stars. The various types of peculiar star are given their own special classification schemes. The Harvard R and N types are now combined into one **carbon star** class, the designations for which include a temperature type and a carbon band strength, as for example in C2,4. A similar classification is used for the **S stars**. A class of **L stars** has been added for cool stars emitting in the infrared, and brown dwarfs. White dwarfs are usually classified on a Harvard-type scheme, with D preceding the type.

spectral type A series of divisions, indicating surface temperature and hence color, into which stars are classified according to the nature of their spectrum. SEE **spectral classification**; **A star**, **B star**, **carbon star**, **F star**, **G star**, **K star**, **L star**, **M star**, **O star**, **S star**.

spectrogram A photographic record of the spectrum of an object, produced by a **spectrograph**.

spectrograph A **spectroscope**, fitted with a camera or an electronic detector such as a CCD, used to obtain a permanent record of the spectrum of a celestial object. Astronomical spectrographs are generally designed for use with a specific telescope. On telescopes with a **coudé** system or **Nasmyth focus** a large spectrograph can be mounted in a permanent position. Spectrographs generally operate in a band of wavelengths from the near-infrared to the near-ultraviolet. There are various designs for different purposes. *High-dispersion* instruments such as the **spectroheliograph** spread the spectral lines widely so that a narrow band of wavelengths can be studied in detail.

spectroheliograph A type of **spectrograph** used to photograph the Sun in the light of one particular wavelength only. A second slit placed in front of the photographic plate or detector images a narrow strip of the Sun's disk at a chosen wavelength. By moving the entrance slit and the second slit in tandem, the whole disk is scanned and an image of it, a *spectroheliogram*, is built up. The spectroheliograph was invented independently by

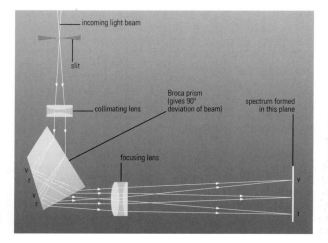

incoming light beam

slit

collimating lens

Broca prism
(gives 90°
deviation of beam)

focusing lens

spectrum formed
in this plane

v
r

v
r

v

r

◀ **spectroscope**
Shown here is one type of traditional spectroscope, using a prism and lenses. The study of spectra, known as spectroscopy, is the key technique by which the physical properties of astronomical bodies are revealed.

▶ spectroscopic binary Redshift is used to ascertain the relative motions of stars in close binary pairs.

George Ellery **Hale** and Henri Deslandres in the 1890s. The term is also used for a number of modern instruments of different design but working on the same principle.

spectrohelioscope A **spectroheliograph** adapted for visual use. The two slits are oscillated rapidly enough for the motion to be undetectable, so that by persistent vision the observer sees a steady, monochromatic image of the Sun's disk.

spectrometer A spectroscope equipped to measure accurately the positions and intensities of spectral lines, by means of a device such as a **Fabry–Pérot interferometer**. Spectrometers are also loosely referred to as **spectrographs**. Since spectra are now routinely recorded on CCDs, from which information in digital form can be analyzed by computer, the distinction between modern spectrometers and spectrographs lies in their application rather than their design.

spectrophotometer An instrument that scans the lines in a spectrum recorded by a spectrograph and measures their intensities. It outputs a graph of intensity against wavelength called a *line profile*.

spectroscope An instrument for producing spectra for the analysis of electromagnetic radiation (light or other wavelengths). *Spectroscopy* is the use of such an instrument to probe the chemical composition and physical conditions (such as

temperature and motion) of an object. From its origins in the pioneering work by William **Huggins**, Angelo **Secchi**, and others in the mid-19th century, spectroscopy developed into the most important investigative technique available to astronomers.

In the arrangement shown in the diagram opposite, a focused beam of light or other radiation is passed through a narrow slit, to prevent images of different wavelengths from overlapping and to ensure a sharply defined spectrum, and then through a lens to collimate it (make it parallel). The parallel beam is then split into its component wavelengths by a prism (often not the simple triangular form), and focused by another lens to produce the spectrum. In modern instruments a **diffraction grating** is usually used instead of a prism, and concave mirrors replace the collimating and focusing lenses.

Astronomical spectroscopes are known as **spectrographs** or **spectrometers**. Strictly, a spectrograph is a spectroscope equipped with a camera for recording a permanent record of a spectrum, whereas a spectrometer incorporates devices for accurately measuring the wavelengths and intensities of the spectral lines.

spectroscopic binary A **binary star** system whose two components are too close for them to be resolved visually as separate objects, but whose binary nature can be deduced from the periodic **Doppler shift** of the absorption lines in the combined spectrum. As the diagram shows, when the

215

two stars have reached positions in their orbits around each other where one is moving away from us and the other toward us, the spectral lines of the one are slightly redshifted, and those of the other slightly blueshifted, from their mean positions.

spectroscopic parallax A method of determining the distance of a star from its spectral type (SEE **spectral classification**) and apparent magnitude. Analysis of a star's spectrum reveals its spectral type. If the absolute magnitude of a star of that type is known, the star's distance can be calculated by comparing its absolute and apparent magnitudes. This is the most common method of determining stellar distances. However, the scale must be calibrated against nearby stars whose distances can be measured directly by **trigonometric parallax**.

spectrum (plural **spectra**) The distribution of intensity of **electromagnetic radiation** with wavelength. When a beam of light from a source is deflected by a prism or by another dispersing agent such as a diffraction grating, the radiation is fanned out into a spectrum because different wavelengths are deviated by different degrees.

Spectra are of various types. A *continuous spectrum* is an unbroken distribution of radiation over a broad range of wavelengths. White light, for example, is split into a continuous "rainbow" band of colors from red to violet. A *line spectrum* contains lines corresponding only to certain wavelengths or frequencies which are characteristic of the chemical elements present in the source. An *emission line spectrum* contains lines emitted at particular wavelengths, and is produced by substances in an incandescent state at low pressure. An *absorption line spectrum* contains dark lines at particular wavelengths superimposed on a continuous spectrum. It is produced when radiation from a hot source which is emitting a continuous spectrum passes through a layer of cooler gaseous material. The spectra of most stars are continuous spectra crossed by absorp-

tion lines, and sometimes by emission lines. Emission line spectra are typical of luminous nebulae. *Band spectra* contain absorption bands, wider than the spectral lines of atoms, and these are the spectral signatures of molecules.

The intensity of an emission line in a spectrum is a measure of the energy of the radiation at that wavelength. The width of the line indicates the wavelength band over which it is distributed. The intensity of the emission is greatest at the center of the line and decreases toward the edges; the reverse is true for an absorption line. These variations in the appearance of a line are known as the *line contours*. The pressure of the gas which is producing the spectral line can affect its width; this effect is known as *pressure broadening*. SEE ALSO **flash spectrum, hydrogen spectrum, Zeeman effect**.

speculum SEE **mirror**.

spherical aberration SEE **aberration** (2).

spheroid The surface formed by rotating an ellipse about one of its axes. If the minor axis is chosen, the result is an *oblate spheroid*. This is the shape of fluid bodies such as stars and gas planets like Jupiter rotating at a sufficient speed, and also of solid but non-rigid bodies like the Earth.

Spica The star Alpha Virginis, magnitude 0.98, distance 262 light years, luminosity 2200 times that of the Sun. It is a main-sequence star of type B1 and a spectroscopic binary.

spicule A narrow jet, like a tiny prominence, seen protruding from the Sun's chromosphere when it is viewed edge-on. Spicules are short-lived (about 5–10 minutes) and congregate at the edges of **supergranulation** cells. They are possibly caused by chromospheric gases shooting into the lower corona.

spiral galaxy A **galaxy** consisting of a nucleus of stars from which spiral arms emerge, winding around the nucleus and forming a flattened, disk-shaped region. The arms contain gas, dust, and young stars, while the nucleus contains old stars.

sporadic meteor A meteor which is not part of a recognized **meteor shower**. Its path cannot be traced back to a known **radiant**. The Earth is constantly sweeping up meteoroids, and 3–8 sporadic meteors should be visible each night under ideal conditions. SEE ALSO **radar astronomy**.

Spörer's law The appearance of sunspots at lower latitudes over the course of the 11-year **solar cycle**. The average position drifts from midlatitudes (about 30–40°) toward the equator as the cycle progresses from one minimum to the next. The phenomenon was first studied in detail by German astronomer Gustav Spörer (1822–95), after whom it was later named. Despite it being called a "law," there is no mathematical formulation of the statement. SEE ALSO **butterfly diagram**.

spring equinox SEE **vernal equinox**.

spring tide SEE **tides**.

SS Cygni star An alternative name for a **U Geminorum star**.

SS433 An unusual binary star, number 433 in a catalog of stars with bright hydrogen emission lines compiled by American astronomers Bruce Stephenson and Nicholas Sanduleak. SS433 lies at the center of a supernova remnant in Aquila called W50. What makes SS433 puzzling is the variable **Doppler shift** in its emission lines. SS433 is thought to consist of a hot, massive star orbited every 13.1 days by a neutron star, which was presumably formed by the supernova that created W50. A stellar wind from the hot star produces one set of emission lines. Matter is transferred from the massive star to the neutron star via an **accretion disk**, but some of this material is ejected at high speed along two jets. The jets sweep around the sky every 164 days, producing the variable Doppler shift in SS433's spectral features.

S star A giant star whose surface temperature lies in the range for M stars, but whose spectrum contains absorption bands of zirconium oxide as well as the titanium oxide characteristic of the M stars. The over-abundance of zirconium, and also of carbon and many other heavy elements, is a result of convective mixing which brings to the surface products from nuclear reactions in the interior. As a result, many S stars show spectral lines of the radioactive element technetium. M stars with weak zirconium oxide bands are assigned the intermediate spectral type MS. S stars often show hydrogen emission lines in their spectra, and many are long-period variables.

standard epoch A fixed point in time, or **epoch**, to which star positions are referred in order to remove the effects of precession and proper motion, together with a reference system in which the positions are measured. The standard epoch currently used for coordinates in catalogs and star charts is designated 2000.0. Its fixed points are noon on January 1, 2000, and the mean equator and equinox (i.e. ignoring minor variations in the positions of the celestial equator and the equinoxes).

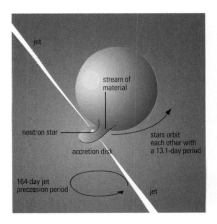

▲ **SS433** Shown here is a theoretical model of the unusual binary pair SS433, which is thought to consist of a hot, massive star orbited by a neutron star.

star A self-luminous ball of gas whose radiant energy is produced by nuclear reactions, mainly the conversion of hydrogen into helium. The temperatures and luminosities of stars are prescribed by their masses (according to the **mass–luminosity relation**). The most massive stars are about 100 solar masses (i.e. a hundred times heavier than the Sun). Above this mass the star is not stable and is likely to break up. Large stars can be very luminous and hot, and therefore appear blue. Medium-sized stars like the Sun are yellow, while small stars are a dull red. The smallest stars contain less than one-twentieth of a solar mass; below this the temperature does not become high enough for nuclear reactions to take place, and the object becomes instead a **brown dwarf**. The colors, and hence the temperatures, of stars give rise to the scheme of **spectral classification**. Most stars are born in pairs or groups (SEE **binary star**, **star cluster**). SEE ALSO **stellar evolution**, **stellar populations**.

starburst galaxy A **galaxy** that appears much brighter at infrared wavelengths than would be expected from its visual brightness as a result of a burst of star formation. Starburst galaxies were first detected in 1983 by the Infrared Astronomical Satellite, IRAS, which found some galaxies that emitted up to 50 times more energy as infrared radiation than they do as visible light; by contrast, the Andromeda Galaxy, a normal spiral, emits only about 4% of its total energy in the infrared. Hot young stars emit strongly in the ultraviolet. This radiation is absorbed by dust clouds and re-emitted as infrared. Some starburst galaxies are near other galaxies, and it may be that the gravitational influence of a neighbor is responsible for triggering the starburst. However, many starburst galaxies have no close companions, so this cannot be the only explanation.

star cluster An assembly of stars moving as a whole through space. Star clusters

STARS: THE BRIGHTEST (first magnitude or brighter)				
Star	Name	Apparent magnitude	Spectral type	Distance (light years)
α Canis Majoris	Sirius	−1.44 v	A0	8.60
α Carinae	Canopus	−0.62 v	F0	313
α Centauri	Rigil Kentaurus	−0.28**	G2+K1	4.39
α Boötis	Arcturus	−0.05 v	K2	36.7
α Lyrae	Vega	0.03 v	A0	25.3
α Aurigae	Capella	0.08 v	M1	42.2
β Orionis	Rigel	0.18 v	B8	773
α Canis Minoris	Procyon	0.40	F5	11.4
α Eridani	Achernar	0.45 v	B3	144
α Orionis	Betelgeuse	0.45 v	M2	427
β Centauri	Hadar	0.61 v	B1	525
α Aquilae	Altair	0.76 v	A7	16.8
α Crucis	Acrux	0.77**	B0.5+B1	321
α Tauri	Aldebaran	0.87	K5	65.1
α Virginis	Spica	0.98 v	B1	262
α Scorpii	Antares	1.05 v **	M1+B2.5	604
β Geminorum	Pollux	1.16	K0	33.7
α Piscis Austrini	Fomalhaut	1.16	A3	25.1
β Crucis	Mimosa	1.25 v	B0.5	353
α Cygni	Deneb	1.25 v	A2	3230
α Leonis	Regulus	1.36	B7	77.5
ε Canis Majoris	Adhara	1.50	B2	431
v variable ** combined magnitude of double star				

are classified as globular or open. A *globular cluster* is one in which an extremely large number of stars (10^5 to 10^7) are packed closely into a roughly spherical space typically 50 to 150 light years across. The stars are strongly concentrated toward the center. Globular clusters contain very old (**Population II**) stars and generally lie in the halo of our Galaxy. The compactness and large total masses of globular clusters make them dynamically very stable systems that remain largely unchanged over many thousands of millions of years. There is, however, a slow but steady loss of stars, mostly of low mass, that are swung out of the cluster through encounters with other stars; this is called *cluster evaporation*.

Open clusters are much less rich than globulars, and the stars are not as concentrated toward the center; they are also irregularly shaped. Open clusters contain young (**Population I**) stars and are found in or near the central plane of the Galaxy. They are susceptible to disruption, not only through evaporation but also through tidal interaction with the Galaxy as a whole or encounters with interstellar clouds. Only the richest survive for more than 10^9 years, while the smallest and least tightly bound last no more than a few million years. As all stars in an open cluster are moving together, it is possible to find the distances of nearby clusters by the **moving cluster** method.

STARS: THE NEAREST (closer than 12 light years)				
Star	Apparent magnitude	Spectral type	Absolute magnitude	Distance (light years)
Sun	−26.78	G2	4.82	—
Proxima Centauri	11.01 v	M5	15.45	4.22
α Centauri A	−0.01	G2	4.34	4.39
α Centauri B	1.35	K1	5.70	4.39
Barnard's Star	9.54	M4	13.24	5.94
Wolf 359	13.46 v	M6.5	16.57	7.8
Lalande 21185	7.49	M2	10.46	8.31
Sirius A	−1.44	A0	1.45	8.60
Sirius B	8.44	DA2	11.34	8.60
UV Ceti A	12.56 v	M5.5	15.42	8.7
UV Ceti B	12.96 v	M6.5	15.81	8.7
Ross 154	10.37	M3.5	13.00	9.69
Ross 248	12.27	M5.5	14.77	10.3
ε Eridani	3.72	K2	6.18	10.50
HD 217987	7.35	M2	9.76	10.73
Ross 128	11.12 v	M4.5	13.50	10.89
L789–6	12.32	M5	14.63	11.2
61 Cygni A	5.20 v	K5	7.49	11.36
Procyon A	0.40	F5	2.68	11.41
Procyon B	10.7	DF	13.0	11.41
61 Cygni B	6.05 v	K7	8.33	11.42
HD 173740	9.70	K5	11.97	11.47
HD 173739	8.94	K5	11.18	11.64
GX Andromedae A	8.09 v	M1	10.33	11.64
GX Andromedae B	11.10	M4	13.35	11.64
G51–15	14.81	M6.5	17.01	11.8
ε Indi	4.69	K5	6.89	11.83
τ Ceti	3.49	G8	5.68	11.90
L372–58	13.03	M4.5	15.21	11.9
v variable				

star diagonal SEE **diagonal** (1).

Stardust A NASA space probe to Comet Wild 2, launched in 1999. It will fly by the comet in 2004, making measurements and taking images. By using a collector covered with a special aerogel, it will also capture dust particles from the coma which will be ejected to Earth in a capsule when Stardust returns in 2006.

Star of Bethlehem The Biblical "star seen in the east" which guided the Wise Men, mentioned only by St Matthew. It has been ascribed to various astronomical phenomena, including comets, novae, meteors, and the conjunction of two or three planets. The search for an astronomical explanation is complicated by uncertainty over the exact date of Christ's birth. The three separate conjunctions of Jupiter and Saturn in Pisces in 7 BC are widely favored. However, the single most remarkable astronomical event about that time was the apparent merging in Leo of Venus and Jupiter, the two brightest planets, in June of 2 BC, which would have created a brilliant temporary "star."

star streaming The overall effect of the proper motions of stars (corrected for the Sun's motion) not being random, giving rise to two preferred directions of motion in exactly opposite directions. One vertex is in the constellation Orion, the other in Scutum. Star streaming results from the motion of stars in their orbits around the Galaxy. It was discovered by the Dutch astronomer Jacobus Kapteyn in 1904.

star trailing In **astrophotography**, the appearance of stellar images as curved lines, or trails, on photographs taken by a stationary camera – one that is not driven to compensate for the rotation of the Earth. The length of the trails depends on the exposure time. A camera aimed at the celestial pole will record star trails as concentric arcs centered on the pole.

stationary point The point in the apparent path of a planet when it appears motionless – SEE article and diagram at **retrograde motion** (2).

steady-state theory A cosmological theory put forward by Hermann **Bondi** and Thomas **Gold** in 1948, and further developed by Fred **Hoyle** and others. According to this theory the Universe has always existed; it had no beginning and will continue for ever. Although the Universe is expanding, it maintains its average density – its steady state – through the continuous creation of new matter. The theory was rejected because it cannot naturally explain the **cosmic microwave background** or the observation that the appearance of the Universe has changed with time.

stellar association SEE **OB association, T association**.

stellar evolution The development of a star over its lifetime, which can range from thousands of years to thousands of millions of years, depending on its mass. The theory of stellar evolution is based on mathematical models of stellar interiors. Its account of the evolution of stars can be checked by observations of stars at each of the predicted stages.

Stars form when a cloud of gas and dust collapses under its own gravity. As the cloud collapses its temperature rises because the energy of infalling atoms is released as heat when they collide. At this stage it is termed a *protostar*. A protostar will continue to contract under gravity until its center is hot enough (about 15 million K) for nuclear reactions to commence, with the release of energy (SEE **carbon–nitrogen cycle**, **proton–proton reaction**). The outward flow of radiation from deep within halts the collapse (SEE **radiation pressure**). As this is happening, the star's outer layers are still falling in. **T Tauri stars** are low-mass stars in this final stage of collapse.

Once nuclear fusion has commenced in the core of a star, the star adopts a stable structure with its gravity balanced by the radiation from its center. This is known as the main-sequence phase of a star's life

(SEE **Hertzsprung–Russell diagram**). It is the longest phase in the life of a star. The Sun has a main-sequence lifetime of some 10,000 million years, of which half has expired. Less massive stars are redder and fainter than the Sun, and stay on the main sequence for longer. More massive stars are bluer and brighter and have a shorter lifetime. As a star ages on the main sequence, nuclear fusion changes the composition of its core from mainly hydrogen to mainly helium.

Eventually, the hydrogen supply in the core of the star runs out. With the central energy source removed, the core collapses under gravity and heats itself further until hydrogen fusion is able to take place in a spherical shell surrounding the core. As this change occurs the outer layers of the star expand considerably and the star becomes a **red giant** or, in the most massive stars, a **supergiant**. While it is a red giant, the star's core temperature will reach 100 million K, hot enough for the fusion of helium to carbon to begin (the **triple-alpha process**). In low-mass stars the onset of helium fusion is sudden (the *helium flash*), but in high-mass stars it gets under way more gradually. After this stage high-mass stars can fluctuate in size, becoming **long-period variable** stars like Mira.

When all of the helium fuel in the core has been converted into carbon, the core of the star will again collapse and heat up. In a low-mass star the central temperature will not rise far enough to initiate carbon fusion. The red giant will lose its outer layers as a **planetary nebula**, after which only a **white dwarf** remains. In a high-mass star the contraction of the carbon core will lead to further episodes of nuclear fusion involving heavier elements, during which the star will remain a supergiant. Once fusion has converted all of the core into elements with atomic weights close to that of iron, no further fusion is possible. At this point the core will collapse explosively, and the star will throw off its outer layers in a **supernova** explosion. The explosion takes place not at the center but on the exterior of a dense core. The explosion throws off several solar masses of gas at speeds of many thousands of kilometers per second, and also pushes inward on the core with immense force. The effect of this implosion is to crush the already extraordinarily dense core still further, forming a **neutron star** or a **black hole**, both effectively endpoints of stellar evolution.

stellar interferometer SEE **optical interferometer**.

stellar nomenclature Some stars have proper names given to them by Greek, Roman, and Arab astronomers, such as Sirius, Capella, and Aldebaran. In addition to any proper name, the stars within each constellation are identified by a letter or number followed by the name of the constellation in the genitive (possessive) form (SEE the table of **constellations** on pages 48–9). The brightest stars are indicated by Greek letters, as in α (Alpha) Lyrae or ε (Epsilon) Eridani. These letters are known as *Bayer letters* because they were assigned by Johann **Bayer** in his star atlas *Uranometria* of 1603. Stars that do not have Bayer letters are prefixed by a number, as in 61 Cygni. These are known as *Flamsteed numbers* because they relate to the stars charted in John Flamsteed's *Historia coelestis Britannica* of 1725. Fainter stars are referred to by their designation in any of a number of other catalogs. **Variable stars** have their own system of nomenclature.

stellar populations A classification of stars according to their age and location in the Galaxy. There are two principal populations. *Population I* consists of relatively young stars which are located in the plane of the Galaxy, in its spiral arms. *Population II* consists of relatively old stars dispersed throughout the entire Galaxy but prominently visible in its center and halo. There is also a gradation of stars intermediate between these two extremes.

The distribution of populations in our Galaxy is explained in terms of the collapse of the Galaxy during its formation. The oldest population of stars was made during the infall, which is why Population II is distributed throughout the galactic halo and central regions of the Galaxy. Stars in these regions, including those in globular clusters, continue to have elliptical orbits around the Galaxy. After the Galaxy had developed its flat rotating disk and spiral arms, stars of Population I formed, moving in circular orbits. Between these two extremes lie intermediate populations showing progressively flatter distributions, representing successive stages in the collapse of the Galaxy.

Chemical elements produced in stars are recycled back into the interstellar material by stellar winds and supernova explosions, so Population I stars have greater concentrations of heavier elements than Population II. The Sun's content of heavy elements puts it among the younger Population I stars. The chemical history of the Galaxy suggests that there has existed a Population III, created when the Galaxy first formed, which have now disappeared. These stars would have manufactured the smattering of heavy elements present in Population II before dying to form compact objects like neutron stars.

stellar wind A stream of charged particles, mostly protons and electrons, from the surface of a star. The strength of the wind depends on the type of star, and its velocity can range from a few hundred to several thousand kilometers per second. Young stars evolving toward the main sequence have powerful stellar winds, up to a thousand times stronger than the Sun's **solar wind**. Old stars evolving into red giants also have strong stellar winds.

Stephan's Quintet A group of peculiar galaxies in the constellation Pegasus discovered in 1877 by the French astronomer Edouard Stephan (1837–1923). Four of the objects (NGC 7317, 7318A, 7318B and 7319) have nearly the same redshift ($z = 0.02$), but a fifth galaxy, NGC 7320, has a much lower redshift ($z = 0.0003$). Presumably it merely happens to lie in the same line of sight. Some astronomers now refer to the group, omitting NGC 7320, as *Stephan's Quartet*.

Steward Observatory The observatory of the University of Arizona, Tucson. Its main telescope is a 2.3-m (90-inch) reflector, sited on Kitt Peak.

stony-iron meteorite A meteorite consisting of approximately equal proportions of silicates (stony material) and metals, mostly nickel–iron. There are two main subtypes: Pallasites and Mesosiderites. *Pallasites*, which consist of olivine (a magnesium–iron silicate) mixed with nickel–iron, may have originated near the core/mantle interface of a planetary body. *Mesosiderites* are a much coarser combination of chunks of various silicates and nickel–iron, and could have been produced by impacts on a planetary surface.

stony meteorite A **meteorite** that consists mostly of silicates (stony material), with only a small amount of chemically uncombined metals (typically 5% nickel–iron). There are two subtypes: **chondrites** and **achondrites**.

Stratospheric Observatory for Infrared Astronomy (SOFIA) A joint venture between NASA and the German Space Agency to install a 2.5-m (100-inch) telescope aboard a Boeing 747-SP aircraft. It will fly at altitudes above 13 km (8 miles), where it will have uninterrupted access to wavelengths of 0.3 to 1600 μm, from the visible to the microwave region. In this part of the spectrum, water vapor impedes ground-based observations. SOFIA, like its predecessor, the Kuiper Airborne Observatory, is to be based at the Ames Research Center, and is expected to be ready in 2004.

strewnfield SEE **meteorite, tektites**.

Strömgren, Bengt George Daniel (1908–87) Swedish-born Danish astronomer. In 1940 he suggested the idea of what is now called a Strömgren sphere (SEE **H II region**) to explain how some nebulae shine. Strömgren also worked on photoelectric photometry, and the internal structure of stars.

Struve Russian–German family that produced four generations of astronomers. They included:

Struve, (Friedrich Georg) Wilhelm von (1793–1864), emigrated from Germany to Russia in 1833. In 1824 he had begun to observe double stars with a telescope made by Joseph von **Fraunhofer**, and in 1837 he published the first good catalog of double stars, containing nearly 2500 doubles he had discovered himself. In 1840 he made the third good measurement of the parallax of a star (Vega).

Struve, Otto (Wilhelm) (1819–1905), son of Wilhelm. He studied double stars, and determined an accurate value for the rate of **precession**, taking into account the motion of the Sun with respect to nearby stars.

Struve, Otto (1897–1963), a grandson of the above, emigrated and became a naturalized American. A leading observer and astrophysicist of his time, he applied spectroscopy to the study of binary stars, stellar rotation, and discovered interstellar hydrogen and calcium regions.

Subaru Telescope An 8.3-m (327-inch) reflecting telescope for use at visible and infrared wavelengths, sited at **Mauna Kea**. It commenced operation in 1999. Subaru is the Japanese name for the Pleiades.

subdwarf A star that is less luminous by 1 to 2 magnitudes than main-sequence stars of the same **spectral type**. Subdwarfs are mainly of types F, G, and K, and lie below the main sequence on the **Hertzsprung–Russell diagram**. Most are **Population II stars**. They are placed in luminosity class VI, although an alternative designation is to prefix their spectral type with the letters "sd."

subgiant A star of smaller radius and lower luminosity than a normal giant star of the same **spectral type**. They are mainly of types G and K, and lie between the main sequence and the giants on the **Hertzsprung–Russell diagram**. Subgiants are placed in luminosity class IV.

submillimeter-wave astronomy Observations of **electromagnetic radiation** at wavelengths of less than a millimeter; in practice, between 0.3 and 3 mm. The techniques for detecting submillimeter radiation are similar to those for radio wavelengths, but as it is absorbed by water vapor, submillimeter telescopes have to be sited at high altitudes. Important targets at these wavelengths are molecular clouds and regions of star formation.

subsolar point The point on the surface of the Earth or other body in the Solar System at which, at any given moment, the Sun is directly overhead. Similar points are defined for other pairs of bodies.

substellar object SEE **brown dwarf, extrasolar planet**.

summer solstice SEE **solstice**.

Summer Triangle The prominent triangle that is formed by the first-magnitude stars Altair (in Aquila), Vega (in Lyra) and Deneb (in Cygnus). The triangle is overhead on summer nights in northern temperate latitudes.

▲ **stellar wind** The Bubble Nebula is caused by the strong stellar wind from the star at the bottom of the image. When young stars undergo a phase called the T Tauri phase their extremely strong stellar winds blow away much of the nebula in which they were formed.

Sun The star at the center of the Solar System, around which all other Solar System bodies revolve in their orbits. The apparent daily motion of the Sun across the sky and its annual motion along the **ecliptic** are caused by the Earth rotating on its axis and revolving in its orbit. The Sun does, however, have its own motion in the Galaxy of about 20 km/s ($12\frac{1}{2}$ mile/s) toward the point in the sky known as the **apex**. The Sun's light is occasionally blocked by the Moon (which, like the Sun, has an apparent diameter of about half a degree) in a **solar eclipse**. The amount of sunlight normally reaching the Earth is quantified as the **solar constant**. Data for the Sun are given in the table.

The Sun is a dwarf **G star**, on the **main sequence** of the Hertzsprung–Russell diagram, and is thus a typical, average star. It consists of about 70% hydrogen (by weight) and 28% helium, with the remainder mostly oxygen and carbon. Its temperature, pressure, and density increase toward the center, where the values are about 15×10^6 K, 10^{11} bar, and 150 g/cm^3, respectively. As with all stars, the Sun's energy is generated by nuclear fusion reactions that take place under the extreme conditions in the core, chiefly the **proton–proton reaction**, which converts hydrogen into helium. The mass converted into energy is 4.3 million tons per second, but even at this enormous rate the loss amounts to only 0.07% of the total mass in 10^{10} years.

The Sun's core is about 400,000 km (250,000 miles) across. Energy released from the core passes up through the *radiative zone*, which is about 300,000 km (nearly 200,000 miles) thick, by a process of successive absorptions and re-emissions. It then passes through the 200,000-km (125,000-mile) thick *convective zone*, transported by rising and falling cells of gas, to the surface, the **photosphere** (meaning "sphere of light"), from where it is radiated into space.

Most of the Sun's visible activity takes place in the 500-km (300-mile) thick photosphere. The pattern of convective cells that transport energy from below is visible as the **granulation** (SEE ALSO **supergranulation**). **Sunspots** are darker, cooler regions of the photosphere where the local magnetic field is enhanced. Their passage across the Sun's disk reveals the Sun's **differential rotation**, which might indicate that its core is spinning more quickly than its outer regions. Associated with sunspots are **flares** – sudden, violent releases of energy and material that lead to **aurorae** in the Earth's atmosphere. Other phenomena occurring in the photosphere are **faculae** and **plages**. There is a periodicity in level of solar activity known as the **solar cycle**.

Above the photosphere lies the **chromosphere** (meaning "sphere of color," so-called because of its rosy tint when seen during a total solar eclipse), which consists of hot gases and extends for thousands of kilometers. It is generally in a state of turbulence, the temperature rising extremely rapidly with height. This is the realm of **spicules** and **prominences**, huge eruptions of material from the Sun's limb. Extending outward from the chromosphere for millions of kilometers is the extremely tenuous solar **corona**. A continuous emission from the corona of particles and radiation is known as the **solar**

2001/03/29 09:36 UT

▲ **Sun** Although at a stable part of its life cycle, the Sun is not unchanging and undergoes a roughly 11-year cycle, during which its magnetic field "winds up" and then declines. Sunspots are the most obvious result of these changes.

wind. There are also large-scale eruptions of material from the corona, known as **coronal mass ejections**. The solar wind and the Sun's magnetic field dominate a region of space called the **heliosphere**, which extends to the boundaries of the Solar System.

The **continuous spectrum** of the Sun, with its dark **Fraunhofer lines**, results from reactions in the photosphere and has yielded much information. The **Zeeman effect** is a splitting of lines of the solar spectrum which indicates the presence of a magnetic field.

The Sun must never be viewed directly through any optical instrument. There are two ways for amateurs to observe the Sun's disk. A *full-aperture filter* has a special metallic coating which cuts out harmful radiation before the Sun's light enters the telescope. Alternatively the Sun's image can be projected on to a white screen. Temporary or permanent blindness can result from direct viewing or from the use of unsuitable filters. Professional instruments for solar work include the **coronagraph** and the **spectroheliograph**.

There have been many probes (such as **Ulysses**) and satellites (such as the **Solar and Heliospheric Observatory**, **Yohkoh**, and the **Transition Region and Coronal Explorer**) launched to study the Sun.

sundial A simple timekeeping device in which a shadow cast by the Sun falls on a dial graduated in hours. The shadow is cast by a short pillar called a *style* or *gnomon* standing out from the dial, which may be mounted vertically or horizontally. A sundial shows **apparent solar time**.

sundog SEE **parhelion**.

sungrazer A **comet** which at perihelion passes through the Sun's corona. Many of the bright long-period comets of the 19th century were sungrazers. Often, such comets do not survive their close passage of the Sun.

sunrise The moment when the Sun's upper limb first appears above the horizon.

SUN: DATA	
Distance from the Earth	
mean (the astronomical unit)	149.6 × 10⁶ km
maximum (at aphelion)	152.1 × 10⁶ km
minimum (at perihelion)	147.1 × 10⁶ km
Diameter	1.391 × 10⁶ km
Density (mean)	1.409 g/cm³
Mass	1.989 × 10³⁰ kg
Volume	1.412 × 10¹⁸ km³
Period of axial rotation	
at equator	25.4d
at poles	about 35d
Inclination of axis of rotation to pole of ecliptic	7° 15′
Surface gravity (Earth = 1)	28
Spectral type	G2V
Luminosity	3.86 × 10²⁶ W
Magnitude (mean visual)	−26.86 (apparent) +4.71 (absolute)
Rotational velocity (mean)	1.9 km/s
Escape velocity	617 km/s
Temperature of surface	5780 K
Temperature of core	15 million K

It is defined as the moment when the Sun's **zenith distance** is 90° 50′, and decreasing. This value is arrived at by making allowance for the Sun's semidiameter (16′) and for atmospheric refraction (34′).

sunset The moment when the Sun's upper limb disappears below the horizon. Similarly to **sunrise**, it is defined as the moment when the Sun's zenith distance is 90° 50′, and increasing.

sunspot A region in the Sun's **photosphere** which is cooler than its surroundings and therefore appears darker. Sunspots consists of a dark central region, the *umbra*, and a gray outer region, the *penumbra*. They vary in size from around 1000 to 50,000 km (600 to 30,000 miles), and occasionally up to about 200,000 km (125,000 miles), at which size they are visible to the naked eye. Their duration varies from a few hours to a few weeks, or months for the very biggest. Sunspots are seen to move across the face of the Sun as it rotates. The number of spots visible depends on the stage of the 11-year **solar cycle**.

Sunspots occur where there is a local strengthening of the Sun's magnetic field, which cools the area to around 4000 K. The stronger field is believed to suppress the convection of hotter gases from lower levels. Spots are usually found in pairs of opposite magnetic polarity, between latitudes 30° and 40° north or south of the equator, although larger groups are not uncommon (SEE ALSO **following**, **preceding**).

The spectrum of a sunspot differs from that of the photosphere as a result of the spot's lower temperature. The motion of gases in the penumbra was detected by a Doppler shift of spectral lines (known as the **Evershed effect**). The **Zeeman effect** in the spectra of some sunspots demonstrates the presence of strong magnetic fields, the polarity of which reverses for each solar cycle.

supercluster SEE **cluster of galaxies**.

supergiant An extremely luminous star of large diameter and low density. Supergiants can be of any spectral type, from O to M. Rigel (type B) and Betelgeuse (type M) are examples. The luminosities of supergiants are several magnitudes greater than those of giant stars of the same spectral type, and so they lie at the top of the **Hertzsprung–Russell diagram**. Red (M-type) supergiants like Betelgeuse have the largest diameters, around 1000 times that of the Sun. Supergiants are assigned to luminosity class I. The brightest are often given the separate class Ia, and the others placed in class Ib.

supergranulation A pattern of convective cells distributed fairly uniformly over the Sun's **photosphere**, and much larger than ordinary photospheric granules (SEE **granulation**). Material has been detected flowing from the center to the edge of the cells, where most of the magnetic flux coming from the photosphere is concentrated, and it is believed that it is the magnetic field at the edges of the cells that leads to the formation of **spicules**.

superior conjunction SEE **conjunction**.

superior planet Any of the planets Mars, Jupiter, Saturn, Uranus, Neptune and Pluto, which orbit outside the Earth's orbit.

superluminal Describing the motion of an object which appears to be moving faster than the speed of light. Parts of some radio sources appear to be moving apart at superluminal velocities. This has been explained as a trick of the geometry of the source as we see it, for such velocities are impossible according to **Einstein**'s Special Theory of Relativity.

supernova (plural **supernovae**) A stellar explosion in which virtually an entire star is disrupted. For a week or so, a supernova may outshine all the other stars in its galaxy. The luminosity (absolute magnitude up to about −19) is some 23 magnitudes (1000 million times) brighter than the Sun, and the energy released in the explosion is the same as is released over the star's entire previous life. A supernova is about 1000 times brighter than a **nova**. The diagram shows the light curve of Supernova 1989B, in the galaxy M66. Supernovae are designated by the year in which they are observed, plus one or more letters to indicate the order of discovery in that year.

Analysis of ancient records identifies several supernovae in our own Galaxy, before the invention of the telescope. One was observed in 1054 by Chinese astronomers. The **Crab Nebula** and its associated pulsar are the remains of this event. In 1572 Tycho Brahe and others saw a new star as bright as Venus in the constellation Cassiopeia (SEE **Tycho's Star**). The last naked-eye supernova in our Galaxy was seen in 1604 in Ophiuchus, and was observed for over a year (SEE **Kepler's Star**). There is probably about one supernova every 30 years in a galaxy like our own, but most of them are concealed by dust.

Supernovae are of two main types, classified by their light curves and spectra. *Type II* supernovae have hydrogen lines in their spectra, whereas *Type I* supernovae

▶ **supernova** The light curve of a supernova shows a sudden brightening as the outer atmosphere is thrown off after the stellar core has collapsed. This peak is followed by a steady decline for about two months, after which this fading slows further.

do not. Type I supernovae have been found in all kinds of galaxies, but Type II have never been found in ellipticals. Type II supernovae are thought to be caused by the explosion of stars greater than about 8 solar masses at the end of their life. When the nuclear fuel inside the star gives out, the core collapses to form a **neutron star** or **black hole**. The collapse of the star's interior releases energy which is picked up by the outer layers, and these layers are ejected into space at about 5000 km/s (3000 mile/s). Type I supernovae are believed to be formed by the explosion of the white dwarf component of a binary star. Hydrogen from the companion leaks on to the white dwarf and drives it over the critical mass of 1.4 solar masses, causing it to explode.

After a couple of years the supernova has expanded so much that it becomes thin and transparent. For hundreds or thousands of years the ejected material remains visible as a **supernova remnant**.

Supernova 1987A A bright supernova which flared up in the Large Magellanic Cloud in February 1987, reaching magnitude 2.8 – the first naked-eye supernova since Kepler's Star of 1604. A blue supergiant designated Sanduleak −69° 202 was identified as the star which had exploded.

supernova remnant (SNR) A gaseous emission nebula, the expanding shell of matter thrown off into space during the outburst of a **supernova**. These remnants are often strong radio and X-ray sources. The ejected layers of the supernova collide with the surrounding interstellar gas and

heat up to perhaps a million degrees K, emitting X-rays. Electrons spiral in the magnetic fields in the compressed gas and emit **synchrotron radiation** in the form of radio waves. The most intense radio source in the sky, **Cassiopeia A**, is a supernova remnant. Supernova remnants usually appear as hollow shells. However, some (including the Crab Nebula) are filled balls of radio and X-ray emission. Such a remnant is termed a *plerion*, meaning "filled." This implies that they contain electrons produced by an active pulsar, formed in the supernova explosion.

Surveyor A series of seven US spacecraft, five of which made successful soft landings on the Moon between May 1966 and January 1968 in preparation for the Apollo landings. They took photographs and made chemical analyses of the surface.

Swan Nebula SEE **Omega Nebula**.

Swift–Tuttle, Comet A **comet** discovered in 1862 by American astronomers Lewis Swift (1820–1913) and Horace Tuttle (1837–1923), subsequently identified with Comet Kogler of 1737, and observed at its return in 1992. It is the second-brightest short-period comet, Halley's Comet being the brightest. The large number of jets seen emerging from the nucleus in 1992 suggest that the comet's changing period (at present 130 years) may be caused by **non-gravitational forces**. It is the parent comet of the **Perseid** meteor shower; this relationship was discovered by Giovanni Schiaparelli, and was the first comet–meteor connection to be established.

Sycorax An outer **satellite** of Uranus in an eccentric, retrograde orbit discovered by Brett J. Gladman and Philip D. Nicholson in 1997.

symbiotic variable A **binary star** consisting of two stars of widely differing surface temperatures, such as a cool red giant and a hot dwarf. Gas from the cool star falls on to, and heats, the smaller star. Symbiotic variables are also known as *Z Andromedae stars*, after their prototype, which has a range from magnitude 8.3 to 12.4.

synchronous orbit An orbit in which a satellite's period of revolution is the same as the primary's period of axial rotation. From the planet's surface the satellite appears to hover over one point, neither rising nor setting. An example is **Chiron**'s orbit around Pluto.

synchronous rotation The axial rotation of a celestial body in the same period as its period of revolution. Consequently, it always presents the same face toward the body about which it revolves, as in the case of the Moon orbiting the Earth. Tidal friction (SEE **tides**) has locked the Moon into this condition, which is also known as *captured rotation*. Other examples are the regular satellites of Jupiter and Saturn.

synchrotron radiation The electromagnetic radiation emitted by charged particles (usually electrons) that are accelerated by a strong magnetic field to speeds which are a significant fraction of the speed of light. The higher the energy of the particles, the shorter the wavelength of the radiation they emit. Synchrotron radiation is so called because it was first observed in particle accelerators called snychrotrons. Cosmic sources include **radio galaxies** and supernova remnants such as the **Crab Nebula**.

synodic month The period between two identical phases of the Moon; it is the same as the duration of one **lunation**. Its length is 29.53059 mean solar days.

synodic period The period of apparent revolution of one body about another as observed from the Earth, for example from one opposition or conjunction to the next. COMPARE **sidereal period**.

synthesized aperture SEE **aperture synthesis**.

Syrtis Major A dark, triangular feature near the Martian equator, and the planet's most conspicuous feature in telescopic views from the Earth. It was first recorded by Christiaan **Huygens** in 1659. It is a sloping, cratered area, officially named Syrtis Major Planum.

Systems I and II Regions of **Jupiter**'s upper atmosphere with slightly differing speeds of rotation.

SY Ursae Majoris star SEE **U Geminorum star**.

syzygy The approximate alignment of three celestial bodies; in particular, the alignment of the Earth, the Sun, and the Moon or another planet. Thus syzygy occurs at full Moon and new Moon, and at planetary oppositions and conjunctions.

T

TAI ABBREVIATION FOR **International Atomic Time** (from its name in French).

tail A long train of dust or gas particles expelled from the nucleus of a **comet**.

tangential velocity The component of a star's velocity at right angles to the line of sight. It is also known as *transverse velocity*. It can be found from measurements of the annual **proper motion** of the star and its distance in parsecs. If the **radial velocity** is also known, then the **space velocity** of the star can be calculated.

Tarantula Nebula An emission nebula faintly visible to the naked eye on the southeastern edge of the Large Magellanic Cloud; also known as NGC 2070 or 30 Doradus. It is 1000 light years across and 500,000 times the Sun's mass, bigger and brighter than any nebula in our Galaxy or any other nearby galaxy. It has a complex filamentary structure, and has a dense cluster of stars, designated R136, at its center.

T association A region of recent and active star formation consisting of low-mass stars known as **T Tauri stars**. The members of T associations offer a series of snapshots of stars like our Sun at various stages through its infancy. They are still surrounded by the dusty, obscuring material of the cloud from which they are forming, and consequently they are often brightest in the infrared. COMPARE **OB association**.

Taurus A constellation of the zodiac, representing the bull into which Zeus changed himself to carry off Princess Europa. It contains two of the most famous open clusters in the sky, the **Pleiades** and the **Hyades**, and also the **Crab Nebula**. Its brightest star is **Aldebaran**. Beta (Elnath), magnitude 1.7, and Zeta, magnitude 3.0, mark the tips of the bull's horns. Lambda is an eclipsing binary (range 3.4 to 3.9, period 3.95 days). There is also the prototype of the **T Tauri stars**.

tectonics Name given to the various processes by which the surface of a planetary body is deformed as a result of heating from within. Tectonic processes operate on bodies that have undergone **differentiation** and so possess a surface crust with a molten layer below. Large-scale movements of the molten layer affect the crust, producing faults and folds, and lifting mountains. Several satellites (e.g. **Ariel** and **Europa**) have surface features which appear to indicate tectonic activity. *Plate tectonics* operates, as far as we know uniquely, on the Earth, whose crust consists of a number of *plates* which move with respect to one another at a few centimeters per year, supported on the mantle below. Where these plates meet there are earthquakes, volcanoes, and other geological activity, producing features such as mid-oceanic ridges and mountain chains.

tektites Small, glassy objects, typically centimeter-sized, found scattered across certain specific areas (*strewnfields*) of the Earth's surface. They consist mainly of sil-

▲ **Tarantula Nebula** Located in the Large Magellanic Cloud, the Tarantula Nebula is much the brightest of the external galaxies visible from Earth.

ica, with small quantities of metallic oxides, and are from 600,000 to 65 million years old. Their chemical and physical characteristics and their ages are specific to the areas in which they are found, and have nothing in common with the geological characteristics of their surroundings. All tektites have clearly solidified rapidly from a temperature of around 1500–2000 K. Their origin is uncertain, though their flight markings, which are characteristic of solidification while flying through the air at high speed, suggest that they originated as terrestrial rock was vaporized by the impact of large meteorites. Tektites are named for the location of the strewnfield: for example, *Australites* are from Australia, and *Indochinites* from Southeast Asia.

telescope An instrument for collecting and magnifying light or other electromagnetic radiation from a distant object. There are two basic types of optical telescope: the **refracting telescope**, in which the light-collecting element is a lens, and the **reflecting telescope**, in which it is a mirror; **catadioptric** systems are hybrids of the two. The image formed by the primary (main) lens or mirror is magnified by an **eyepiece**. The eyepiece may be replaced by one of several devices: for example, a camera, photographic plate, or **CCD** for **astrophotography** (SEE ALSO **Schmidt camera**); a **photometer** for measuring brightness; or a **spectrometer** for examining spectra. A telescope is supported (SEE **mounting**) in such a way that its weight is supported and it is easily aimed at celestial objects.

The **focal ratio** of a telescope is equal to the focal length of the primary lens or mirror divided by its diameter. Refractors have longer focal ratios than reflectors, which, aperture for aperture, are thus shorter. The most compact designs are short-focus instruments like the **Schmidt–Cassegrain telescope**. The image produced by an astronomical telescope is inverted, as opposed to that formed in a terrestrial telescope, which is erect (upright).

The refractor installed in the Yerkes Observatory in 1897, which has an objective 1 m (40 inches) in diameter, is still the largest of its kind. Large telescopes built since then are all reflectors, because large mirrors are easier to make and support than large lenses. The world's largest reflectors are the **Keck Telescopes**, which each have a composite mirror 10 m (396 inches) across. With the development of spacecraft, it has become possible to send instruments such as the **Hubble Space Telescope** into space above the Earth's atmosphere, and so make observations unaffected by atmospheric absorption. SEE ALSO **drive**, **radio telescope**.

Telescopium A faint southern constellation representing a telescope, introduced by **Lacaille**. Alpha, its brightest star, is magnitude 3.5, but it contains little of interest.

Telesto A small **satellite** of Saturn, discovered by Bradford Smith and others in 1980 between the two Voyager encounters. Telesto is irregular in shape, measuring 30 × 16 km (19 × 10 miles). It is a **coorbital satellite** with Tethys and **Calypso**.

telluric line An **absorption line** in the spectrum of a celestial object produced by molecules such as oxygen and water in the Earth's atmosphere.

terminator The boundary between the illuminated and unilluminated hemispheres of a planet or satellite as viewed from Earth. Mercury, Venus, Mars, and the Moon show phases and therefore have their terminators are visible. The mountainous lunar surface gives the Moon a visibly irregular terminator.

terrestrial planets The planets Mercury, Venus, Earth and Mars, so called because they have rather similar characteristics in respect of size, density, and few or no satellites.

Tethys A medium-sized **satellite** of Saturn, diameter 1058 km (657 miles), discovered by Giovanni Cassini in 1684. The whole of Tethys's surface is densely cratered, but evidence of past geological activity is provided by some less heavily cratered areas which must have undergone resurfacing. The two dominant features are Odysseus, a crater 440 km (270 miles) in diameter, and Ithaca Chasma, a 100-km (60-mile) wide trough running three-quarters of the way around the satellite; their formation may be linked. Tethys shares its orbit with two coorbital satellites, **Calypso** and **Telesto**.

Thalassa One of the small inner **satellites** of Neptune discovered in 1989 during the Voyager 2 mission.

Thales of Miletus (*c.* 625 – *c.* 550 BC) Greek philosopher. He provided the first cosmogony – a theory of the Universe's origin – in which everything developed from a primordial mass of water. He knew of the **saros**, and is credited with having used it to predict the eclipse of the Sun which occurred probably on May 28, 585 BC, during a battle between the Lydians and the Medes.

Thebe One of the small inner **satellites** of Jupiter discovered in 1979 during the Voyager missions. It is irregular, measuring about 110×90 km (70×55 miles).

theory of everything (TOE) SEE **grand unified theory**.

third contact In a total **solar eclipse**, the moment when the trailing edge of the Moon's disk begins to uncover the Sun's disk, and totality ends. In a **lunar eclipse**, it is the moment when the Moon begins to leave the umbra of the Earth's shadow. The term is also used for the corresponding stage in eclipses involving other bodies.

three-body problem A fundamental problem in celestial mechanics: to determine the motions of three bodies under the influence only of their mutual gravitational attractions. There is no exact general solution, only solutions for special cases, but highly accurate approximations can be

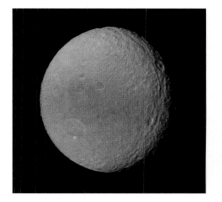

▲ **Tethys** Ithaca Chasma can be seen to the right in this image of Saturn's satellite Tethys. Younger craters are superimposed on the giant trough, indicating that it is an ancient feature.

achieved by using modern computers. The three-body problem was examined first by Isaac **Newton**, and subsequently tackled by many astronomers and mathematicians, including Leonhard **Euler**, Joseph Louis **Lagrange** (SEE ALSO **Lagrangian points**), Carl Jacobi, and Henri **Poincaré**. Work on the three-body problem has been stimulated mostly by the need to understand the orbit of the Moon, under the gravitational influence of the Earth and the Sun, and more recently by the need to calculate the orbits of artificial satellites. SEE ALSO **many-body problem**.

tides Distortions induced in a celestial body by the gravitational attraction of one or more others. The gravitational attraction a body experiences is greatest on the side nearest the attracting body, and least on the side farthest away, causing it to elongate slightly in the direction of the attracting body, acquiring a *tidal bulge* on each side. If the deformed body is orbiting the attracting body, different parts of its surface periodically experience tidal distortion as it rotates. These tides most obviously affect a gaseous atmosphere or a fluid ocean, but the solid crust of a planet like the Earth, which is supported by a fluid mantle, is able to "flex" and so also expe-

▲ **Titania** The surface of Titania is heavily cratered and traversed by faults. This Voyager 2 picture shows details down to 7 km (4 miles) across.

riences tidal distortion, but to a much lesser degree. The enormous gravitational forces exerted by Jupiter on its satellite **Io** produce a large amount of flexure and interior heating, and this *tidal heating* is responsible for Io being highly volcanically active. Smaller and completely rigid bodies can under certain circumstances be broken up by tidal forces.

The familiar ocean tides on the Earth are raised by the gravitational attraction of the Moon and the Sun, the Moon's influence being about three times the Sun's. There are two high tides and two low tides each day. When the Sun and the Moon are exerting a pull in the same direction (as at new Moon or full Moon) their effects are additive and high tides are higher (*spring tides*). When the pull of the Sun is at right angles to that of the Moon (as at first quarter or last quarter), high tides are lower (*neap tides*). *Tidal friction* caused by the tidal ebb and flow of water over the ocean floor acts as a brake on the Earth's rotation, as a result of which the length of the day is increasing by about one millisecond per century. The angular momentum which is thus being lost is transferred to the Moon (SEE **secular acceleration**).

time A scale for measuring the duration of events by reference to some regularly recurring phenomenon. From remote antiquity the basis of time has been the passage of the Earth once around its orbit, giving the unit of one **year**, and the rotation of the Earth on its axis, giving the unit of one **day**. However, the rotation of the Earth is not constant, and so it cannot be used as an accurate and invariable standard of time. Modern time-scales are now based on a continuous count of time units defined by atomic resonators (**International Atomic Time**). The basis of civil timekeeping is **Greenwich Mean Time** (known technically as **Universal Time**). In addition, astronomers reckon time by the stars: **sidereal time**. SEE ALSO **apparent solar time, equation of time, local time, mean solar time, time zone**.

time zone One of 24 divisions of the Earth's surface, each 15° of longitude wide, within which the time of day is reckoned to be the same. At a conference held in Washington in 1884 the meridian of Greenwich was adopted as the zero of longitude, and zones of longitude were established. Standard time in each successive zone westward is one hour behind that in the preceding zone. Large territories such as the USA and Russia span several time zones. The adoption of time zones results in a discrepancy at longitude 180° (the International Date Line), which is resolved by omitting one day from the calendar when crossing from west to east, or repeating one day if the crossing is from east to west.

Titan The largest **satellite** of Saturn, and the second largest in the Solar System, discovered by Christiaan Huygens in 1655. Titan is unique among planetary satellites in having a substantial atmosphere. At 1.88 g/cm^3 Titan is the densest of Saturn's large satellites, and is composed of rock and water ice in roughly equal proportions. The space probe Voyager 1 found no gaps in an opaque, reddish cloud layer 200 km (125 miles) above the surface. The atmosphere consists mostly of nitrogen, with some methane and traces of other hydrocarbon compounds, and exerts a surface pressure 1.5 times that on the Earth. The surface temperature is 95 K, at which methane can exist as solid, liquid, or gas, so methane may play the role that water does on Earth, forming, clouds, rain, lakes, and even snow. The surface may therefore have been shaped by the same types of erosion that have operated on Earth. More will be known in 2004 when the probe called Huygens is due to be landed on Titan by the **Cassini** spacecraft.

Titania The largest of the five main **satellites** of Uranus, discovered by William **Herschel** in 1787. It is 1580 km (980 miles) in diameter, and has a density of 1.68 g/cm³, indicating that it consists of a mixture of rock and water ice. Titania is Oberon's twin in size, but its surface resembles that of **Ariel**, with similar features and a similar geological history. The most prominent trough is named Messina Chasma, and is 1500 km (950 miles) long.

Tombaugh, Clyde William

(1906–97) American astronomer. In 1930, nearly a year into a search based on predictions by Percival **Lowell**, he discovered the planet **Pluto**. He continued to search for other planets for more than ten years, discovering in the process star clusters, clusters of galaxies, a comet, and hundreds of asteroids. After World War II he developed telescopic cameras for tracking rockets after launch.

topocentric Pertaining to observations made from a point on the surface of the Earth, e.g. topocentric coordinates as opposed to geocentric coordinates (which are measured from the center of the Earth).

total eclipse SEE **lunar eclipse, solar eclipse**.

Toutatis Asteroid no. 4197, discovered by Christian Pollas in 1989. Radar observations suggest that it consists of two bodies, 4 km (2.5 miles) and 2.5 km (1.5 miles) in diameter, in close contact. Toutatis is an **Apollo asteroid**, and one of the largest **near-Earth asteroids**.

TRACE ABBREVIATION FOR **Transition Region and Coronal Explorer**.

transit (1) The passage of a celestial body across the observer's **meridian**. Such transits, observed with a **transit instrument**, were formerly important in the measurement of time and for navigational purposes.

transit (2) The passage of a body directly between the Earth and the Sun. The planets Mercury and Venus do so on occasion, when they appear as a black spot crossing the solar disk (SEE ALSO **black drop**). Accurate measurements of such transits helped to establish the scale of the Solar System.

transit (3) The passage of a planetary satellite across the planet's disk. Jupiter's four main satellites frequently transit its disk, and Saturn's inner satellites are occasionally seen in transit. Observations of transits of Jupiter's satellites by Ole **Römer** established the enormous but finite speed of light.

transit (4) The passage of a surface or atmospheric feature of a body across its **central meridian** as it rotates, the timing of which gives a means of measuring the body's rotation period.

▲ **Toutatis** These four images of the asteroid Toutatis were constructed using radar information from NASA's Goldstone and Arecibo radar telescopes. Toutatis appears to consist of two irregularly shaped, cratered objects.

transit instrument A telescope that is mounted on a horizontal east–west axis and movable only in the vertical plane of the meridian. Transit instruments were once used to time the transits (1) of bodies across the local meridian and measure their altitudes. Such measurements were

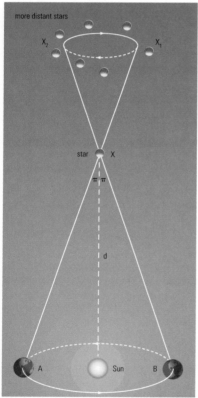

▲ **trigonometric parallax** If a nearby object (X) is observed from opposite sides of the Earth's orbit (A and B), it will appear in different positions against the stellar background (X_1 and X_2). As the distance between A and B is known, the distance, d, to X can be calculated.

used for precision timekeeping, but have been superseded by devices such as the **photographic zenith tube**.

Transition Region and Coronal Explorer (TRACE) A NASA satellite launched in 1998 to study the Sun's **corona** and the transition region – the border between the chromosphere and the corona. The primary mirror of its 300-mm (12-inch) Cassegrain telescope has different coatings on its four quadrants to enable TRACE to make images at different ultraviolet wavelengths.

trans-Neptunian object SEE **Kuiper Belt**.

transverse velocity SEE **tangential velocity**.

Trapezium Popular name for the multiple star Theta Orionis in the Orion Nebula. Its four main components, magnitudes 5.1, 6.7, 6.7 and 8.0, are arranged in the shape of a trapezium. The stars are very hot, and light up the nebula.

Triangulum A small but easily found constellation between Andromeda and Aries. Its brightest star, Beta (magnitude 3.0), forms a well-marked triangle with Alpha (3.4) and Gamma (4.0). The main feature of interest is the spiral galaxy known as the Triangulum Galaxy, or M33, 2.7 million light years away in our **Local Group**, visible with binoculars under clear, dark skies.

Triangulum Australe A southern constellation representing a triangle, marked by the stars Alpha (magnitude 1.9), Beta, and Gamma (both 2.9). The open cluster NGC 6025 is visible with binoculars.

Trifid Nebula An emission nebula in the constellation Sagittarius, apparently divided into three main sectors by dark lanes of dust. It is also known as M20 or NGC 6514. At its center is an 8th-magnitude double star. The Trifid lies 6700 light years away.

trigonometric parallax A means of determining the distance of a star using the principle of triangulation. As the Earth revolves around the Sun, nearby stars show a small change in position with respect to more distant stars. This change in position can be measured from photographs taken six months apart, i.e. when the Earth is on opposite sides of the Sun. As the diagram shows, the angular displacement is a direct measure of the star's parallax (π). Since the baseline (the distance from the Sun to the Earth) is known, the distance d of the star can be determined by trigonometry.

The first star to have its distance measured by this method was 61 Cygni, by Friedrich **Bessel** in 1838. Parallax is measured in arc seconds. The inverse of the parallax is the distance in parsecs. Alpha Centauri is the star with the largest known parallax, 0.752″, which corresponds to a distance of 1.3 parsecs. Typical errors in measuring parallax photographically are about 0.01″, but by using many plates this error can be reduced to about 0.004″. The parallax of a star at 25 parsecs is 0.04″, so the uncertainty is around 10%. The most accurate parallax determinations have been made by the **Hipparcos** satellite.

triple-alpha process A nuclear reaction in which three alpha particles (helium nuclei) are transformed into carbon with the release of energy. It takes place after all the hydrogen in a star's core has been exhausted. The core contracts and its temperature rises until it exceeds 100 million K, when the triple-alpha reaction begins. In each reaction two helium nuclei combine to from a beryllium nucleus which, in turn, captures a further helium nucleus to form a carbon nucleus. Additional helium captures can produce oxygen, neon, and a number of heavier elements. The reaction is believed to be the dominant energy-producing process in red giants. It is also known as the *Salpeter process* after the US physicist Edwin Salpeter (1924–), who described it.

▲ **Triton** Voyager 2 imaged Triton in 1989, and this mosaic shows some of the varied terrains. The broad area at the bottom of the picture is the southern polar ice cap, which probably consists mainly of nitrogen ice.

triplet A compound lens consisting of an assembly of three component lenses, which may be air-spaced or cemented.

Triton The largest **satellite** of Neptune, discovered in 1846 by William Lassell. It is in an inclined, retrograde orbit and has almost certainly been captured by Neptune, probably from the **Kuiper Belt**. Its diameter is 2705 km (1680 miles), and its density is 2.07 g/cm^3, consistent with a composition largely of three parts rock to one of water ice. Triton is the coldest and most distant world so far visited: Voyager 2's instruments measured a surface temperature of 35 K ($-238°$C). Its surface is complex and varied, and includes *cantaloupe terrain* – a dimpled landscape crossed by broad fissures and resembling the skin of a melon. There is much evidence of very recent volcanic activity in a variety of features formed by the upwelling of fluid material. In the south polar region, which is covered with a thin **polar cap** of nitrogen ice, dark eruptive plumes send fine dark material rising to altitudes of 8 km (5 miles) before being carried downwind for 100 km (60 miles) or more in the tenuous nitrogen atmosphere.

Trojan asteroid A member of one of two groups of **asteroids** sharing Jupiter's orbit. One group lies ahead of Jupiter, and the other behind, oscillating about the L_4

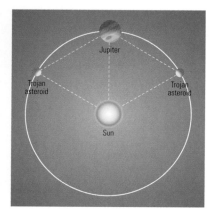

▲ **Trojan asteroid** The asteroids known as the Trojans orbit at the Lagrangian points in front of or behind Jupiter. They oscillate around these points over a period of 150–200 years.

and L_5 **Lagrangian points** of Jupiter's orbit around the Sun. The Trojans ahead of Jupiter, about two-thirds of the total, are known as the *Achilles group*, after the first one to be found, **Achilles** in 1906; those behind form the *Patroclus group*, named for the second Trojan to be discovered, also in 1906. About 1600 were known by early 2003; six Martian Trojans and one Neptunian have been found.

tropical month One revolution of the Moon around the Earth relative to the vernal **equinox**. It equals 27.32158 days.

tropical year The time taken for the Earth to orbit the Sun relative to the vernal equinox. Its length is 365.24219 mean solar days. Because of the precession of the vernal equinox, which moves along the ecliptic in a direction opposite to the Sun's motion, the tropical year is about 20 minutes shorter than the sidereal year. It is sometimes also known as the *solar year*.

Tropic of Cancer The parallel of latitude on Earth, $23\frac{1}{2}°$ north, that marks the most northerly declination reached by the Sun at the summer **solstice**, on or about June 21.

Tropic of Capricorn The parallel of latitude on Earth, $23\frac{1}{2}°$ south, that marks the most southerly declination reached by the Sun at the winter **solstice**, on or about December 22.

true anomaly SEE **anomaly**.

Trumpler classification The classification of open star clusters devised by Swiss–American astronomer Robert Trumpler (1886–1956) on the basis of three characteristics: the number of stars in the cluster, the concentration of stars toward the center of the cluster, and the range of brightness within each cluster.

T Tauri star A very young star still settling on to the main sequence and belonging to a class of irregular variables, named for the prototype, T Tauri. T Tauri stars are found in nebulae or young clusters, and are characterized by high-velocity infall or outflow of gas as they adjust to the onset of nuclear reactions. Emission lines in their spectra indicate an extended atmosphere of gas. Most known T Tauri stars are less massive than the Sun. Heavier stars either pass through the T Tauri stage while they are still obscured in the nebulae from which they were born, or have a different appearance at the same stage of life. SEE ALSO **Hertzsprung–Russell diagram, T association**.

Tucana A far southern constellation representing a toucan. Its overall faintness is redeemed by the presence of the Small Magellanic Cloud and the superb globular cluster 47 Tucanae (NGC 104). At magnitude 4 and half a degree across, 47 Tucanae is inferior only to Omega Centauri. It lies 15,000 light years away. NGC 362 is another globular, visible in binoculars.

Tully–Fisher relation A relationship that links the absolute brightness of a galaxy to its speed of rotation as revealed by the width of its 21 cm hydrogen line. It is one of the most important methods for estimating distances to galaxies. It was formulated by Brent Tully (1943–) and Richard Fisher (1943–) in 1977.

Tunguska Event A huge aerial explosion on June 30, 1908, just north of the Stony Tunguska River in Siberia, which devastated the surrounding forest. The explosion, which was heard more than 800 km (500 miles) away and was recorded seismographically all around the world, was preceded by a fireball as bright as the Sun. Trees up to 40 km (25 miles) away were felled, and in the central 1000 square km (400 square miles) trees and animals were incinerated.

The absence of impact craters or meteoritic fragments puzzled earlier investigators, but the best current theory is that a **near-Earth asteroid** roughly 50 m (150 ft) in diameter entered the atmosphere obliquely, and shattered and vaporized at an altitude of about 8 km (5 miles), blanketing the area with dust. Dust particles preserved in resin from trees in the area have the composition of known stony meteorites, and the damage pattern is consistent with the impactor having had an asteroidal density. An incoming comet, which would be less dense, would vaporize too high up to cause much damage on the ground; a former theory had part of Encke's Comet as the impactor.

tuning fork diagram A diagram of **galaxy** types, originated by Edwin Hubble. The "handle" of the fork consists of elliptical galaxies (E), numbered according to their degree of elongation, while the two prongs of the fork are made up of ordinary (S) and barred (SB) spirals, designated a, b, or c according to how tightly wound the spiral arms are. The classification was originally interpreted as a sequence of evolution, with ellipticals developing into spirals, but this is now known to be incorrect. SEE the diagram at **Hubble classification**.

twenty-one centimeter line The emission line of neutral hydrogen in interstellar clouds. It lies in the radio spectrum at a wavelength of about 21 cm; the frequency is 1420 MHz. Its existence was predicted by Hendrik van de **Hulst** in 1944 and discovered by him and others in 1951. SEE ALSO **H I region**.

twilight The intermediate period during which the illumination of the sky gradually increases before sunrise, and decreases after sunset. The phenomenon is caused by the scattering of sunlight by molecules of air and particles of dust in the Earth's atmosphere. The duration of twilight depends on the steepness of the Sun's apparent path with respect to the horizon, so that twilight lasts longer at higher latitudes. The following three forms of twilight are distinguished. *Civil twilight* ends, or begins, when the center of the Sun's disk is 6° below the horizon, and is regarded as a period during which normal daytime activities cease to be possible. *Nautical twilight* ends, or begins, when the center of the Sun's disk is 12° below the horizon, and the marine horizon is no longer visible. *Astronomical twilight* begins in the morning and ends in the evening when the center of the Sun's disk is 18° below the horizon, and is the time when the faintest stars can be seen with the naked eye.

Tycho A large lunar crater, 85 km (53 miles) in diameter, in the Moon's southern uplands. It is the center of the most conspicuous system of lunar **rays**, stretching across much of the Moon's nearside, which suggests that Tycho may be a relatively young crater.

Tycho Brahe SEE **Brahe, Tycho**.

Tychonian system A planetary theory put forward by Tycho **Brahe**.

Tycho's Star A supernova in Cassiopeia in 1572 that was observed and described by Tycho **Brahe**. It was brighter than Venus at maximum and was visible during daytime. Its remnant has been detected as a radio and X-ray source. Its light curve identifies it as a Type I supernova.

U

UBV system A system of three-color photometry devised by Harold Johnson and William Morgan of Yerkes Observatory. Stellar magnitudes are determined at three different wavelength bands through three color filters: ultraviolet (U) peaking at 360 nm; blue (B) peaking at 440 nm; and yellow (V for visual) peaking at 550 nm. SEE ALSO **color index**.

U Geminorum star A member of a class of dwarf novae which show sudden outbursts of between 2 and 6 magnitudes, followed by a slower return to minimum where they remain until the next outburst. Periods range from 10 days to several years. They are also known as *SS Cygni stars*. The best known examples are SS Cygni, whose light curve is shown in the diagram, and U Geminorum. All are close binaries, comprising a subgiant or dwarf star of type K or M which has filled its Roche lobe, and a white dwarf surrounded by an **accretion disk** of infalling matter. Almost all members show rapid irregular flickering of about half a magnitude at minimum. As they are all binaries, some show eclipses at minimum, such as U Geminorum itself. The

SY Ursae Majoris subclass has normal maxima and "supermaxima" that last about five times as long and are twice as bright. It has been suggested that they evolve from **W Ursae Majoris stars**. They are closely related to **Z Camelopardalis stars**.

Uhuru The first X-ray astronomy satellite, launched by the US in December 1970. Many X-ray sources bear numbers prefixed with the letter U, being their designation in the Uhuru catalog.

ultraviolet astronomy The study of **electromagnetic radiation** from space with wavelengths between those of the visible spectrum and X-rays, i.e. from about 350 nm down to about 90 nm. Apart from the longest wavelengths (called the *near ultraviolet*), ultraviolet radiation does not penetrate the atmosphere, so observations have to be made from rockets and satellites. The first ultraviolet satellites to study the Sun and stars were the Orbiting Solar Observatory and Orbiting Astronomical Observatory series in the 1960s. A major observatory satellite, the **International Ultraviolet Explorer** (IUE), was

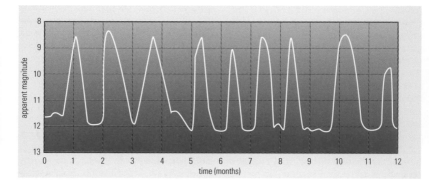

▲ **U Geminorum star** The light curve of a typical U Geminorum star shows repeated outbursts at intervals ranging from approximately ten days to several years. During outburst, the star's brightness climbs rapidly, with a slower decline. Outbursts occur as a result of mass transfer to an accretion disk around the smaller, more massive star in these close binary systems.

launched in 1978. Coverage was extended to the shortest ultraviolet wavelengths by **Rosat** in 1990 and the **Extreme Ultraviolet Explorer** (EUVE) in 1992.

Unlike other wavelengths such as radio, infrared, and X-rays, ultraviolet astronomy has discovered few new sources. Instead, the main application is in spectroscopy. Many of the abundant atoms and ions in celestial sources have their strongest spectral lines in the ultraviolet. Observations at ultraviolet wavelengths have been significant for studies of the composition and motions of interstellar gas. Hot stars of spectral types O and B emit strongly in the ultraviolet, and they excite surrounding nebulae by their ultraviolet emissions. In the Solar System, ultraviolet spectroscopy is used to study planetary atmospheres and the gas in comets. Further off in the Universe, ultraviolet astronomy is important in the study of luminous galaxies and quasars, which contain hot stars and large quantities of gas.

Ulugh Beg (1394–1449) Mongol ruler and astronomer. He established an observatory at Samarkand, and compiled the first star catalog to surpass those of Hipparchus and **Ptolemy** in precision.

Ulysses A European Space Agency probe launched in October 1990 to study the polar regions of the Sun. It was thrown out of the ecliptic by a fly-by of Jupiter in February 1992, and looped over the south pole of the Sun in September 1994, flying over the north pole of the Sun a year later. In the late 1990s, Ulysses studied the evolution of **coronal mass ejections** from a position back near the orbit of Jupiter.

umbra (1) The inner part of the shadow cast by a celestial body illuminated by an extended source such as the Sun. An observer in the umbral region of a shadow sees all of the source obscured, as, for example, in a total **solar eclipse**. SEE ALSO **lunar eclipse**. COMPARE **penumbra** (1).

umbra (2) The dark central area of a sunspot. COMPARE **penumbra** (2).

▲ **Umbriel** This pole-on view of Umbriel was obtained by Voyager 2 on January 24, 1986. The smallest details that can be resolved are 10 km (6 miles) across, and the heavily cratered surface is clearly visible.

Umbriel One of the five main **satellites** of Uranus, discovered by William **Lassell** in 1851. Its diameter is 1170 km (727 miles), and its density of 1.51 g/cm^3 is consistent with a composition of mostly water ice and rock. Umbriel is a geologically inactive, heavily cratered satellite with a predominantly dark surface. The only bright spot is the floor of the 140-km (85-mile) diameter crater named Wunda.

United States Naval Observatory The US Government observatory with headquarters at Washington, DC. Its main telescope there is a 0.66-m (26-inch) refractor. The Observatory has an observing station at Flagstaff, Arizona, with a 1.55-m (61-inch) reflector for astrometric work.

Universal Time (UT) Standard time used for scientific purposes throughout the world, popularly known as **Greenwich Mean Time** (GMT). There are several versions of Universal Time. *UT0* is the observed rotation of the Earth relative to the stars. *UT1* is UT0 corrected for a slight wandering of the Earth's geographical poles, which affects the longitude of the place of observation. *Coordinated Universal Time* (UTC; the abbreviation is of the term in French) is a regular system of time based on the caesium atomic clock, and is used for radio time signals. As the Earth's rotation is gradually slowing, the time shown by an atomic clock gradually diverges from UT1. To keep atomic time in step with UT1 to the nearest second, time signals are retarded when necessary by one **leap second** at midnight on June 30 and December 31. UTC is what is generally known as GMT.

Universe All of space and everything contained in it. SEE **cosmology**.

upper culmination SEE **culmination**.

Uranus The seventh major planet from the Sun, and the second-smallest of the four giant planets. With a mean magnitude of 5.5, Uranus is visible to the naked eye under good conditions. Its apparent diameter varies from 3.3 to 4.1 arc seconds, and through a telescope it appears as a small, featureless, greenish blue disk. In size, mass, atmosphere and color, Uranus resembles Neptune. Like all the giant planets, it possesses a ring system and a retinue of satellites. The main data for Uranus are given in the table.

▲ **Uranus** Cloud features on Uranus are fairly elusive. This heavily processed Voyager 2 image from 1986 shows some structure in the planet's atmosphere at lower right.

Uranus was the first planet to be discovered since ancient times. Although it had been observed on several occasions (and once had even been cataloged as a star), its non-stellar nature was first recognized by William **Herschel**, who observed its disk on March 13, 1781, during a telescopic survey of faint naked-eye stars. He first took it to be a comet, but as observations accumulated it became clear that the object was in a circular, planetary orbit and was indeed a planet.

Like Pluto, Uranus's axis of rotation is steeply inclined, and lies close to the ecliptic plane. Its poles therefore spend 42 years in sunlight, followed by 42 years in darkness. Highly exaggerated seasonal variations are therefore experienced by both the planet and its satellites. The cause of the tilt is not known; a possibility is a collision during the final stages of its **accretion**.

The fly-by of the Voyager 2 probe in 1986 provided most of our current knowledge of the planet. The upper atmosphere is about 83% molecular hydrogen, 15% helium, and the other 2% mostly methane, whose strong absorption of red light gives Uranus its predominant hue. Strong enhancement of Voyager images was needed to show anything other than the featureless disk visible from Earth. Faint banding and a haze over the Sun-facing pole, possibly of hydrocarbons produced by the effect of solar ultraviolet radiation on methane, were discernible, along with traces of methane clouds.

Near-infrared images obtained from the ground and from the Hubble Space Telescope have since revealed dark spots and bright clouds similar to those visible when Voyager flew past **Neptune**. It may be that Uranus's "weather" was unusually calm at the time of the Voyager visit; the Hubble Space Telescope imaged about 20 clouds in 1998, circling the planet at 500 km/h (300 mile/h). However, the planet has little internal heat, and certainly insufficient to drive a dynamic meteorology like Neptune's. **Differential rotation** operates on Uranus, atmospheric features taking 16 to 17 hours to rotate, compared with the internal rotation period of over 17 hours.

Wind speeds in the atmosphere vary from about 150 to 600 km/h (90 to 370 mile/h).

Uranus's interior structure remains conjectural. There may be an iron–silicate core of about ten Earth masses, surrounded by a deep "mantle" of perhaps six Earth masses, consisting of water, ammonia and methane ices. The mantle may be liquid, or "mushy" – a mixture of solids, liquids and gases in proportions that vary with depth. It is in this region that the planet's magnetic field, which is of comparable strength to Saturn's, originates. The magnetic axis is tilted by 59° to the axis of rotation, and is displaced by nearly 8000 km (5000 miles) from the planet's center. (In this respect Uranus is also similar to Neptune.) The magnetic field gives rise to an appreciable **magnetosphere**.

The five largest **satellites** were known before the Voyager encounter, which led to the discovery of 11 more. These 16 are all regular satellites orbiting in or close to Uranus's equatorial plane. They are all darkish bodies composed of ice and rock. Five more have since been found, orbiting in distant orbits.

The main components of Uranus's ring system were discovered from the Kuiper Airborne Observatory in 1977 when the planet occulted a star (SEE **occultation**). Others were imaged by the Voyager spacecraft. The brightest and outermost is the Epsilon Ring, while the innermost is a very diffuse sheet of material. Between the two are nine narrow, darkish rings. SEE ALSO **ring, planetary**.

Ursa Major A famous northern constellation, the Great Bear, whose main pattern is known as the Big Dipper or Plough. This feature consists of seven stars: Alpha (Dubhe), magnitude 1.8; Beta (Merak), 2.3; Gamma (Phekda), 2.4; Delta (Megrez), 3.3; Epsilon (Alioth), 1.8; Zeta (**Mizar**, a famous double), 2.2; and Eta (Alkaid), 1.9. Dubhe and Merak are the "pointers" to the pole star. Five of the Big

URANUS: DATA	
Globe	
Diameter (equatorial)	51,118 km
Diameter (polar)	49,947 km
Density	1.29 g/cm^3
Mass (Earth = 1)	14.53
Volume (Earth = 1)	62.18
Sidereal period of axial rotation (equatorial)	17h 14m (retrograde)
Escape velocity	21.3 km/s
Albedo	0.51
Inclination of equator to orbit	97° 52′
Temperature at cloud-tops	55 K
Surface gravity (Earth = 1)	0.79
Orbit	
Semimajor axis	19.22 AU = 2871 × 10^6 km
Eccentricity	0.046
Inclination to ecliptic	0° 46′
Sidereal period of revolution	84.01y
Mean orbital velocity	6.81 km/s
Satellites	21

Dipper stars make up a **moving cluster**; the exceptions are Dubhe and Alkaid. There are many galaxies in Ursa Major, including several Messier objects (M81, M82, M101, M108 and M109) as well as the **Owl Nebula**.

Ursa Minor The constellation that contains the north celestial pole. Its brightest star is Alpha (**Polaris**), the pole star, magnitude 2.0. The constellation's seven main stars make a pattern rather like a faint and distorted Big Dipper, and is known as the Little Dipper.

UT ABBREVIATION FOR **Universal Time**.

UV Ceti star Another name for a **flare star**. UV Ceti itself is a binary consisting of 13th-magnitude red dwarf components, which exhibits large flares every few hours. It is among the closest stars to us, lying 8.7 light years away.

V

Valhalla A large impact basin on Jupiter's satellite **Callisto**, surrounded by a series of concentric rings representing "ripples" from the impact.

Valles Marineris SEE Mariner Valley.

Van Allen Belts Two zones of plasma (high-energy charged particles) in the Earth's **magnetosphere**. The concentric, tyre-shaped belts trap charged particles which then spiral around magnetic field lines, back and forth between the two magnetic poles. The diagram, which is roughly to scale, shows the form of the belts. The outer belt contains mainly electrons from the **solar wind**, and the inner belt mainly protons from the same source. Within the inner belt is a radiation belt consisting of particles produced by interactions between the solar wind and heavier **cosmic ray** particles. The activity of the Van Allen Belts varies with the level of solar activity (SEE **solar cycle**). Because the Earth's magnetic axis is offset from its rotational axis, the belts are not a uniform distance from the Earth's surface. The inner belt comes very close to the surface over the South Atlantic Ocean; this is called the *South Atlantic Anomaly*. The two main belts were discovered by American physicist James Van Allen (1914–) and his collaborators during investigations using Geiger counters on board the first American Explorer satellites in 1958. The cosmic ray belt was found by Russian Cosmos satellites in 1991.

variable star A star whose brightness varies with time. *Intrinsic variables* are stars that vary because of some inherent feature, such as pulsations in size or events in the atmosphere. In *extrinsic variables*, external factors such as eclipses or obscuring dust affect the amount of light reaching us from the star. There are three main classes of variable. **Eclipsing binaries** are extrinsic variables in which members of a binary system periodically pass in front of each other as seen from Earth. **Pulsating stars** are intrinsic variables which expand and contract in size, either regularly or irregularly. **Cepheid variables** and **RR Lyrae stars** are examples of pulsating stars with regular cycles of variation. Less regular variation is found among the **long-period variables** (or Mira stars), **semiregular variables**, and **irregular variables**. Eruptive (or cataclysmic) variables are a mixed class, including **supernovae, novae, dwarf novae, flare stars, R Coronae Borealis stars**, and **shell stars**. Each main group also has its own subclassification.

Variable stars have their own peculiar naming system. The letter R is assigned to the first variable to be discovered in a particular constellation (unless it already has

◄ Van Allen belts
Earth is surrounded by radiation belts, which contain energetic particles trapped by the magnetic field. The inner belt dips low over the South Atlantic, where its particle population can present a hazard to satellites.

an existing designation, such as a **Bayer letter** or Flamsteed number). Further discoveries are denoted by S to Z. Then come RR to RZ, then SS to SZ, and so on down to ZZ. After that come AA to AZ, BB to BZ, and so on down to QZ, but the letter J is never used. At this point the 334th variable has been reached. Subsequent variables are numbered V335, V336, and so on.

variation A perturbation in the Moon's motion caused by the Sun's changing gravitational pull on the Moon as the Moon orbits the Earth.

Vega The star Alpha Lyrae, fifth-brightest in the sky, magnitude 0.03, distance 25 light years, luminosity 50 times that of the Sun. It is the standard star of spectral type A0 and luminosity class V in the **UBV system**. The Infrared Astronomy Satellite, IRAS, found a disk of gas and dust extending out to 85 AU from it, possibly a **proplyd** – a planetary system in the process of formation.

Vega probes Two space probes launched by the Soviet Union in December 1984 which dropped balloons into the atmosphere of Venus in June 1985 and encountered Halley's Comet in March 1986. They flew past the comet's nucleus at distances of about 8000 km (5000 miles), photographing it and analyzing the dust and gas given off.

Veil Nebula The brightest part of the **Cygnus Loop**, a large supernova remnant in Cygnus.

Vela A constellation representing the sails of the dismembered constellation Argo Navis. Its leading star is Gamma, magnitude 1.8 but slightly variable; this is the brightest **Wolf–Rayet star**, with a luminosity 40,000 times that of the Sun. Vela is a rich region; the stars Delta (magnitude 1.9) and Kappa (2.5) make up the **False Cross** with Iota and Epsilon Carinae. Other features of the constellation are the **Gum Nebula** and the **Vela Pulsar**.

▲ **Venus** Centered on longitude 180°, this image is a composite view of Venus prepared from Magellan Imaging Radar results in 1991.

Vela Pulsar A pulsar with a period of 89 milliseconds, discovered in 1968 in the constellation Vela; known as PSR 0833-45. The optical counterpart, discovered in 1977, flashes with the same period.

Venera A series of space probes launched by the Soviet Union to the planet Venus between 1961 and 1983. Venera 7 in 1970 was the first craft to land successfully on Venus, although it transmitted for only 23 minutes before being destroyed by the extreme temperatures and pressures at the surface. Venera 8 in 1972 proved more resilient and transmitted for 50 minutes. Veneras 9 and 10 in 1975 took the first photographs of the surface. Veneras 11 and 12 landed in 1978 and sent back information about the atmosphere, but no pictures. Veneras 13 and 14 landed in 1982, taking color photographs of the surface and sampling the rocks. Veneras 15 and 16, both radar mappers, went into orbit around the planet in October 1983 and operated until the following July.

Venus The second major planet from the Sun, and the second-largest of the four inner planets. Visible around dawn or dusk as the so-called *morning star* or *evening star*, it is the most conspicuous celestial object, after the Sun and Moon. At its brightest

VENUS: DATA	
Globe	
Diameter	12,104 km
Density	5.20 g/cm^3
Mass (Earth = 1)	0.8149
Volume (Earth =1)	0.8568
Sidereal period of axial rotation	243d 0h 30m (retrograde)
Escape velocity	10.4 km/s
Albedo	0.76
Inclination of equator to orbit	177° 20′
Surface temperature	750 K
Surface gravity (Earth =1)	0.90
Orbit	
Semimajor axis	0.723 AU = 108.2 × 10^6 km
Eccentricity	0.007
Inclination to ecliptic	3° 24′
Sidereal period of revolution	224.701d
Mean orbital velocity	35.02 km/s
Satellites	0

(magnitude −4.7) it is even visible to the naked eye in the daytime when the Sun is high in a clear sky. At greatest elongation Venus is 45° to 47° from the Sun, and has an apparent diameter of 25 arc seconds. Like the Moon and Mercury, it is seen to go through **phases**. **Transits** of Venus are rare, and occur in pairs 8 years apart, with over a century between each pair. The next pair are on June 7, 2004, and June 5, 2012. There is no known satellite. The main data for Venus are given in the table.

A telescope shows the planet's dazzling yellowish white cloud cover, with faint markings. In ultraviolet images these markings show clearly as a Y-shaped feature, enabling the clouds' rotation period of just under 4 days (nearly 60 times as fast as the rotation period of the planet itself) to be measured. The **ashen light**, the **cusp caps**, and the **Schröter effect** may be watched for by amateur observers; their causes have been long debated, but they are most likely real atmospheric phenomena rather than optical effects, as has sometimes been claimed.

Venus's very high surface temperature was indicated by measurements at radio wavelengths in 1958, but space probes were needed to reveal more about the surface. The most successful were the long Soviet **Venera** series, the two US **Pioneer Venus** probes, and, most recently, **Magellan**. After early failures, Venera craft were built to withstand the extreme surface conditions – a temperature of 750 K and a pressure of over 90 bars – for long enough to transmit measurements and pictures. Magellan has produced highly detailed radar mappings of the surface.

A gently undulating plain covers two-thirds of Venus. Highlands account for a further quarter, and depressions and chasms the remainder. The two principal highland regions are Ishtar Terra in the northern hemisphere, and Aphrodite Terra, largely in the southern. On Ishtar Terra is the Maxwell Montes mountain range, rising to 11 km (7 miles) above the average level, and the highest point on Venus. The plains are cratered; the largest crater is Mead, 280 km (170 miles) in diameter, but there are no craters smaller than a few kilometers across as smaller objects would have been destroyed by their passage through the dense atmosphere. However, most of the surface features are volcanic in origin. There are large shield volcanoes like Rhea Mons in the Beta Regio highlands, and calderas like Colette, extensive lava flows from which have covered the plateau Lakshmi Planum. The volcanic peak Maat Mons, at 8.5 km (5.3 miles) high the second-highest point on Venus, is surrounded by lava flows which have been estimated at no more than 10 years old. If this is so, then Venus is almost certainly still volcanically active. There are many other types of near-circular volcanic feature, which have been given names such as *arachnoids*, *coronae*, and *ovoids*. With a handful of exceptions, the planet's surface features are named for famous women, some from legend, from all the world's cultures.

The atmosphere consists of 96% carbon dioxide and $3\frac{1}{2}$% nitrogen, with traces of helium, argon, neon, and krypton. Its great density has produced the very high surface temperature by an extreme **greenhouse effect**. At various altitudes there are haze layers of sulfuric acid, sulfur dioxide, water vapor and sulfur; and ultraviolet light from the Sun causes some carbon dioxide to break down into carbon monoxide and oxygen. Pioneer Venus orbiter measurements showed that the sulfur dioxide level in the cloud-tops dropped tenfold between 1979 and 1986, which may reflect changes in the level of volcanic activity. A chemically active atmosphere is probably the only agent for "weathering" of surface features. Wind speeds in the atmosphere, which are 350 km/h (220 mile/h) at the cloud-tops, fall to 10 km/h (6 mile/h) at the surface.

The internal structure is largely unknown. Venera measurements indicate a low-density crust. There may be a silicate mantle and iron–nickel core, but that is just guesswork based on analogy with the similar-sized Earth. Certainly there is no appreciable magnetic field, although the solar wind does interact strongly with Venus's **ionosphere** to produce a well-defined bow shock – a **magnetosphere** of sorts.

vernal equinox The point at which the Sun crosses the celestial equator from south to north. It is also known as the *spring equinox* and the *First Point of Aries*. It is the zero point of the celestial coordinate called right ascension. The vernal equinox moves westward by about one-seventh of a second of arc daily because of the effect of **precession**.

Very Large Array (VLA) The most complex radio telescope on a single site in the world. It consists of 27 dishes of 25 m (82 ft) in diameter, movable along the arms of a giant Y on the plains near Socorro, New Mexico. Each arm can be up to 21 km (13 miles) long, using the technique of **aperture synthesis** to produce a virtual dish some 34 km (21 miles) wide.

Very Large Telescope (VLT) A group of four 8.2-m (232-inch) telescopes at Paranal Observatory, Chile, built by the **European Southern Observatory**. They were completed in 1998, 1999, 2000, and 2001. With all four telescopes linked together, the VLT has a light-gathering power equivalent to that of a single 16-m (630-inch) mirror. The four main instruments can be linked with several smaller ones to form an **optical interferometer**.

Very Long Baseline Array (VLBA) An array of ten 25-m (82-ft) radio telescopes spread across the continental USA and Hawaii, giving a baseline for **very long baseline interferometry** 8000 km (5000 miles) long. Resolutions of 0.001 arc second are possible. The instrument began operation in 1993. SEE ALSO **radio interferometer**.

very long baseline interferometry (VLBI) A technique in radio astronomy which extends the concept of **aperture**

▶ **Very Large Array**
One of the world's premier radio observatories, the Very Large Array is located at an elevation of 2120 km (6970 ft) on the plains of San Augustin, near Socorro.

synthesis by combining signals from radio telescopes which may be thousands of kilometers apart to achieve an accuracy of positional measurements of 0.001 arc second. An example is the **Very Long Baseline Array**. By combining signals from radio astronomy satellites and ground-based radio telescopes, baselines of around 40,000 km (25,000 miles) have been achieved.

Vesta Asteroid no. 4, diameter 501 km (311 miles), discovered in 1807 by Heinrich Olbers. At magnitude 6.4 it is the brightest asteroid and the only one ever visible (under ideal conditions) to the naked eye. Vesta's spectrum suggests that its surface, unique among the larger asteroids, was once molten (SEE **achondrite**). There is a huge, 460-km (285-mile) impact basin at the south pole.

vignetting A defect in an optical instrument in which light is unevenly distributed across the field of view, being typically darker near the edges than at the center. The most common cause is an obstacle in the light path.

Viking Two American space probes to Mars, launched in 1975. Each was a combined orbiter and lander. Viking 1 went into orbit around Mars in June 1976. The lander spacecraft was separated from the orbiter and landed on July 20, 1976. Viking 2 went into orbit in August 1975 and its lander reached the surface on September 3, 1976. Both landers photographed their surroundings, reported the weather at their locations, and studied the planet's soil. Experiments conducted by the landers found no indications of life at the landing sites. The orbiters photographed the surface of the planet and its two small moons, Deimos and Phobos.

Virgo The largest constellation of the zodiac, lying on the celestial equator, and representing the goddess of justice. Its brightest star is **Spica**. Gamma is a splendid binary with equal components of magnitude 3.6, period 169 years; they are closest in the year 2005. Virgo is one of the richest areas for faint galaxies since it includes the **Virgo Cluster**. Closer to us than this cluster is the **Sombrero Galaxy**.

Virgo Cluster A rich concentration of galaxies in the direction of the constellation Virgo, although some lie across the border in Coma Berenices. The center of the cluster is about 45 million light years away, and is marked by the giant elliptical galaxy M87, also known as the radio source Virgo A, which is ejecting a jet of gas. In all there are 16 Messier objects in the cluster; the total number of galaxies in the cluster is about 3000.

visible spectrum The range of wavelengths in the electromagnetic spectrum perceptible by the human eye. It varies slightly from one person to another, but extends typically from 385 nm (violet) to 700 nm (red). SEE ALSO **electromagnetic radiation**.

visual binary A binary system that can be observed as a double star with a telescope, as distinct from a spectroscopic binary. SEE **binary**.

visual magnitude (symbol m v) The **apparent magnitude** of a celestial object in the color region to which the human eye is most sensitive, i.e. about 560 nm.

VLA ABBREVIATION FOR **Very Large Array**.

VLBA ABBREVIATION FOR **Very Long Baseline Array**.

VLBI ABBREVIATION FOR **very long baseline interferometry**.

VLT ABBREVIATION FOR **Very Large Telescope**.

Volans A small, faint southern constellation adjacent to Carina, representing a flying fish. Its brightest stars are of 4th magnitude.

volatile A substance that solidifies at a low temperature. Substances such as water and carbon dioxide, familiar as gases on Earth, are examples. Bodies in the Solar System that formed nearer the Sun have a lower volatile content than those that formed farther away. Volatiles in comets, which formed in the outer reaches of the Solar System, evaporate as the gas tail when the comet approaches the Sun.

volcanism The eruption of molten material at the surface of a planetary body. Volcanism in which molten silicate rock erupts, cools, and becomes new surface material has taken place on Venus and Mars as well as the Earth, and similar material produced the lunar maria (SEE **mare**) by volcanic flooding. In the outer Solar System a mixture of water ice and ammonia ice can behave as lava, and has produced, for example, the volcanic floodplains on **Ariel**. Sulfur volcanism operates on **Io**, the most volcanically active world known. On **Triton**, dark particles are carried aloft by nitrogen geysers termed *plumes*.

Voyager Two American space probes to the outer planets. Both were launched in 1977. They passed Jupiter in March and July 1979, photographed the planet and satellites, and discovered Jupiter's ring system. Voyager 1 reached Saturn in November 1980, followed by Voyager 2 in August 1981, photographing the rings in detail, as well as the moons. Voyager 1's path was bent out of the ecliptic by its approach to Titan, but Voyager 2 proceeded to Uranus, reaching it in January 1986, and Neptune, in August 1989. Both Voyager probes are now on their way out of the Solar System, their last task to locate the heliopause, the boundary of the **heliosphere**.

Vulcan A supposed planet within the orbit of **Mercury**. It was originally invoked to explain, by its gravitational effect, the residual in the advance of Mercury's perihelion, since accounted for by Einstein's General Theory of Relativity. In 1859 a French physician, Edmond Lescarbault, observed a small body in transit across the Sun's disk and made subsequent careful observations of it. Urbain **Le Verrier** examined Lescarbault's observations closely, and was convinced of his integrity and the planet's existence, but no further trace of it has ever been found.

Vulpecula A dim constellation next to Cygnus, representing a fox. Its brightest stars are of 4th magnitude. It contains the **Dumbbell Nebula**, reputedly the easiest planetary nebula to see in the sky.

▲ **Viking** Compiled from images obtained by the Viking orbiters, this composite shows the Amphitrites Patera region of Mars. Radial ridges extend northward for about 400 km (250 miles) from this old volcano.

W

weakly interacting massive particle SEE **WIMP**.

Weizsäcker, Carl Friedrich von
(1912–) German theoretical physicist and astrophysicist. In 1938 he and Hans **Bethe** independently proposed a detailed theory for the production of energy in the Sun and other stars (SEE **carbon–nitrogen cycle**) in which hydrogen is converted into helium by nuclear fusion. In 1944 he revived the *nebular hypothesis* of the origin of the Solar System originally proposed by Immanuel Kant and Pierre Simon de **Laplace**.

West, Comet
A bright comet discovered by Danish astronomer Richard West in 1975. The following year it became a prominent naked-eye object with a fan-shaped tail. The comet was extremely active, and the nucleus broke into at least four fragments as it passed within 30 million km (20 million miles) of the Sun.

▲ **The Whirlpool Galaxy** The magnificent Whirlpool Galaxy takes its name from its pronounced spiral structure. The small companion galaxy at top, NGC 5195, appears to be interacting with the Whirlpool.

Westerbork Radio Observatory
A radio astronomy observatory located near Groningen in the Netherlands, operated by the Netherlands Foundation for Research in Astronomy. Its major instrument is an **aperture synthesis** telescope consisting of 14 dishes 25 m (82 ft) in diameter, along an east–west baseline 2.7 km (1.7 miles) long.

Whipple, Fred Lawrence (1906–)
American astronomer. He is best known for his "dirty snowball" theory of comets, proposed in 1949 and shown to be correct in 1986 when space probes were sent to Halley's Comet. Whipple discovered several comets and worked on cometary orbits; he also studied planetary nebulae, stellar evolution, and the Earth's upper atmosphere.

Whirlpool Galaxy
A well-defined spiral galaxy of type Sc which appears face-on to us and has very prominent arms; also known as M51 and NGC 5194. It lies in Canes Venatici, not far from the end of the tail of Ursa Major, and was the first galaxy in which spiral structure was noted, by Lord **Rosse** in 1845. A small companion galaxy, NGC 5195, appears to be connected to it by an extension of one of the spiral arms. The Whirlpool Galaxy is 37 million light years away.

white dwarf
A type of star about the size of the Earth, but with a mass about that of the Sun. As a result, its density is enormously greater than that of any terrestrial material (0.1 to 100 tons/cm^3). This is because the normal atomic structure is broken down completely, with electrons and nuclei packed tightly together (the state known as **degenerate matter**). A white dwarf cannot have a mass of more than about 1.4 solar masses (the **Chandrasekhar limit**). For larger masses, gravity would always overwhelm the pressure of the electrons and the star would collapse under its own weight, forming a **neutron**

star or black hole. White dwarfs are of low luminosity and gradually cool down to become cold, dark objects. They represent the final stage in the evolution of low-mass stars after they have lost their outer layers. The first white dwarf to be discovered was the companion of **Sirius**.

Widmanstätten pattern The distinctive pattern on a sectioned and polished surface of an **iron meteorite** that has been etched with acid. It is named for Count Aloys Joseph von Widmanstätten (1754–1849), who observed it in 1808. The pattern is only seen in meteoritic material, and consists of intersecting plates of the iron–nickel minerals kamacite and taenite.

Wild Duck Cluster An open cluster in the constellation Scutum, also known as M11 and NGC 6705. Its wedge shape resembles wild ducks in flight. The cluster lies 6200 light years away.

Wilkinson Microwave Anisotropy Probe (WMAP) A NASA spacecraft launched in 2001 to study the small-scale variations (anisotropy) in the **cosmic microwave background** (CMB). It is positioned at the Earth's L_2 **Lagrangian point**, from where it can map the CMB at high resolution. It is named for David Wilkinson, a member of the mission's science team.

William Herschel Telescope A reflecting telescope with a 4.2-m (165-inch) mirror at the **Roque de los Muchachos Observatory** in the Canary Islands. It was opened in 1988 and is owned and operated by the **Royal Observatory, Edinburgh**.

Wilson, Robert Woodrow (1936–) American physicist. With Arno **Penzias**, he detected in 1964 the **cosmic microwave background** radiation that provides the strongest evidence for the **Big Bang**. They shared the 1978 Nobel Prize for Physics for their discovery.

Wilson effect The foreshortening of a **sunspot** when it is near the Sun's limb, accompanied by a widening of the pen-

▲ **white dwarf** This close-up view from the Hubble Space Telescope shows part of the globular cluster M4 in Scorpius. While most of the objects seen here are old, red stars, several white dwarfs (circled in blue) have also been found.

umbra on the side nearest the limb, and a narrowing on the side farthest from the limb. The phenomenon was discovered by the Scottish astronomer Alexander Wilson (1714–86), who took it as an effect of perspective, indicating that sunspots are saucer-shaped depressions. However, not all spots show the Wilson effect. The cause is now thought to be that sunspots are more transparent than the surrounding photosphere, and the umbra more transparent than the penumbra.

WIMP Abbreviation for "weakly interacting massive particle," a hypothetical atomic particle whose existence has been invoked to account for the **missing mass** of the Universe.

winter solstice SEE solstice.

WMAP ABBREVIATION FOR **Wilkinson Microwave Anisotropy Probe**.

Wolf, Maximilian Franz Joseph Cornelius ("Max") (1863–1932) German astronomer. He was a pioneer of photographic methods in astronomy. In 1891 he made the first photographic discovery of an asteroid (Brucia, no. 323); in all he made 232 asteroid discoveries. He detected several new nebulae, including the **North America Nebula**, the periodic comet (1884 III) which bears his name, the nebulosity around the Pleiades, numerous dark nebulae, and the first cluster of galaxies (in Coma Berenices) to be identified as such, as well as the **Virgo Cluster**.

Wolf number SEE **relative sunspot number**.

Wolf–Rayet star (WR star) A star whose spectrum contains bright emission lines rather than dark absorption ones. WR stars are divided into two kinds. In the WN type, emission lines from nitrogen dominate the spectrum. In the WC type, emission lines of carbon and oxygen predominate. Both types have strong lines of helium and a few have moderate or weak lines of hydrogen as well. All the emission lines are broad. WR stars are very hot, with surface temperatures between 25,000 and 50,000 K, luminosities between 100,000 and 1 million times the Sun's, and masses from 10 to 50 solar masses. They have strong **stellar winds** that carry away 3 solar masses per million years. Many central stars of planetary nebulae are WR stars. The type of WR star that emerges after the original hydrogen-rich envelopes have been lost depends on how far the star's evolution had progressed. Wolf–Rayet stars are named for their French discoverers, Charles Wolf (1827–1918) and Georges Rayet (1839–1906).

wormhole A hypothetical tunnel in **spacetime**. In 1916 Albert Einstein discovered that one of the solutions to equations describing a black hole described a connection between different regions of spacetime. In principle, this would make time travel possible. Recent cosmological theories suggest that on very small scales spacetime has a foam-like structure, pervaded by wormholes. These wormholes could lead to the formation of "baby universes," which would have profound implications for our understanding of the Big Bang.

wrinkle ridge A winding ridge on the surface of a lunar mare, having sloping sides rising to a typical height of 200 m (650 ft), and several hundred kilometers long. They are often associated with **rilles**, and may have originated when the lava that formed the maria cooled and contracted. Wrinkle ridges are also found on Venus.

W Ursae Majoris star A member of a class of eclipsing binary stars, almost in contact as they orbit one another. They differ from **Beta Lyrae stars** in that the two components are smaller, less luminous stars of nearly identical brightness.

W Virginis star A member of a group of pulsating variable stars superficially similar to Cepheids. They are giant stars, typically in the spectral range G0 to M0 and with absolute magnitudes from -1 to -4, on the instability strip in the **Hertzsprung–Russell diagram**. Their masses can be as low as 0.5 solar masses, suggesting that they have evolved from low-mass main-sequence stars. Light curves of W Virginis stars can be distinguished from those of classical Cepheids by their less regular shape and a double-peaked maximum. Their periods range from about 1 to about 100 days, and their **period–luminosity law** is distinctly different from that of normal Cepheids; however, they can still be used as distance estimators for galactic and extragalactic objects. Some belong to Population I and some to Population II. Occasionally W Virginis stars show small period changes but in one case, RU Camelopardalis, the variations suddenly stopped in 1964 for about three years. It is not understood how pulsations can stop, or start, in such a short time.

X Y Z

▲ **X-ray astronomy** This Chandra X-ray Observatory image shows a young supernova remnant in Centaurus. The rapidly expanding shell of gas contains large amounts of oxygen, neon, magnesium, sulfur and silicon. It has a pulsar at its center.

X-ray astronomy The study of **electromagnetic radiation** of short wavelength (about 0.01 to 10 nm), shorter than ultraviolet radiation but longer than gamma-rays. The Earth's atmosphere is opaque to X-rays, so this radiation must be studied by means of instruments aboard rockets or satellites. For astronomical purposes X-rays are usually described according to the energy associated with the X-ray photons; this energy is expressed in **electron-volts**.

X-ray astronomy began in 1949 with the discovery, by a team at the United States Naval Research Laboratory using captured German rockets, that the Sun emits X-rays. By the time that the first X-ray satellites were launched, the general properties of the Sun's X-ray emission had been established. Other celestial X-ray sources were discovered by rocket flights in the 1960s. The first all-sky survey at X-ray wavelengths was made by the US satellite Uhuru, launched in 1970. Uhuru, and later the UK satellite Ariel V, discovered the existence of **X-ray binaries**, in which

one component is a neutron star or a black hole, as well as X-ray emission from supernova remnants, active galaxies, quasars, and hot gas pervading clusters of galaxies. An **X-ray telescope** aboard the Einstein satellite, launched in 1978, produced the first true images of X-ray sources, rather than simply counting X-rays as previous instruments had done. Objects in nearby galaxies such as M31 and the Magellanic Clouds were resolved and studied for the first time. It was also discovered that almost all types of ordinary star emit X-rays. Subsequent X-ray satellites have included **Rosat, Yohkoh**, the **Chandra X-ray Observatory** and **Newton**.

X-ray binary A binary system consisting of a normal star and a collapsed star – a **neutron star**, **black hole**, or, in less intense sources, a **white dwarf**. The two components are very close together. In some cases, where the normal star is a giant or supergiant, a stellar wind blows directly on to the compact companion, causing

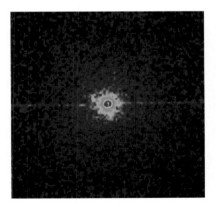

▲ **X-ray binary** The strong X-ray source Cygnus X-3 is a close binary system in which matter from a normal star is being drawn into a neutron star or black hole. The X-ray emission from Cygnus X-3 varies regularly with a 4.8 hour period, as the compact star circles its companion star.

the X-ray emission. In other cases, expansion of the normal star in the course of its evolution results in a flow of gas toward the collapsed star. This matter forms an **accretion disk** of hot material which emits X-rays. SEE ALSO **X-ray burster**.

X-ray burster A source of intense flashes of X-rays. These X-ray bursts have rise times of about 1 second, fall times of about 60 seconds, and total luminosities equivalent to one week's energy output of the Sun. The intervals between bursts are irregular, ranging from hours to days, while many sources undergo burst-inactive phases that can last for weeks or even months. Burst sources are associated with old Population II objects and emit no pulsations. They are thought to be binaries containing an old neutron star with no magnetic field, which explains the absence of pulsations. The neutron star is paired with an old low-mass star, and individual bursts come from thermonuclear explosions after material accreted from the companion star has exceeded a critical mass on the neutron star's surface. *X-ray transients* are nova-like outbursts of X-rays believed to come from similar binaries in which the mass transfer is very uneven.

X-ray Multi-Mirror Mission (XMM) SEE **Newton**.

X-ray telescope An instrument for imaging X-ray sources. All X-ray telescopes are satellite-borne because X-rays do not penetrate the Earth's atmosphere. Unlike less energetic electromagnetic radiation, X-rays cannot be focused by reflection from a conventional concave mirror (as in a **reflecting telescope**) because, for all but the shallowest angles of incidence, they penetrate the surface. In *grazing incidence* X-ray telescopes the focusing ele-

ment is a pair of coaxial surfaces, one paraboloidal and the other hyperboloidal, off which incoming X-rays are reflected at a very low "grazing" angle toward a focus. The detecting element is often a **CCD** adapted for X-ray wavelengths. An alternative instrument is the *microchannel plate detector*. The incident radiation falls on a plate which is made up of many fine tubes, rather like a short, wide fiber-optic bundle. The plate is charged, so that the radiation generates electrons which are accelerated down the tubes and together form an image that can be read off.

X-ray transient SEE **X-ray burster**.

Yagi antenna A basic form of antenna used in simple **radio telescopes**. It consists of several parallel elements mounted on a straight member, and often forms the basis of cheap arrays used in **aperture synthesis**; it is also a familiar form of TV aerial. It was developed by the Japanese engineer Hidetsugu Yagi (1886–1976).

year The time taken by the Earth to complete one revolution around the Sun. Various years, defined according to the choice of reference point, are given in the table. The *civil year* (**calendar** year) averages 365.2425 **mean solar days**. SEE ALSO the entries for each type of year listed in the table.

Yerkes Observatory The observatory of the University of Chicago, at Williams Bay, Wisconsin. It was founded by George Ellery **Hale**. Its main instrument is a 1-m (40-inch) refractor, opened in 1897 and still the largest in the world.

Yohkoh A satellite launched in 1991 to study the Sun, particularly solar flares, at X-ray and gamma-ray wavelengths. The mission was a Japanese–British–American collaboration; Yohkoh is Japanese for "sunbeam." The satellite returned many X-ray images of the Sun, the first obtained from orbit since Skylab missions in 1973. Its instruments revealed much about the rapid changes that occur in the corona, and how flares originate and evolve.

TYPES OF YEAR	
anomalistic year	365.25964 mean solar days
eclipse year	346.62003 mean solar days
sidereal year	365.25636 mean solar days
tropical year	365.24219 mean solar days

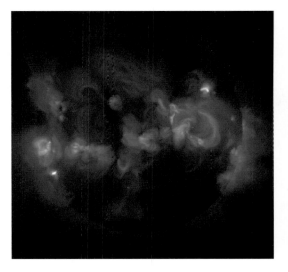

► **Yohkoh** This Yohkoh image of the active Sun at X-ray wavelengths shows hot coronal plasma in magnetic loops above active regions. Cooler, dark regions above the poles are described as coronal holes; these extend to lower solar latitudes at sunspot minimum.

Young, Charles Augustus (1834–1908) American solar astronomer. He made some of the earliest spectroscopic studies of the corona, and discovered the **flash spectrum** of the chromosphere during a total solar eclipse in 1870. He also used spectral line measurements to determine the Sun's rotation period.

Z Andromedae star SEE **symbiotic variable**.

Z Camelopardalis star A member of a small subgroup of **dwarf novae** similar to **U Geminorum stars**, except that they experience occasional "standstills," remaining more or less constant at some intermediate brightness. The occurrence of the standstills, and their duration, ranging from a few days to many months, are unpredictable.

Zeeman effect The splitting of a spectral line into several components by a strong magnetic field. In circumstances where these components cannot be resolved, the effect is apparent as a widening of the original line. The Zeeman effect occurs in the spectra of sunspots and stars. It demonstrates the existence of magnetic fields in celestial bodies and, since the field

strength depends on the degree of splitting, it allows the fields to be measured. The phenomenon was predicted by Hendrik Lorentz, and discovered in 1896 by Pieter Zeeman (1865–1943). The two Dutch physicists shared the 1902 Nobel Prize for Physics.

Zelenchukskaya Observatory An observatory, also known as the Special Astrophysical Observatory, on Mount Pastukhov in the Caucasus Mountains of southern Russia, the site of the 6-m (236-inch) Large Azimuthal Telescope, which was opened in 1975. Also at Zelenchukskaya is the RATAN-600 radio telescope, which consists of a ring of reflecting panels 600 m (2000 ft) in diameter.

zenith The point on the **celestial sphere** that is vertically above the observer, 90° from the horizon. The point directly opposite, beneath the observer's feet, is called the *nadir*.

zenithal hourly rate (ZHR) The number of meteors that would be seen at the maximum of a particular **meteor shower** under good conditions if the radiant were immediately overhead. The ZHR is obtained by applying a number of correction factors to the observed hourly rate of meteors in the shower. The observed hourly rate is always less than the ZHR. There are short-term and long-terms variations in showers' ZHRs as the associated **meteor streams** evolve.

zenith distance The angular distance of a celestial body from the zenith. It is equal to 90° minus the altitude of the body above the horizon.

zero-age main sequence (ZAMS) The **main sequence** on the Hertzsprung–Russell diagram as defined by stars of zero age, i.e. before they have undergone any substantial evolution.

ZHR ABBREVIATION FOR **zenithal hourly rate**.

zodiac A belt on the celestial sphere, about 8° either side of the ecliptic, which forms the background for the motions of the Sun, Moon, and planets (except Pluto). The zodiac is divided into 12 *signs*, each 30° long, which are named for the constellations they contained at the time of the ancient Greeks: Aries, Taurus, Gemini, Cancer, Leo, Virgo, Libra, Scorpio, Sagittarius, Capricorn, Aquarius and Pisces. Because of **precession** these signs no longer coincide with the constellations of the same name. The ecliptic also passes through the constellation Ophiuchus, which is not a zodiacal sign.

zodiacal light A cone of faint light, usually fainter than the Milky Way, visible at all seasons in the tropics in the absence of moonlight. It stretches along the ecliptic from the western horizon after evening twilight, or from the eastern horizon before morning twilight. At the point in the sky opposite the Sun, the zodiacal light brightens a little; this is the **gegenschein**. The spectrum of the zodiacal light resembles the Sun's, indicating that the phenomenon results from the scattering of sunlight by particles in the plane of the ecliptic. These particles constitute the *zodiacal dust cloud*, and probably originate partly from matter ejected by the Sun, and partly from the decay of comets and asteroids.

zone of avoidance A region of the sky along the plane of the Milky Way where almost no external galaxies can be seen because of absorption by interstellar gas and dust in our Galaxy. It is between 10° and 40° wide.

Zwicky, Fritz (1898–1974) Swiss–American astronomer, born in Bulgaria. In 1934 he and Walter **Baade** suggested that what is left after a supernova explosion is a neutron star; this was confirmed in 1968 with the discovery of the pulsar in the **Crab Nebula**. In the mid-1960s, using the newly developed **Schmidt camera**, Zwicky discovered and examined many supernovae in other galaxies, and began a long study of **clusters of galaxies**.

Appendix

THE GREEK ALPHABET

α	A	alpha	ν	N	nu
β	B	beta	ξ	Ξ	xi
γ	Γ	gamma	o	O	omicron
δ	Δ	delta	π	Π	pi
ε	E	epsilon	ρ	P	rho
ζ	Z	zeta	σ	Σ	sigma
η	H	eta	τ	T	tau
θ	Θ	theta	υ	Y	upsilon
ι	I	iota	φ	Φ	phi
κ	K	kappa	χ	X	chi
λ	Λ	lambda	ψ	Ψ	psi
μ	M	mu	ω	Ω	omega

CONVERSION FACTORS

Distances

1 nm = 10 Å
1 inch = 25.4 mm
1 mm = 0.03937 inch
1ft = 0.3048 m
1 m = 39.37 inches = 3.2808 ft
1 mile = 1.6093 km
1 km = 0.6214 mile
1 km/s = 2237 mile/h
1 pc = 3.0857 $\times 10^{13}$ km = 3.2616 l.y. = 206,265 AU
1 l.y. = 9.4607 $\times 10^{12}$ km = 0.3066 pc = 63,240 AU

Temperatures (to the nearest degree)

°C to °F :	×1.8, +32
°C to K :	+273
°F to °C :	−32, ÷1.8
°F to K :	÷1.8, +255
K to °C :	−273
K to °F :	×1.8, −460

Note: To convert temperature *differences*, rather than points on the temperature scale, ignore the additive or subtractive figure and just multiply or divide.

SYMBOLS FOR UNITS, CONSTANTS AND QUANTITIES

a	semimajor axis	L	luminosity	T	temperature (absolute), epoch (time of perihelion passage)
Å	angstrom unit	L_n	Lagrangian points (n = 1 to 5)		
AU	astronomical unit	l.y.	light year		
c	speed of light	m	meter, minute	T_{eff}	effective temperature
d	distance	m	apparent magnitude, mass	v	velocity
e	eccentricity	m_{bol}	bolometric magnitude	W	watt
E	energy	m_{pg}	photographic magnitude	y	year
eV	electron-volt	m_{pv}	photovisual magnitude	z	redshift
f	following	m_v	visual magnitude		
F	focal length, force	M	absolute magnitude, mass (stellar)	α	constant of aberration, right ascension
g	acceleration due to gravity	N	newton	δ	declination
G	gauss	p	preceeding	λ	wavelength
G	gravitational constant	P	orbital period	μ	proper motion
h	hour	pc	parsec	ν	frequency
h	Planck constant	q	perihelion distance	π	parallax
H_0	Hubble constant	q_0	deceleration parameter	ω	longitude of perihelion
Hz	hertz	Q	aphelion distance	Ω	observed/critical density ratio, longitude of ascending node
i	inclination	r	radius, distance		
IC	*Index Catalogue*	R	Roche limit	°	degree
Jy	jansky	s	second	′	arc minute
k	Boltzmann constant	t	time	″	arc second
K	degrees kelvin				

255

MULTIPLES AND SUBMULTIPLES USED WITH SI UNITS

Multiple	Prefix	Symbol	Submultiple	Prefix	Symbol
10^3	kilo-	k	10^{-3}	milli-	m
10^6	mega-	M	10^{-6}	micro-	μ
10^9	giga-	G	10^{-9}	nano-	n
10^{12}	tera-	T	10^{-12}	pico-	p
10^{15}	peta-	P	10^{-15}	femto-	f
10^{18}	exa-	E	10^{-18}	atto-	a

Acknowledgements

Abbreviations
AURA – Association of Universities for Research in Astronomy, Inc.
CXC – Chandra X-ray Observatory Center, Harvard-Smithsonian Center for Astrophysics
DLR – Deutschen Zentrum für Luft- und Raumfahrt
ESA – European Space Agency
ESO – European Southern Observatory
GSFC – Goddard Space Flight Center
JHU – Johns Hopkins University
JPL – Jet Propulsion Laboratory
JSC – Johnson Space Center
MPE – Max-Planck-Institut für extraterrestrische Physik, Garching
NASA – National Aeronautics and Space Administration
NOAO – National Optical Astronomy Observatory
NRAO – National Radio Astronomy Observatory
NSF – National Science Foundation
SOHO – Solar and Heliospheric Observatory. SOHO is a project of international cooperation between ESA and NASA
STScI – Space Telescope Science Institute
ST-ECF – Space Telescope European Coordinating Facility
USGS – US Geological Survey

5 Nigel Sharp/NOAO/AURA/NSF; **6** NOAO/AURA/NSF; **7** ESO; **8** NASA/JPL; **13** NASA/David R. Scott; **14** National Astronomy and Ionosphere Center/Cornell Univ./NSF; **15** NASA/JPL; **25** A. Dupree Harvard-Smithsonian Center for Astrophysics/NASA/STScI; **26** Big Bear Solar Observatory; **30** ESO; **37** NASA/JPL/Univ. Arizona; **39** NOAO/AURA/NSF; **43** NASA/JPL/USGS; **50** Courtesy of the TRACE consortium, Stanford-Lockheed Institute for Space Research/NASA; **51** DMR/COBE/NASA/Two-Year Sky Map; **54** ESO; **57** NASA; **59** NASA/JPL; **62** Bill Schoening/NOAO/AURA/NSF; **63** ESO; **69** NOAO/AURA/NSF; **70** NASA/JPL/USGS; **72** NASA/JHU Applied Physics Laboratory; **73** NASA/JPL/DLR; **76** Courtesy SOHO/MDI Consortium; **77** SEC/NOAA – National Oceanic and Atmospheric Administration/US Department of Commerce; **78** Courtesy SOHO EIT Consortium; **79** Nigel Sharp/NOAO/AURA/NSF; **83 top** NASA/JPL/DLR; **83 bottom** NASA/JPL/Univ. Arizona; **84** ESO; **85** NASA/JPL/USGS; **86** Gemini Observatory/NSF/Univ. Hawaii Institute for Astronomy; **87** Nigel Sharp/NOAO/AURA/NSF; **88** ESA/MPE; **89 top** NOAO/AURA/NSF; **89 bottom** Michael Rich, Kenneth Mighell and James D. Neill (Columbia Univ.)/Wendy Freedman (Carnegie Observatories)/NASA; **90** A. Fruchter and the ERO Team (STScI, ST-ECF)/NASA; **91** ESO; **92** NASA/JPL/Cornell Univ.; **97** (X-ray) NASA/UMass/D. Wang et al (optical) NASA/HST/D. Wang; **98** NOAO/AURA/NSF; **99** ESO; **102** ESA; **103** T.A.Rector (NOAO/AURA/NSF) and Hubble Heritage Team (STScI/AURA/NASA); **105** NASA/JSC; **108** NASA/JPL/USGS; **109 top** NASA/JPL; **109 bottom** NASA/GSFC/Langley Research Center/JPL/Multi-angle Imaging SpectroRadiometer Team; **110** Infrared Processing and Analysis Center of California Institute of Technology/JPL; **111** ESA/ISOCAM/ISOGAL Team; **112** NASA; **113** ESA; **115** NASA/JPL/Lunar and Planetary Laboratory; **116** Ian Morison/Jodrell Bank Observatory; **117** NASA/JPL/Univ. Arizona; **121** NOAO/AURA/NSF; **124** NASA/JPL; **125** NOAO/AURA/NSF; **131** NASA/JSC; **136** ESO; **137** NASA/JPL/USGS; **138** NASA/JPL/USGS; **139** NASA/JPL/Malin Space Science Systems; **140** NASA/JHU Applied Physics Team; **141** NASA/JPL; **146** NASA/JPL; **147** NASA/JPL/USGS; **149** NASA/JPL/USGS; **154** NASA/JPL; **156** NASA/North Carolina School of Science and Mathematics/C. Olbert et al; **161** NASA/JPL; **162** Todd Boroson/NOAO/AURA/NSF; **165** Mark McCaughrean/ESO; **167** Mir 27 Crew © Centre National d'Études Spatiales, France; **170** NASA/JPL/Malin Space Science Systems; **171** ESO; **172** Bill Schoening/NOAO/AURA/NSF; **173** NASA/ESA/Hubble Heritage Team (STScI/AURA); **175** Dr R. Albrecht ESA/ESO/ST-ECF/NASA; **180** NASA/JPL; **181** NASA/JPL; **182** NASA/CXC/Smithsonian Astrophysical Observatory; **183** John Bahcall, Institute for Advanced Study, Princeton/Mike Disney (Univ. Wales)/NASA; **184** NASA/JPL; **186** NRAO/Associated Universities, Inc.; **188** NASA/JPL/USGS; **190** NASA/Hubble Heritage Team (STScI/AURA); **193** NASA/JPL; **202** NASA/Hubble Heritage Team (STScI/AURA)/R. G. French (Wellesley College)/J. Cuzzi and J. Lissauer (NASA/Ames Reseach Center)/L. Dones (Southwest Research Institute, Boulder, Colorado); **207** NASA/AURA/NSF; **223** NASA/Donald Walter (S. Carolina State Univ.)/ Paul Scowen and Brian Moore (Arizona State Univ.); **224** Courtesy SOHO/MDI Consortium; **229** ESO; **231** NASA/JPL; **232** NASA/JPL; **233** NASA/NSSDC; **235** NASA/JPL/USGS; **239** NASA/JPL; **240** NASA/JPL; **243** NASA/JPL; **245** NRAO/AU/NSF; **247** NASA/JPL/USGS; **248** T.A. Rector and Monica Ramirez/NOAO/AURA/NSF; **249** Harvey Richer (Univ. British Columbia)/STScI/NASA; **251 top** NASA/CXC/Rutgers/J. Hughes et al; **251 bottom** NASA/SRON/MPE; **253** Institute of Space and Astronautical Science, Japan.

Artwork © Philip's